AF601554

Green Energy and Technology

Climate change, environmental impact and the limited natural resources urge scientific research and novel technical solutions. The monograph series Green Energy and Technology serves as a publishing platform for scientific and technological approaches to "green"—i.e. environmentally friendly and sustainable—technologies. While a focus lies on energy and power supply, it also covers "green" solutions in industrial engineering and engineering design. Green Energy and Technology addresses researchers, advanced students, technical consultants as well as decision makers in industries and politics. Hence, the level of presentation spans from instructional to highly technical.

Indexed in Scopus.

Indexed in Ei Compendex.

More information about this series at https://link.springer.com/bookseries/8059

Antonio Urbina

Sustainable Solar Electricity

Springer

Antonio Urbina
Institute for Advanced Materials
and Mathematics (INAMAT2)
and Department of Sciences
Public University of Navarra (UPNA)
Pamplona, Spain

ISSN 1865-3529 ISSN 1865-3537 (electronic)
Green Energy and Technology
ISBN 978-3-030-91770-8 ISBN 978-3-030-91771-5 (eBook)
https://doi.org/10.1007/978-3-030-91771-5

This Springer imprint is published by the registered company Springer Nature Switzerland AG
The registered company address is: Gewerbestrasse 11, 6330 Cham, Switzerland

To Marijose,
for everything,
including the little thing(s).

Foreword

Over the last thirty years (my time as a researcher of photovoltaics), the solar photovoltaic (PV) industry has grown at an astonishing rate from an installed global capacity of less than 100 to over 800,000 MW_p. Solar PV has become the fastest-growing energy technology and the primary future source of electricity in most scenarios for low-carbon development. The "coming of age" of renewable electricity has been one of few good news stories in our efforts to mitigate climate change. Thanks to solar power and wind, the decarbonisation of electricity is no longer seen as a major challenge and has become a tool that can assist the decarbonisation of other sectors (transport, buildings, industry). The question of how to harness solar electricity affordably is more or less solved. But we still need to establish how to achieve the energy transition sustainably.

In this book, Antonio Urbina presents a lucid account of the principles and technology of solar photovoltaics, alongside an introduction to the concept of sustainability and to the metrics that quantify sustainability. He shows how, starting from the production process, the environmental impacts, resource requirements and energy balance can be quantified, making these impact assessments a natural extension to the design of PV technology. He also places photovoltaic technology in its larger global context by addressing economic aspects and the international regulatory and policy framework, in a detailed, up to date and informative manner. To my knowledge, this is the first book of its kind and I find it timely for three reasons.

First, while energy technologies are commonly compared in terms of cost and performance metrics (such as power conversion efficiency for a solar cell), cost is not enough to distinguish options in terms of their effectiveness at reducing carbon emissions. A solar module with a shorter energy payback time and a smaller life-cycle averaged emissions intensity will be more effective than an alternative at mitigating CO_2 emissions when it replaces a higher carbon technology. Life Cycle Assessment (LCA) techniques allow PV developers to evaluate the module designs and production processes that optimise those metrics. The best options may not be those of the lowest cost or the highest conversion efficiency. Identifying these priorities at the design stage allows more efficient use of resources.

Second, as the energy transition advances, it becomes more necessary to consider solutions from a system-level perspective. For solar PV, that means not only considering the output of a module but also the effectiveness of integrated systems, such as solar PV integrated with hydrogen generation for fuel supply or solar PV with desalination for clean water supply. To evaluate different technical solutions to the same demand in terms of their energy balance, resource costs or emissions impact, a means of comparing quite different technologies is needed. Life Cycle Assessment provides that and can be applied as part of the selection of technologies, avoiding lock-in to solutions that are less effective in terms of energy or emissions balance.

Third, the technological revolution that lies ahead of us (if we are to avert the worst consequences of climate change) will be as great as the last industrial revolution, but much more rapid. Rapid change brings risks of social, economic, environmental and geopolitical impacts as well as emissions impacts. Before choosing pathways, it would be wise to evaluate them in terms of their overall sustainability. This book provides the basic knowledge to formulate and evaluate these questions.

Antonio Urbina is well qualified to write this work, having researched the science of PV materials, evaluated solar PV systems and pioneered the application of LCA and sustainability assessment to emerging PV technologies. From this experience, base he shows how to make sustainability a central part of technology evaluation. Although the book presents LCA and sustainability analysis in the context of solar electricity, the methodologies are very readily transferrable, and increasingly relevant, to other energy, and non-energy, technologies.

London, UK
November 2021

Jenny Nelson

Acknowledgements

I started working on photovoltaics at two levels in the mid 90s during the final years of my Ph.D. (which was focussed on the Quantum Hall Effect, a very different issue, but which shares with photovoltaics the use of advanced semiconducting devices). The first level was a very practical approach: the use of small photovoltaic solar home systems for rural electrification in developing countries, an interest which started with a course delivered by the Instituto de Energía Solar (Madrid) and I must acknowledge the enthusiasm on the subject put by the researchers that delivered the course: Dr. Pablo Díaz, Dr. Estefanía Caamaño and Dr. Miguel A. Egido, which taught me the fundamentals of practical PV system design. The second level was the deepening of the theoretical understanding provided by the books of Prof. Jenny Nelson (Imperial College London) and Prof. Eduardo Lorenzo (Instituto de Energía Solar, Madrid), and I must acknowledge the authors not only for writing the books, but also for facilitating always friendly communications and discussions on photovoltaic technology and its practical deployment. The acknowledgement to Prof. Jenny Nelson must be extended to her invitation for a research stay at Imperial College, and the subsequent research collaboration that we have kept since then and which continues to this day, also including other colleagues at Imperial College which I acknowledge: Prof. Ji-Seon Kim, Prof. James Durrant, Dr. Sachetan Tudhalar, Dr. Christopher Emmott and Dr. Wing Chung Tsoi (now at Swansea University).

Regarding my research work in organic and hybrid photovoltaic technologies with a special focus on stability studies, I acknowledge Prof. Frederik Krebs (CEO of Infinity PV, Denmark) and Prof. Mónica Lira-Cantú (Institut Catalá de Nanociència i Nanotecnolog a) for his and her constant support and fruitful collaboration, and Prof. Ana Rosa Lagunas (Centro Nacional de Energas Renovables, CENER, Spain) for helping me to bridge the gap between academic research and the complex world of standardization, certification and industrial applications of photovoltaic technology. It has also been very important the work of Dr. Lucía Serrano (Universidad Rey Juan Carlos, Madrid), Dr. Nieves Espinosa (Joint Research Centre, European Commission), Dr. Rafael García-Valverde (Infinity PV, Denmark), Dr. Carlos Toledo (ENEA, Italy) and Dr. Rodolfo García (Universidad Politécnica de Cartagena, Spain), who have been fundamental contributors to the research of our group on Life Cycle Assessment of photovoltaic technologies, during and after their respective Ph.D. thesis work, which was completed under my supervision a few years ago. This research work was carried out in the context of projects in collaboration with Dr. José Abad, Dr. Antonio J. Fernández-Romero, Dr. Javier Padilla (UPCT), Prof. Jaime Colchero (Universidad de Murcia), Prof. Ana Cros and Prof. Nuria Garro (both at Universidad de Valencia), Prof. Wolfgang Maser and Prof. Ana Benito (both at Instituto de Carboquímica ICB-CSIC, Zaragoza); to all of them I acknowledge their support with access to instruments and materials that have been used to fabricate and characterize organic and hybrid solar cells in the context of several collaborative projects and the discussions during seminars (and coffee breaks) during many fruitful years.

Financial support must be acknowledged to Agencia Estatal de Investigación (Ministerio de Ciencia e Innovación, Spain), grant PID2019-104272RB-C55, and to Fundación Séneca (Spain), grant 19882-GERM-15, both including European Commission FEDER funds.

Contents

Acronyms

AC	Alternating Current
AM	Air Mass
a-Si	Amorphous Silicon
BGS	British Geological Survey
BHJ	Bulk Heterojunction
BIPV	Building Integrated Photovoltaics
CdTe	Cadmium Telluride
CIGS	Copper Indium Gallium (di)Selenide, chalcopyrite structure
c-Si	Crystalline Silicon
CSR	Corporate Social Responsibility
CSS	Closed Space Sublimation
CZ	Czochralski
CZTS	Copper Zinc Tin (di)Selenide, kesterite structure
DC	Direct Current
EC	European Commission
ED	Electro Deposition
EPBT	Energy Payback Time
EPR	Extended Producer Responsibility
ETL	Electron Transporting Layer
EVA	Ethylene-Vinyl-Acetate
FAPI	Formamidinium Lead Iodide, perovskite structure
FF	Fill factor
FTO	Fluor Tin Oxide
FU	Functional Unit
FZ	Floating Zone
GHG	Greenhouse Gases
GRR	Ground Requirement Ratio
HIT	Heterojunction with an Intrinsic Thin layer
HTL	Hole Transporting Layer
I_{sc}	Short circuit current
IBC	Interdigitated Back contact Cell

IEA	International Energy Agency
IEC	International Electrotechnical Commission
III-V	Elements of groups III and V of the periodic table
IPCC	Intergovernmental Panel on Climate Change
IRENA	International Renewable Energy Agency
ISO	International Organization for Standardization
ITO	Indium Tin Oxide
I-V	Current–voltage characteristic curve of a solar cell or module
JRC	Joint Research Centre (European Commission)
LCA	Life Cycle Assessment
LCIA	Life Cycle Impact Assessment
LCOE	Levelized Cost of Energy (or Electricity)
LEC	Liquid Encapsulated Czochralski
LPE	Liquid Phase Epitaxy
MAPI	Methyl Ammonium Lead Iodide, perovskite structure
MBE	Molecular Beam Epitaxy
mc-Si	Multi-crystalline Silicon
MOCVD	Metal-Organic Chemical Vapour Deposition
mono-Si	Mono-crystalline Silicon
MOVPE	Metal-Organic Vapour Phase Epitaxy
mpp	Maximum power point (in a I–V or P–V curve)
NREL	National Renewable Energy Laboratory (USA)
OPV	Organic Photovoltaics
P_{mpp}	Power at maximum power point
P3HT	Poly-(3-Hexyl-Thiophene-2,5-diyl)
PANI	Poly-Aniline
PAR	Photosynthetically Active Radiation
PCBM	Phenyl-C_{61}-Butyric acid Methyl ester
PCE	Power Conversion Efficiency
PECVD	Plasma Enhanced Chemical Vapour Deposition
PEDOT	Poly-3,4-Ethylene-Dioxy-Thiophene
PERC	Passivated Emitter and Rear Cell
PET	Poly-Ethylene Terephthalate
PPV	Poly-(p-Phenylene-Vinylene)
PR	Performance Ratio
PVD	Physical Vapour Deposition
PVF	Poly-Vinyl Fluoride
PVPS	Photovoltaic Power Systems Programme (IEA)
RFS	Radio Frequency Sputtering
sc-Si	Single-crystalline Silicon
SLS	Soda Lime Silica
UNFCCC	United Nations Framework Convention on Climate Change
USGS	United States Geological Survey
V_{oc}	Open circuit Voltage
VALCOE	Value-Adjusted Levelized Cost of Energy (or Electricity)

Part I
Introduction

Part I is an introductory part which describes the main concepts regarding photovoltaic technology and life cycle assessment. The book contents are built upon the combination of both areas of knowledge, and it is, therefore, important from the beginning to clarify the purpose and the scope of the study. This part also emphasizes the importance of the problem that the energy transition is facing: a huge amount of photovoltaic systems has been already deployed and many more are planned for the near future; many of these systems will have to be revamped, replaced or extended with new modules, and the old ones will need to be recycled or landfilled. In Chap. 1, the working scenarios proposed by the International Energy Agency are presented and the implications for photovoltaic capacity growth will be analysed in detail. In Chap. 2, the main components of photovoltaic systems are presented, ranging from cells to modules and then to whole systems; this chapter describes each component, its principles of work and the equations governing its main output (but not going into details of the physics behind semiconductor photogeneration and transport dynamics); the objective of this chapter is to define the main parameters used to evaluate photovoltaic (PV) cells, modules and system performance and to classify the "product" parts (a classification which is used for the Life Cycle Assessment (LCA) study of the different technologies). The "product" from the LCA perspective is the final PV system, which includes different steps: cells, modules and whole system (with Balance of System (BoS), components). In Chap. 3, the Life Cycle Assessment methodology is presented, with a special focus on its application to energy systems in general and photovoltaic systems in particular and also the inclusion of social and economic considerations for a broader LCA approach (methodologies still under discussion in the scientific community).

Chapter 1
Scenarios for Solar Electricity at the TeraWatt Scale

A world shock has occurred in 2020, and it has strongly affected the energy sector. According to the preliminary estimations included in the most recent report from the International Energy Agency (World Energy Outlook 2020, [9]), the global energy demand dropped by 5% in 2020, and energy-related CO_{2eq} emissions dropped by 7%. This shock in the demand side, concentrated in a single year, is higher in terms of energy demand reduction than the shock in the supply side that started in October 1973 due to an oil export embargo proclaimed by the Organization of the Petroleum Exporting Countries (OPEC) that lead to a sudden rise in oil prices. The impact of the oil crisis was long lasting, it reshaped the energy landscape worldwide and triggered the first steps to unlock the "carbon lock-in" and initiate an energy transition that is now fully fledged [1].

After a sudden shock, a well-established paradigm can be shifted if the policy response is clearly defined and enough investment is provided, initially in research activities and later in demonstration projects. The initial efforts triggered by the oil crisis put in place technological advancements that supported the early stages of the energy transition a few decades ago. In 2020, an external shock, the catastrophic COVID-19 pandemia led to public policies designed with strong investment efforts to reactivate the economy, and this "new deal" has created the opportunity to accelerate the energy transition with renewable mature technologies that are cost-competitive. Wind and photovoltaic technologies are already the cheapest source of electricity in many parts of the world. This combination of shock, new investment and technological readiness could definitely move the world from the carbon lock-in to a renewables lock-in. Large investments have been announced worldwide to reactivate the economy, and a good share of this investment is oriented to reinforce the energy transition and to mitigate climate change. It is a great opportunity that will require new ambitious policies and a worldwide coordination of a good regulatory framework to support this move towards a more sustainable energy landscape.

A. Urbina, *Sustainable Solar Electricity*, Green Energy and Technology,
https://doi.org/10.1007/978-3-030-91771-5_1

But despite the brilliant perspective, this transition is still in its early stages. In 2019, total primary energy consumption in the world was 583.9 Exajoules,[1] with an annual growth rate of 1.6% averaged for the past ten years. The renewables contribution to the total primary energy consumption was 66.64 Exajoules (11.3%, of which 6.4% from hydropower), while oil continues to hold the highest share (33.1%), followed by coal (27.0%), natural gas (24.2%) and then renewables (11.3%) that have already surpassed nuclear (4.3%). Electricity generation in 2019 was 27004.7 TWh[2] (average annual growth in the past ten years was 2.7%, almost doubling the primary energy average annual growth, a clear indicator of the "electrification" of the global energy consumption), and renewable electricity generation was 7027.7 TWh (26.1%, including hydropower, its main contributor, with 4222.2 TWh equivalent to 15.6% followed by wind 1429.6 TWh, 5.2% and solar photovoltaic with 724.1TWh, 2.7%) [2]. According to the International Energy Agency Photovoltaic Power Systems Programme, world final electricity consumption was 24,700 TWh in 2019, with a share of renewable energy in the global electricity production of 28%, including 810 TWh produced from solar photovoltaic systems; thus, the solar electricity production share was 3.3% [11].

In 2020, due to the world reduced energy demand and the increment in photovoltaic power installed capacity, around 3.7% of world electricity production has been supplied by photovoltaic systems and the avoided emissions have been 875 Mt of CO_{2eq} (a calculation by the IEA-PVPS based on the emissions that would have been generated from the same amount of electricity produced by the different grid mixes in all countries and taking into consideration life cycle emissions of PV systems). This world average hides a large variation among countries, where a group of seven countries are in the range of 10% and another seven have already surpassed 5%. In this group, it is important to emphasize that the two most populated countries in the world have already reached 6.5% (India) and 6.2% (China) share of its electricity supply from photovoltaic systems [10].

Despite the progress in rural electrification, still 733 million people are lacking access to electricity, three quarters in sub-Saharan Africa (580 million), and another 100 million people cannot afford electricity although they have access to the grid [5]. Either to substitute electricity from non-renewable sources or to supply new demand, the contribution of photovoltaic systems has been growing steadily since many years ago and has now become the fastest growing technology in terms of annual installed capacity. The share of world electricity supply from photovoltaics is going to increase significantly in the coming decades in all scenarios that are proposed by different institutions. The rate of growth and the cumulative capacity depend strongly on the assumptions for these scenarios, and in all of them, photovoltaic technology share is very high, in some cases the top of the list of annual installed capacity during several years. This fact emphasizes the urgent need of a detailed evaluation of the

---

[1] 1 Exajoule (EJ) = 10^{18} Joules; another broadly used unit for primary energy is tonnes of oil equivalent (toe), 1 toe = 4.1868×10^{10} Joules.

[2] 1 TWh = 10^{12} Wh = 3.6×10^{15} Joules.

sustainability of solar electricity massive deployment. This is the purpose of this book.

In this introductory chapter, an overview of the world photovoltaic energy status and trends are presented. After showing the rough numbers of installed capacity and its most recent evolution, the world energy supply and demand in future scenarios proposed by the International Energy Agency are analysed, and the implications for the growth of photovoltaic installed capacity are commented. The analysis of the sustainability of the photovoltaic electricity generation is the subject of the whole book, and the methodological tools both for the calculation of the electricity that can be generated with different photovoltaic technologies and its environmental and economical impacts are the framework to organize the book into three parts and twelve chapters:

Part I. Introduction. It is an introductory part which describes the main concepts regarding photovoltaic technology and life cycle assessment. The book contents are built upon the combination of both areas of knowledge, and it is, therefore, important from the beginning to clarify the purpose and the scope of the study. This part also emphasizes the importance of the problem that the energy transition is facing: a huge amount of photovoltaic systems has been already deployed and many more are planned for the near future; many of these systems will have to be revamped, replaced or extended with new modules, and the old ones will need to be recycled or landfilled. In Chapter 1, the scenarios proposed by the International Energy Agency are presented and the implications for photovoltaic capacity growth will be analysed in detail. In Chap. 2, the main components of photovoltaic systems are presented, ranging from cells to modules and then to whole systems; this chapter describes each component, its principles of work and the equations governing its main output (but not going into details of the physics behind photogeneration and charge transport in semiconducting materials); the objective of this chapter is to define the main parameters used to evaluate PV cells, modules and system performance and to classify the "product" parts (a classification which is used for the Life Cycle Assessment (LCA) study of the different technologies). The "product" from the LCA perspective is the final PV system, which includes different steps: cells, modules and whole system (with Balance of System (BoS), components). In Chap. 3, the Life Cycle Assessment methodology is presented, with a special focus on its application to energy systems in general and photovoltaic systems in particular and also the inclusion of social and economic considerations for a broader LCA approach (methodologies still under discussion in the scientific community).

Part II. Life cycle assessment of solar electricity. The Life Cycle Assessment (LCA) of the photovoltaic systems (the *product*) and the electricity produced by them (the *service*) requires a very clear statement of the scope and the functional unit (FU) used for the LCA study. The main part of the book is devoted to the two stages of the whole life cycle of a PV system: first, the PV system manufacture phase (from cradle to gate), starting with raw materials production and ending with the PV module delivery (at the gate of the factory); then, the second stage focuses on the use phase and the end-of-life phase (including recycling and landfilling) and requires additional tools to calculate the electricity produced during the operational phase. Part II starts with a detailed description of the manufacturing process of all PV technologies, either commercial or emerging (Chap. 4), and the requirements of raw materials (Chap. 5); the energy balance of the PV system life cycle (Chap. 6) will be presented and, together, they comprise a life cycle inventory of the PV technologies. Beyond the standard LCA approach, an analysis of the energy payback time (EPBT) has been included; it is a parameter broadly used to assess the sustainability of electricity production but which is strongly dependant on the operational phase of the PV system life, including the geographical location where it is operated, and some authors consider that it is not a reliable parameter. The impact assessment in several LCA categories of the whole inventory (materials and energy) will be presented in Chap. 7 with a special focus on commercial technologies and a section devoted to emerging technologies. The focus will be shifted to end-of-life and recycling issues in Chap. 8 and the final chapter of Part II is devoted to Balance of System components with a more detailed analysis of the use of batteries for energy storage.

Part III. Beyond Life Cycle Assessment: socioeconomics and geopolitics of solar electricity. Finally, Part III goes beyond the standard approach to LCA and includes economic and social assessment of impacts. Economic evaluation of the economic cost of installed capacity and produced electricity is accomplished in this part. Comments on the geopolitics of photovoltaics provide the closing remarks of the whole book. In Chapter 10, the definition of economic parameters used to evaluate the impact of PV systems is provided. Those comprise the levelized cost of electricity (also with the modern definition of IEA, called the "value-adjusted" LCOE). Employment opportunities by sector and by country are analysed, including investigation on socioeconomic networks that range from NGOs or other associations to small, medium or large companies linked to solar electricity. Chapter 11 provides a list of the regulatory framework worldwide, with a presentation of technical standards and regulatory policies, including a comparison between countries and a comment about its evolution. The book ends with Chap. 12 in which solar electricity will be put into the context of globalization, when on the one hand still a large amount of population lacks access to electricity while on the other hand solar electricity is now subject of speculation by investment funds and big multinationals. Climate change mitigation and the related international agreements are the closing subjects of the book.

1.1 Evolution of Installed Photovoltaic Capacity

At the end of 2020, the cumulative installed photovoltaic capacity in the world reached 760.4 GW_{DC}, steadily approaching the landmark of 1 TW that could be reached in two years if annual installed capacity follows the growing trends of the past few years (see Fig. 1.1, reproduced from [8]). Despite the COVID-19 pandemic,

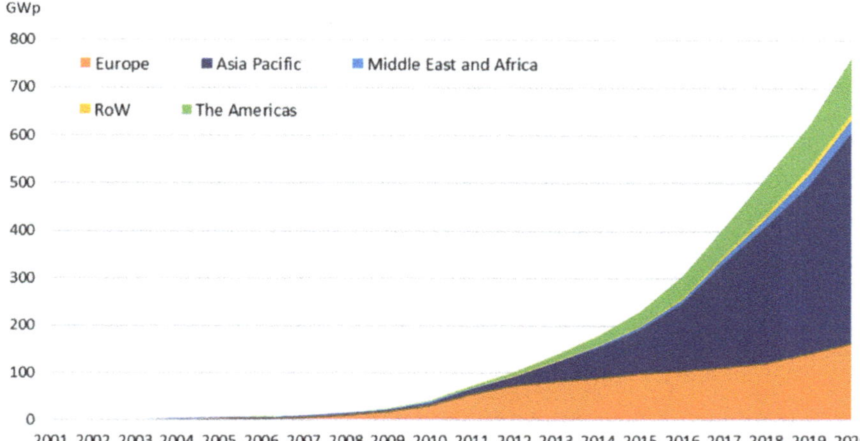

Fig. 1.1 Evolution of cumulative installed capacity (GW_p). *Source* IEA-PVPS (Reproduced with permission from [8])

the annual installed capacity in 2020 was 139.4 GW_{DC}, with at least 20 countries installing more than 1 GW, indicating a sustained annual capacity installation of more than 100 GW/year since 2017 that seems to be accelerating (see Fig. 1.2).

China alone represented 253.4 GW on cumulative installed capacity followed by the European Union (as EU27, 151.3 GW), the USA (93.2 GW), Japan (71.4 GW) and India (47.4 GW). Considering that China installed a third of global new capacity in 2020 and that Vietnam and Korea have seen their highest growth in one year, the trend is clear: Asia is going to be the leading photovoltaic region in the next decade, with Australia also becoming an important actor and reaching the first position in the ranking of PV installed per capita (749 W/capita), surpassing Germany which had been the leader in per capita PV capacity until 2019. The Asia-Pacific region installed 61% of new global PV capacity in 2020. The European Union have been leader for many years, but it seems that the trend is slowing down, with only a few European countries keeping a strong growth (Germany still clearly at the head of installed cumulative capacity with 53.9 GW, followed by Italy and the United Kingdom at some distance). In America, the new USA administration announced a strong investment in new renewable energy infrastructure that could reinforce its already strong position in the photovoltaic market; two countries in Latin America installed more than 1GW (Mexico and Brazil), but others presented a contraction in annual installations (Argentina) or very limited growth (Perú, Chile). Africa and the Middle East, with a large potential for PV (due to its very high annual irradiation), showed a limited growth with new installed capacity in 2020 of only 3% of world total. Still both annually installed and cumulative capacity are mostly concentrated in a few countries, with the rest of the world (ROW in Table 1.1) contributing only 6.8% and 0.3%, respectively. Details of world data for PV annual and cumulative capacity and energy generation can be found in the regular

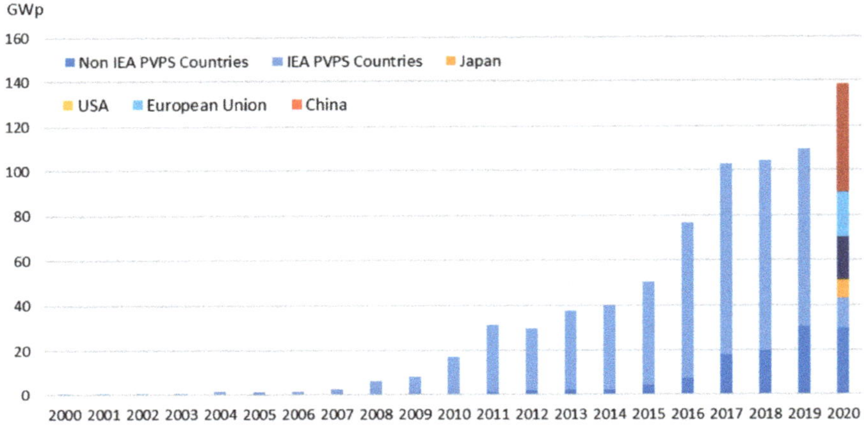

Fig. 1.2 Evolution of annual installed capacity (GW_p) (Reproduced with permission from [10])

reports from the International Energy Agency (IEA) "World Energy Outlook", the IEA-International Renewable Energy (IRENA) "Capacity Statistics and Highlights", the IEA-Photovoltaic Power Systems Programme (PVPS) "Trends in Photovoltaic Applications", the IEA-PVPS "Snapshot of Global PV Markets" and the reports from the World Bank initiatives for off-grid electrification programmes "Energy Sector Management Assistance Programme (ESMAP)" and "Lighting Global".

There are two main categories of photovoltaic system size classification: roof-top or utility scales. Until 2014, the roof-top scale was predominant with more than 50% of annual installed capacity, which kept the cumulative capacity also above 50% for this kind of system; since 2015, the annual installations have been clearly dominated by utility scale (grid-connected PV plants at MW scale), although roof-top systems continued to grow and this application sector has seen an unexpected increase in 2020 due to the very large programme for roof-top systems in Vietnam (and a continuation in Germany and United States were it was already strong): in 2020, around 55GW of new PV systems were roof-top; regarding off-grid systems, further 180 million of roof-top solar home systems have been installed to date providing electricity to 420 million people, and 47 million people are connected to 19,000 photovoltaic powered minigrids in the world (mainly in Asia, with 85% of minigrids, while the future planning is centred in Africa) [3, 4]. Nevertheless, the trend seems to point to a future domination of medium to large size plants. On the other hand, the two broad categories need to be extended to incorporate variations: building integrated photovoltaics (BIPV) complementing the first group of "building attached" (BAPV) roof-top systems (small to medium power systems), or floating systems, agrivoltaics or other utility scale but with very flexible plant design adapted to multiple functionalities of medium to large size plants. Other small groups of applications are still not significant in terms of capacity, but represent targeted markets that could grow significantly in the future: vehicle integrated systems, indoor systems adapted

Table 1.1 Annual installed and cumulative photovoltaic capacity in 2020, with data from IEA-PVPS "Snapshot of Global PV Markets 2021" [10]; the European Union grouped 27 countries in 2020; power is expressed in GW_{DC} and when data are available in GW_{AC} they have been converted for better cross country comparison of data

Annual installed capacity				Cumulative capacity			
		GW_{DC}	%			GW_{DC}	%
1	China	48.2	34.6	1	China	253.4	33.3
(2)	European Union	19.6	14.1	(2)	European Union	151.3	19.9
2	United States	19.2	13.8	2	United States	93.2	12.3
3	Vietnam	11.1	8.0	3	Japan	71.4	9.4
4	Japan	8.2	5.9	4	Germany	53.9	7.1
5	Germany	4.9	3.5	5	India	47.4	6.2
6	India	4.4	3.2	6	Italy	21.7	2.9
7	Australia	4.1	2.9	7	Australia	20.2	2.7
8	Korea	4.1	2.9	8	Vietnam	16.4	2.2
9	Brazil	3.1	2.2	9	Korea	15.9	2.1
10	Netherlands	3.0	2.2	10	UK	13.5	1.8
	ROW	9.5	6.8		ROW	2.1	0.3
	Total:	139.4			Total:	760.4	

to indoor light, portable flexible and low weight systems, cladding systems integrated in paths or roads and a broad range of new system designs in an already old application class dedicated to supply power to signals, lighting or electronic devices. Off-grid systems, mainly for rural electrification in developing countries, represented an important market at the beginning of PV system deployment (80s and 90s), and now its share market is strongly reduced, although in terms of installed capacity, it is still a significant application and have a very large impact in human development in rural livelihoods without previous access to electricity; in 2030, the off-grid PV systems should be extended to provide electricity to 1.2 billion people [3]. The evolution of the broad classes of PV applications can be seen in Fig. 1.3.

The massive deployment of PV capacity is already producing electricity from a renewable source at a lower price than grid electricity in some countries at some time intervals. The produced solar photovoltaic electricity has been growing steadily at a similar pace of installed capacity, in Fig. 1.4, and overview of the aggregated data for different world regions is presented, the data can be downloaded from the IRENA website, and it is updated regularly. Asia is now the leading country in solar electricity production followed by Europe which was surpassed in 2016, North America comes in third position and the rest of the regions are clearly lagging behind, but they are expected to grow significantly in the coming years due to strong cost reductions of PV systems.

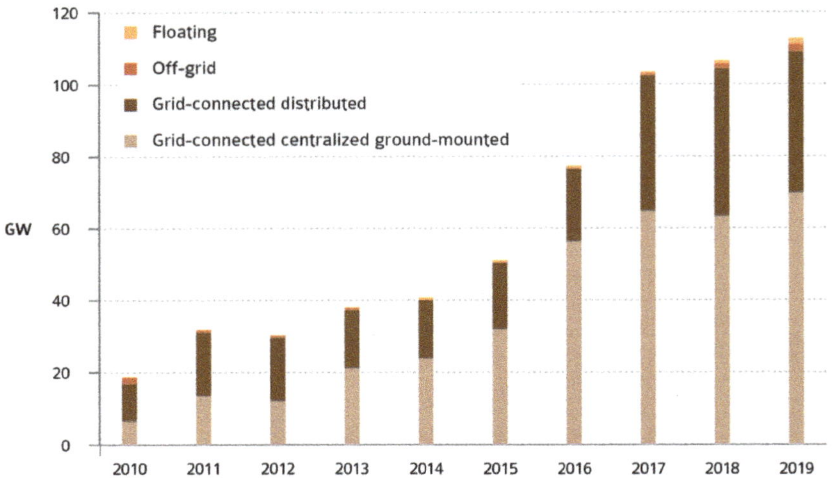

Fig. 1.3 Annual share of centralized, distributed, off-grid and floating installations (GW). *Source* IEA-PVPS Trends in PV Applications 2020 (Reproduced with permission from [11])

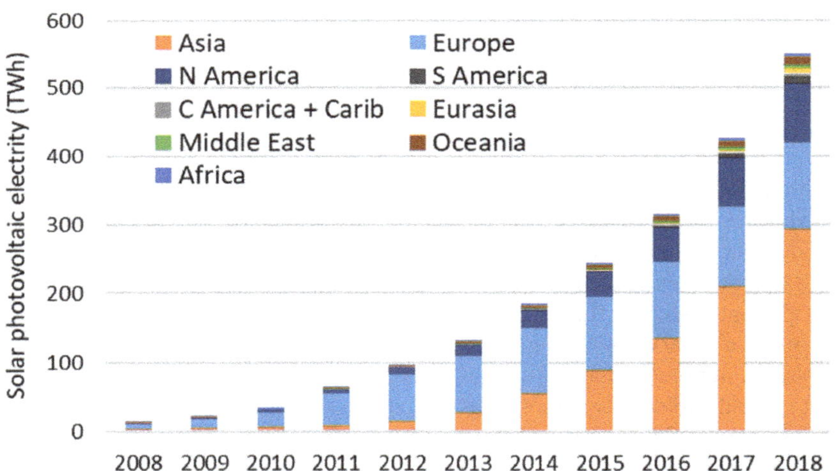

Fig. 1.4 Solar photovoltaic electricity production (TWh) per region during the past eleven years (with most recent real production data from IRENA Renewable Energy Statistics website (last update April 5, 2021, www.irena.org/Statistics/Download-Data)

The availability of cheap electricity from photovoltaics will also contribute to enhance the penetration of other technologies that are energy consumers required in an energy transition aimed at a 100% green electricity. These sectors are hydrogen production and electric vehicles. The developments of PV technologies are acting as a strong driver for the development of other technologies linked to the energy sector and it has created a synergy between the need for efficient storage of electricity produced from photovoltaics at time intervals where demand is lower than supply (similarly for other intermittent sources like wind) and the need for higher electrification of the transport sector. The use of hydrogen as a fuel "vector" and the charging of batteries in electrical vehicles require electricity produced from renewable sources. The link between this renewable intermittent electricity production and the transport sector is pushing the development of technologies for efficient charge storage and green fuel production. This link is still not clear and a strong effort in research and development is currently being carried out.

1.2 Photovoltaics in the Scenarios of the International Energy Agency

The International Energy Agency scenarios are the basis for projections shown in the World Energy Outlook reports, and they are linked to socioeconomic scenarios set up by the United Nations and in particular, the Sustainable Development Goals now used by most countries to set up their own sustainable objectives and to contribute to international cooperation policies [5].

The **Stated Policies Scenario (STEPS)** is a baseline scenario that is built upon the policies announced by each country; the targets related to new renewable energy capacity installations or emission reductions are backed up by detailed technical and economical measures needed for their realization. In particular, the Nationally Determined Contributions (NDC) for emissions reductions that the countries are announcing as part of their commitment with the Paris Agreement are considered in the STEPS scenario only if they are backed by a clear plan of implementation. In contrast, many policies that have been announced with net zero pledges already reaching 70% of global GDP and CO_2 emissions, but still with high level of uncertainty or no technical backing in its energetic policies, are not considered; in general, those lousy undefined pledges are not considered in the STEPS scenario. On the other hand, the STEPS scenario already includes the impact of COVID-19 pandemic in the economic activity of 2020 but considers that the pandemic is brought under control and the economy will recover its pre-crisis levels before the end of 2021. Prior to the crisis, energy demand was projected to grow by 12% between 2019 and 2030, and growth over this period is now estimated at 9% in the STEPS scenario. Additionally, economic policies have already been modified by recovery policies and stimulus packages including additional investments in the energy transition infrastructure towards a low-carbon energy sector. Nevertheless, commitments declared so far, even if successfully fulfilled, will keep global annual emissions in the range of 34–36 Gt CO_2eq between 2020 and 2030, followed by a reduction that would still leave around 22 billion tonnes of CO_2 emissions worldwide in 2050; the continuation of that trend is consistent with a temperature rise in 2100 of around 2.7 °C (with a probability of 50%), well beyond the limits set in the Paris agreement. Furthermore, the

United Nations Framework Convention on Climate Change (UNFCCC) was even more pessimistic and considered that the initial nationally declared commitments (NDC) for greenhouse gases (GHG) emission reductions of 119 countries could lead to a temperature increase in the range of 2.7–3.7 °C, indicating that much greater emission reduction efforts than those associated with the NDCs will be required in the period after 2025 and 2030 to hold the temperature rise below 2 °C above pre-industrial levels [12]. The updates of NDCs by 75 parties (representing about 30% of global GHG emissions) were recently assessed by the UNFCCC, but still are not on track to meet the Paris Agreement; the reality is that far from a reduction, the figures contained in the NDCs will lead in 2025 to GHG emissions around 14.04 Gt CO_{2eq}, that is, 2.0% higher than the 1990 level (13.77 Gt CO_{2eq}), 2.2% higher than the 2010 level (13.74 Gt CO_{2eq}) and 0.5% higher than the 2017 level (13.97 Gt CO_{2eq}). Nevertheless, the long-term mitigation measures announced by many countries for 2050 (still without detailed roadmaps for its fulfilment) are ambitious, and the UNFCCC considers that if implemented, the per-capita emissions by 2050 could be reduced by 87–93% compared to 2017 levels and this is consistent with the objective of a temperature rise in the range of 1.5–2 °C with low overshoot scenarios (with the IPCC models for scenario SR1.5) [13]. In the STEPS, renewables meet 80% of the growth in global electricity demand to 2030, hydropower remains the largest renewable source of electricity, but solar is the main driver of growth as it sets new records for deployment each year after 2022, almost tripling from today's levels and followed by onshore and offshore wind. The modelled change in global energy generation from 2019 to 2040 is expected to be 4813 TWh for photovoltaics in the STEPS scenario, a change in twenty years that is seven times larger than the change occurred in the previous twenty years (664 TWh from 2000 to 2019). This deployment of PV capacity will require a fast development of smart, digital and flexible electricity networks and the requirement of new transmission and distribution lines is 80% larger for the next decade compared to the extension paths seen during the past ten years. Data about population growth in the STEPS is taken from the United Nations and considered that the total population rises from 7.7 billion in 2019 to 10.4 billion in 2070, an average growth of 0.6% per year, with almost three quarters of global increase up to 2070 occurring in Africa, and India accounting for a 10% share in the growth and becoming the most populous country in 2024.

The **Delayed Recovery Scenario (DRS)** is designed with the same policy assumptions as in the STEPS, but considering that a prolonged pandemic causes lasting damage to economic prospects. The global economy returns to its pre-crisis size only in 2023, and the pandemic ushers in a decade with the lowest rate of energy demand growth since the 1930s. Prior to the crisis, energy demand was projected to grow by 12% between 2019 and 2030. Growth over this period is now 9% in the STEPS, and only 4% in the DRS with the consequent slowdown of the economic activity in all end-user sectors and, therefore, in energy demand (with important impacts on transport, for example, where the number of cars in the DRS is 50 million lower than in the STEPS).

The **Sustainable Development Scenario (SDS)**, where a surge in clean energy policies and investment puts the energy system on track to achieve sustainable energy objectives in full, including the Paris Agreement, energy access and air quality goals. The assumptions on population growth, GDP and other socioeconomic parameters are the same as in the STEPS. The SDS scenario is based on a stronger technological development of the energy sector that is modelled by using the Energy Technology Perspectives 2020 Model (ETP) of the International Energy Agency, which explores the evolution in energy supply (using an energy conversion model from primary energy, grouped in fossil, nuclear and renewables to final energy such as electricity, heat, gasoline and diesel) and in the three end-user sectors with the highest energy demand and largest greenhouse gas emissions (using models for industry, transport and buildings). The energy conversion step considers 400 technological options, described in terms of detailed technical and economical parameters including learning curves, thus providing a broad range of possible combinations. Interestingly, the model also considers hydrogen-based fuels (synthetic hydrocarbon fuels from hydrogen and CO_2 or ammonia) and direct air capture of CO_2 from the atmosphere, though a cross-cutting technology option; but although these technological options have been demonstrated at small or medium scale, they are still not deployed commercially, and, therefore, some uncertainty is introduced in the model. The modelled change in global energy generation from 2019 to 2040 is expected to be 8135 TWh for photovoltaics in the SDS scenario. Details of the model can be found in the IEA report "Energy Technologies Perspective 2020 Model" (updated in 2021 from its previous 2016 version, [6]).

The new **Net Zero Emissions by 2050 case (NZE2050)** extends the SDS analysis. The NZE2050 scenario is consistent with around a 50% chance of limiting the long-term average global temperature rise to 1.5 °C, as stated in the Paris Agreement. A rising number of countries and companies are targeting net zero emissions, and all stated policies are considered to come into force although there is still not a clear commitment or detailed plans from governments to do so. The NZE2050 includes the first detailed IEA modelling of what would be needed in the next ten years to put global CO_2 emissions on track for net zero by 2050. Reaching net zero globally by 2050 would demand a set of dramatic additional policies and actions over the next ten years, starting already in 2021 with no new oil and gas fields approved for development and no new coal mines or mine extensions; only new coal plants with carbon capture and storage could be approved beyond 2021.

The NZE2050 scenario considers that total energy supply falls by 7% between 2020 and 2030, reaching a total of 550 exajoules (EJ) and remains at around this level until 2050, this reduction achievement occurs by reducing the energy intensity of GDP growth by 2% annually. Renewable sources will supply 80% of total energy supply by 2050, growing from 20% in 2020. Electrification is one of the key drivers towards a de-carbonization of the energy sector with global electricity demand more than doubling from 2020 to 2050.

Bringing about a 40% reduction in emissions by 2030 requires that low-emission sources provide nearly 75% of global electricity generation in 2030 (up from less than 40% in 2019). Again, hydrogen and CO_2 capture are essential for this horizon; 150 million tonnes of hydrogen should be produced with 650 GW installed capacity of electrolyzers by 2030 (rising to 435 million tonnes and 3000 GW, respectively, in

2045); 4 Gt of CO_2 should be captured by 2035 (rising to 7.6 Gt by 2050). Importantly, the NZE2050 model considers that by 2030 all world population will have access to electricity and clean cooking (at an estimated cost of 40 USD billion) and the cost of energy services for households will be affordable and stable even if an increase in energy consumption is produced. The achievement of the NZE2050 scenario will require of strong policy impulse for emission cuts already in 2030 and a constant technological development (most of the reductions beyond 2030 rely on technologies yet to come); only new international standards, regulations and intense cross-border cooperation could guarantee the needed framework for this ambitious objective. A large investment is required in the electricity generation, energy infrastructure for distribution and end-user sectors. In electricity generation, an initial surge from annual investment of about USD 0.5 trillion (average over the past five years) to USD 1.6 trillion in 2030 should be achieved, then annual investment in renewables in the electricity sector should be around USD 1.3 trillion (slightly more than the highest level ever spent on fossil fuel supply which was USD 1.2 trillion in 2014); after this peak in 2030, investment can be reduced to around 30% by 2050. Similarly, investment in energy infrastructure for distribution (electric vehicle charging stations, hydrogen) and carbon capture, transport and storage should increase from USD 290 billion over the past five years to about USD 880 billion in 2030 and for low-carbon technologies in end-user sectors should rise from USD 530 billion in recent years to USD 1.7 trillion in 2030.

The NZE2050 scenario can be considered as an optimistic path for a more sustainable energy generation, and in particular electricity generation as indicated in Table 1.2; therefore, it is an scenario where photovoltaic electricity will play a substantial role with a large increase both in installed capacity and electricity generation in the coming decades. In this scenario, the TeraWatt scale for PV capacity will be surpassed within two or three years, reaching almost 5 TW in 2030 and surpassing 10 TW in 2040. Beyond this point, new installed capacity will coincide with the decommissioning of several GW of previously installed capacity that would have reached its end of life and recycling could become an important industrial activity.

The contribution of renewable electricity generation is key to achieve the ambitious objective of net zero emissions by 2050. The evolution of total CO_2 emissions in Table 1.2 includes carbon dioxide emissions from the combustion of fossil fuels and non-renewable wastes, from industrial and fuel transformation processes (process emissions) as well as CO_2 removals. The energy transition becomes evident in the evolution of the CO_2 intensity (elec.) shown in Table 1.2, that refers to the CO_2 emissions per each kWh of electricity generation; it will achieve a net zero balance before 2040 and become negative afterwards, with the electricity sector acting as a carbon sink for other sectors. Details of the scenario are provided in the International Energy Agency report "Net Zero by 2050—A Roadmap for the Global Energy Sector" [7].

Table 1.2 Electricity capacity (GW) and generation (TWh) (total, renewables and solar PV) and energy-related CO_2 emissions evolution for the NZE2050 scenario of the International Energy Agency. Data from the International Energy Agency report "Net Zero Emissions by 2050. A Roadmap for the Global Energy Sector" [7]

						Share (%)			CAAGR* (%)	
Electricity		2020	2030	2040	2050	2020	2030	2050	2020–2030	2030–2050
Total capacity	GW	7795	14933	26384	33415	100	100	100	6.7	5
Renewables capacity	GW	2994	10293	20732	26568	38	69	80	13	7.5
Solar PV capacity	GW	737	4956	10980	14458	9	33	43	21	10
Total generation	TWh	26778	37316	56553	71164	100	100	100	3.4	3.3
Renewables generation	TWh	7660	22817	47521	62333	29	61	88	12	7.2
Solar PV generation	TWh	821	6970	17031	23469	3	19	33	24	12
Total CO_2	Mt CO_2	33903	21147	6316	0				–4.6	–55.4
CO_2 (electricity + heat)	Mt CO_2	13504	5816	–81	–369				–8.1	n.a.
CO_2 intensity (elec.)	kg CO_2/kWh	0.438	0.138	–0.001	–0.005				–11	n.a.

[a]CAAGR = compound average annual growth rate

1.3 The TeraWatt Scale of Photovoltaic Deployment: Is There Any Limit?

The energy transition that slowly started after the oil crisis in 1973 has gained momentum and it will change the energy landscape in the coming years. The main driver for this change has been shifted from the fear of a supply risk of fossil fuels, sometimes linked to the frequent claim that fossil fuels were achieving their peak production and become more scarce and more expensive every year. This was not the case so far (although some oil fields have indeed reached their peak). But the main driver now is the need to reduce the demand and consumption of fossil fuels, due to the urgent need to reduce CO_2 emissions and mitigate climate change, the biggest challenge for the twenty-first century. The contribution of renewable energies to the electricity mix and the increasing electrification of the global energy production and consumption for all end-user sectors create a synergy path where photovoltaic could become the main electricity supplier and perhaps the main global primary energy supplier.

In all the International Energy Agency scenarios presented in Sect. 1.2, photovoltaic deployment is going to reach the TeraWatt scale in the coming years, with the most optimistic NZE2050 scenario pointing to 2030 to nearly reach the 5 TW milestone. It seems that at the initial stages of the TeraWatt scale, no insurmountable limiting factor has been pointed out in the reports, although some barriers have been identified and policies have been recommended to overcome them, but:

Is there any limit?

Throughout this book, the different potential insurmountable barriers from the point of view of the sustainability of solar electricity are explored. The reader will find a summary of results that aim to answer this question, but also provide methodological tools related to photovoltaic technology and to sustainability assessment that will allow any researcher to perform his or her own calculations in search for a response.

The main factors that could pose a threat to a massive deployment of photovoltaic technology in the TeraWatt scale are grouped and briefly described below. All issues will be analysed in depth in the corresponding chapters.

The risk of materials supply. A huge amount of photovoltaic modules will have to be manufactured in the coming decades. There are many different photovoltaic technologies based on different materials, but today the PV market is relying in more than 95% on one technology (crystalline silicon), the excessive dependence on one single option could be seen as a weakness. Other technologies require in some cases the use of scarce materials (for example, Indium or Tellurium). This possible risk will be assessed in Chap. 5.

The risk of energy balance. Long ago, it was clearly established that the balance between the energy embedded in a PV module (materials processing and module manufacture) and the energy delivered by the PV module throughout the lifetime of any PV technology is overwhelmingly positive. In a few years of operation (depending on the technology), the energy is "recovered" and there is a net clean energy supply of decades before the module reaches its lifetime. This will be analysed in Chap. 6.

The risk of environmental and health damage. This is an important issue that has been already addressed by many research groups by a detailed Life Cycle Assessment methodology, that is constantly updated and re-evaluated for the commercial technologies and newly developed for the emerging technologies, some of them including materials with potential toxicity risks (for example, cadmium in already commercial CdTe technology, or lead in emerging perovskite technology, just to mention two examples). These results are presented in detail in several chapters throughout the book (Chaps. 7, 8 and 9).

The risk of high economic cost. The cost of PV modules was an important barrier for the deployment of PV systems and several policies were implemented to overcome this barrier. This is no longer the case, and currently in many countries, solar electricity from photovoltaic systems is cheaper than the electricity purchased from the grid. Furthermore, the International Energy Agency considers that solar photovoltaic electricity will become the cheapest source of electricity by mid twenty-first century. This was achieved thanks to an impressive learning curve that is analysed in Chap. 10.

Geopolitical risks. Energy supply from oil was plagued by political risks, and the best examples were the two oil crisis of the 70s. Apparently, renewable energies in general, and specially photovoltaic energy benefit from the ubiquity of the energy source, but the supply chain for manufacture could face some geopolitical risks (materials supply chain, technological dependence, commercial wars, etc...); they are presented and discussed in Chap. 12.

References

1. Aklin M, Urpelainen J (2018) Renewables. The politics of a global energy transition. The MIT Press. https://mitpress.mit.edu/books/renewables
2. BP (2020) Statistical Review of World Energy 2020 (69th edn). Tech. rep., British Petroleum. https://www.bp.com/content/dam/bp/business-sites/en/global/corporate/pdfs/energy-economics/statistical-review/bp-stats-review-2020-full-report.pdf
3. ESMAP (2019) Mini Grids for Half a Billion People: Market Outlook and Handbook for Decision Makers. Tech. Rep. Technical Report 014/19, Energy Sector Management Assistance Program (ESMAP). World Bank. http://hdl.handle.net/10986/31926
4. GOGLA (2020) Global Off-Grid Solar Market Trends Report 2020. Tech. rep., GOGLA—Lighting Global - World Bank. https://www.lightingglobal.org/wp-content/uploads/2020/03/VIVID%20OCA_2020_Off_Grid_Solar_Market_Trends_Report_Full_High.pdf
5. IEA (2020) Sustainable Recovery. Tech. rep., International Energy Agency—World Energy Outlook Special Report, world Energy Outlook Special Report in collaboration with the International Monetary Fund
6. IEA (2021a) Energy Technologies Perspective 2020 Model. Tech. rep., International Energy Agency, Paris. https://www.iea.org/reports/energy-technology-perspectives-2020
7. IEA (2021b) Net Zero by 2050. A Roadmap for the Global Energy Sector. Tech. rep., International Energy Agency, net Zero by 2050 Interactive iea.li/nzeroadmap Net Zero by 2050 Data iea.li/nzedata
8. IEA (2021) The Role of Critical World Energy Outlook Special Report Minerals in Clean Energy Transitions. IEA—World Energy Outlook special report, International Energy Agency
9. IEA (2021d) World Energy Outlook 2020. Tech. rep., International Energy Agency. https://www.iea.org/reports/world-energy-outlook-2020
10. IEA-PVPS (2021) Snapshot of Global PV Markets 2021. Tech. Rep. Report IEA PVPS T1 3 9 : 2021, International Energy Agency—Photovoltaic Power Systems Programme—Task1, iSBN 978-3-907281-17-8
11. Masson G, Kaizuka I (2020) Trends in Photovoltaic Applications 2020. Tech. Rep. Report IEA-PVPS T1-38:2020, International Energy Agency - Photovoltaic Power Systems Programme—Technology Collaboration Programme, iSBN 978-3-907281-01-7
12. UNFCCC (2016) Aggregate effect of the intended nationally determined contributions: an update. Tech. Rep. FCCC/CP/2016/2, United Nations Framework Convention on Climate Change. https://unfccc.int/resource/docs/2016/cop22/eng/02.pdf
13. UNFCCC (2021) Nationally determined contributions under the Paris Agreement. Synthesis report by the secretariat. Tech. Rep. FCCC/PA/CMA/2021/2. https://unfccc.int/documents/268571

Chapter 2
Photovoltaic Technology

2.1 Introduction to the Physics of Solar Cells: Power Conversion from Sun to Electricity

An energy technology can be considered renewable when the source of the supplied work is naturally available or replenished within a certain time frame. The availability of any renewable source is always variable in time, that is, intermittent with different periodicity depending on the technology. Also, the energy density of the renewable source may be low and disperse when compared with non-renewable sources like fossil fuels or radioactive fuels. Those are common characteristics of any renewable technology: wind, geothermal, hydro, tidal, etc…and specially evident for the case of solar photovoltaic technology. The source of photovoltaic energy is the Sun light; it is intermittent in its daily and seasonal cycling; it is low density but universally available on the Earth's surface; it does not require replenishment since the Sun can be considered a permanent source within the human-scale time frame. Although it is not really permanent, since the evolution of a G-type, small to medium size main sequence star like the Sun indicates that it may be through approximately half of its life, and therefore, it will provide light to the Earth for another 4,500 million years before becoming a red giant whose radius will be probably larger than the Earth's orbit.

Solar photovoltaic energy is the technology which converts the Sun light power available on the Earth's surface into useful electricity. It converts an intermittent, low power density resource into a reliable source of electrical work which can be delivered on demand at the required power density. According to this definition, solar photovoltaic is a renewable energy, although it is not a completely clean technology since, like any other energy technology (being it either renewable or not), it requires some input of energy to manufacture the devices that are able to convert the power from the Sun into useful work at the Earth. It is important to distinguish between a renewable energy technology and a clean, greenhouse gas (GHG) emissions-free,

A. Urbina, *Sustainable Solar Electricity*, Green Energy and Technology,
https://doi.org/10.1007/978-3-030-91771-5_2

technology. Within the lifetime of the devices, solar photovoltaic technology will provide much more energy than the one required to its manufacture: the balance is positive from the point of view of generated versus embedded energy, and it is also positive when the GHG emissions associated with electricity production is compared to any other means to produce the same amount of electricity by non-renewable sources. A quantification of this balance is one main conclusion of this book.

An important characteristic of photovoltaic technology is its modularity, that is, its capability to work as an efficient power converter at all scales. A small solar cell is as efficient as a module, or as a generator, or even as a very large plant; in fact, a small laboratory cell is *more* efficient than the larger devices or systems. A device with 100 cm^2 active area is more efficient than a 100 Ha PV plant. Of course, the small cell will provide a few Watts of power, while a solar plant may reach hundreds of Mega Watts (MW, even nowadays a few Giga Watts, GW), but the power conversion efficiency (PCE) when converting the Sun light power into electrical power is better in the small cell. In this chapter, an introduction of the working principles of the solar cell is presented, followed by the "scaling-up" from cell to module with a focus on its material components. Power management once it is converted from light into electricity requires additional elements of the photovoltaic system which are grouped in the so-called "balance of system" components, including electricity storage means.

2.1.1 A Brief History of the Development of the Solar Cell

Three main stages can be proposed to summarize the development of photovoltaic technology. One early stage characterized by slow experimental progress during nineteenth century since the discovery of the photovoltaic effect by Edmund Becquerel in 1839 and culminating with the discovery of the electron by J. J. Thomson in 1897, followed by a second stage coincident with the quantum revolution, from Planck's proposal of the *Light Quanta* in 1900 to the development of quantum solid-state physics, where theory and experiment progressed steadily with two interruptions caused by the First and Second World Wars. Two technological advances culminate this second stage: the discovery of the transistor in 1947 and the first solar cell with power conversion efficiency higher than 5% in 1954. These two stages are summarized in this subsection. The third one is a stage of technological development in pursuit of higher power conversion efficiencies, when experimental advancements in small laboratory solar cells have been quickly applied to commercial photovoltaic modules during fifty years and accelerated since the early 2000s with the onset of organic and hybrid technologies and the massive deployment of installed power capacity of inorganic technologies. This third stage of technological development during the past seventy years is summarized in the final subsection of this chapter.

The first scientifically reported effect of the light on the electrical transport properties of a material was presented by Edmund Becquerel in 1839 [3, 4]. These reports are considered the discovery of the photovoltaic effect. He observed an electrical current passing through a liquid electrolyte (aqueous alkaline, neutral or acidic) when the Sun light illuminated a silver chloride or silver bromide coated platinum electrode and analysed the chemical reactions triggered by the action of light. It took almost forty years for a new report of a photovoltaic effect, in this case on a solid-state selenium sample; in 1876, Adams and Day were studying the photoconductivity of selenium and they observed an increase in photocurrent when the sample was illuminated, but intriguingly, the current was also produced in the absence of a driving voltage: the current was produced by the action of light and not by an applied voltage [1]. They had invented the first solid-state photovoltaic cell: by using two platinum electrodes in the selenium sample, a metal-semiconductor rectifying Schottky barrier contact had been created, although those concepts were not known at that time. The same structures (a metal pressed on a piece of semiconducting material) was used by several scientists with the aim to develop a device which could work as a reliable, calibrated, light sensor: Charles Fritts, by coating the selenium with gold, created the first working solar cell in 1883 with 1% power conversion efficiency [7] which was pushed up to 2% shortly afterwards by Heinrich Hertz with more focus on the photodetector research that he was carrying out and which ultimately lead to the discovery of the photoelectric effect when ultraviolet light was illuminating a metallic plate and produced the effect of discharging the plate [10]. A decade later, and also by illuminating with ultraviolet light, J. J. Thomson discovered that the "cathodic rays" emitted by the metallic plate could be composed of tiny particles, that he called "corpuscles" and were later named electrons [25]. All experimental ingredients of the photovoltaic and photoelectric effects had been discovered by the end of nineteenth century, but the theoretical explanation and the full understanding of the difference between them was only possible after the full development of the quantum theory, which started in the first year of twentieth century (Fig. 2.1).

The discovery of the electron by J. J. Thomson was followed by the revolutionary proposal of Max Planck in 1900, the equation which describes the blackbody radiation in terms of *Light Quanta* [20]. The equation was successful in explaining experimental data about the radiation emitted by a body at temperature T and which had been elusive so far. Planck's equation, written in terms of the light frequency, is

$$B(\nu, T) = \frac{2h\nu^3}{c^2}\frac{1}{e^{\frac{h\nu}{k_B T}} - 1}. \tag{2.1}$$

Equation 2.1 is the spectral distribution of the radiation emitted by the blackbody, that is, the number of *light quanta* at each frequency interval from ν to $\nu + \delta\nu$. Planck was aware that his empirical equation was correct since he had first-hand information from experimental colleagues. He then tried to deduce the equation from first principles, which he did in a second article where the revolutionary proposal of light quanta was made in order to be able to deduce the equation proposed in his first 1900 paper. The light came in packages of energy, each *light quanta* with an energy proportional to its frequency ν [21]:

$$E = h\nu = \frac{hc}{\lambda}, \tag{2.2}$$

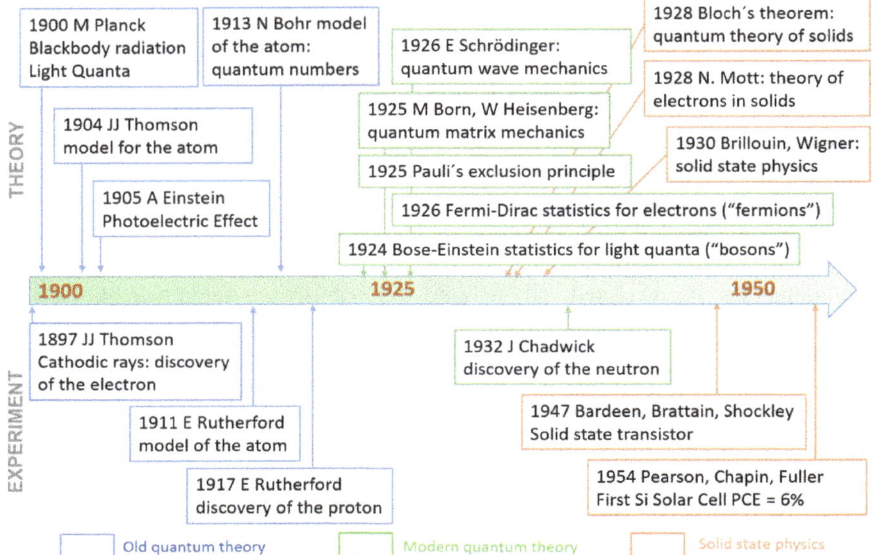

Fig. 2.1 Time frame of the theoretical and experimental developments during the first half of the twentieth century which led from the discovery of the electron and Planck's quantum theory of light to the fabrication of the first solar cell with power conversion efficiency higher than 5%

where h is Planck's constant, $h = 6.62607015 \times 10^{-34}$ Js, the quantum of "action" (energy×time), c is the speed of light, $c = 299792458\ \text{ms}^{-1}$ and λ is its wavelength. Planck's 1900 articles did not have a very strong impact in the first years of the twentieth century. Planck was always trying to keep a connection to classical thermodynamic theory via the concept of entropy and the inclusion of Boltzmann's constant in his equation ($k_B = 1.380649 \times 10^{-23}\ \text{JK}^{-1}$). It was only after Albert Einstein applied the light quanta revolutionary concept to his successful explanation of the photoelectric effect when the old quantum theory started to be broadly accepted [5]. The origin of the old quantum theory is, therefore, linked to photovoltaic technology by two fundamental concepts: first, the blackbody radiation describes the resource which is coming from the Sun, that is, the light and its spectral distribution in terms of the number of photons with given energies at each frequency (or wavelength interval), and second, the *light quanta* and Einstein explanation of the photoelectric effect that describes how ultraviolet light interacts with matter; it explains how the *light quanta* are absorbed by the material: in packages of well-defined energy, later called **photons** [5].

Nevertheless, the photoelectric effect should not be confounded with the photovoltaic effect. In the photoelectric effect, high-energy photons (blue or ultraviolet) are absorbed by a metallic material and electrons are expelled from the material (in air or preferably in a vacuum chamber); its main application are in photomultiplier detectors or photoelectron spectroscopy (UPS, XPS). In the photovoltaic effect, the electrons are not expelled from the material, the photons are absorbed and excite the electrons to higher levels of energy inside the material, and if these electrons can be effectively used to generate a current through an external load, they can supply work to this load; in this sense, the solar cell, driven by the photovoltaic effect, is acting as a current source where the amount of current delivered to the load is controlled by the light arriving at the cell.

The explanation of this process had to wait for the development of the modern quantum theory. At the time of Einstein's 1905 article and the confirmation of the corpuscular nature of both the cathodic rays (electrons) and light (photons), atomic models were being developed and proposed by J. J. Thomson (1904), E. Rutherford (1911) and N. Bohr (1913) in rapid succession, but it was not until the development of modern quantum theory a decade later that the deep understanding of the atom and, therefore, light–matter interaction was possible. First in 1925 with the matrix mechanics (W. Heisenberg, M. Born and P. Jordan) then in 1926 with the wave equation (E. Schrödinger) and finally in 1927 with the relativistic quantum equation of the electron, the discovery of spin and the first proposal for an anti-particle, the positron, was made by Paul Dirac.

For the understanding of the behaviour of electrons and photons with the aim to explain the photovoltaic effect, the equations of modern quantum theory need to be complemented with the statistical description of both kinds of particles. This task was accomplished first by S. N. Bose and A. Einstein for particles with integer spin, called "bosons"; they proposed an equation to describe how these particles occupy states of a given energy. The bosons can condensate in the same energy state, and so do photons (with zero spin) which behave like bosons:

$$f_\gamma(\hbar\omega, T) = \frac{1}{e^{\frac{\hbar\omega-\mu_\gamma}{k_B T}} - 1}, \tag{2.3}$$

where $\hbar\omega$ in the energy of the photon with $\hbar = h/2\pi$ and $\omega = 2\pi\nu$ its angular frequency. The chemical potential of light is μ_γ, which is the average thermodynamical energy of the set of photons at a given absolute temperature T; the link with classical thermodynamics is provided by the energetic term $k_B T$ where k_B is Boltzmann's constant and T the absolute temperature (in Kelvin). This equation, when applied to a body at absolute temperature T which emits electromagnetic radiation (photons), recovers Planck's blackbody radiation, Eq. 2.1.

If the particles have half odd integer spin (s = 1/2, 3/2, etc…), they obey Pauli's exclusion principle and are called "fermions". This principle, proposed by W. Pauli in 1925, indicates that two or more identical fermions cannot occupy the same quantum state (of a given energy). Fermions obey the Fermi–Dirac statistics and electrons, with spin $s = 1/2$, behave like fermions:

$$f_e(E_e, T) = \frac{1}{e^{\frac{E_e-E_F}{k_B T}} + 1}, \tag{2.4}$$

where E_e is the energy of the electron, and E_F is the Fermi energy that indicates the energy level below which all states are fully occupied at T = 0. If T > 0, a small amount of electrons is excited across this Fermi energy and occupy states with $E > E_F$. In intrinsic semiconductors, with well-defined conduction and valence bands, the Fermi level is given by

$$E_F = \frac{E_c - E_v}{2}, \tag{2.5}$$

where E_c is the minimum energy level within the conduction band and E_v is the maximum energy level within the valence band. Both the Bose–Einstein and the Fermi–Dirac statistics recover at high temperatures (and low concentrations of particles) the classical thermodynamic Maxwell–Boltzmann distribution function.

With those statistical ingredients, the development of solid-state physics progressed rapidly. Bloch's theorem (1928) enabled the possibility of solving Schrödinger's equation in crystalline solids and obtaining the wavefunction and eigenenergies of electrons within a solid. When solved for a large number of atoms, the atomic orbitals are very closely spaced in energy (around 10^{22} available states per eV[1]), thus creating some ranges of quasicontinuum energy called "bands"; these bands are separated by ranges of forbidden energy, commonly known as the "energy gap", E_g, for which there is no solution of the wave equation, i.e. there is no wavefunction at this energy, and therefore, there is no available state to accommodate any electron. The combination of the Bloch theorem and the progress in experimental solid-state physics enabled a very rapid progress in the understanding of the behaviour of electrons within solids, with the works of Eugene Wigner and León Brillouin on the atomic structure of materials and Arnold Sommerfeld which developed the first models of electrons in solids (Drude–Sommerfeld model, 1927) and later by Nevill Mott who proposed a full quantum theory for electrons within solids, including metal–insulator transitions and electrons in disordered semiconductors [15, 16].

A detailed description of band calculations and quantum electronic transport in solids is out of the scope of this book and can be found in very good solid-state physics books, like the classical Ashcroft and Mermin book [2] and with more focus on photovoltaic technology, in the excellent books by Jenny Nelson and Peter and Uli Würfel [18, 27]. Nevertheless, the concepts of Fermi energy and energy gap are at the core of semiconducting physics, and an understanding of the underlying physics of photogeneration requires at least a grasp of its physical meaning which is presented in the following subsections.

The final steps of the second stage of the evolution of the solar cell are provided by two inventions. The first one is the fabrication of the first solid-state transistor by John Bardeen, William Shockley and Walter Brattain at Bell Labs in 1947 on a piece of germanium with metallic gold contacts; this experimental device opened the door to solid-state electronics based on semiconducting materials which was rapidly

[1] The electron-volt, eV, is a very convenient energy unit in solid-state physics, it is defined as the energy that an electron acquires when accelerated in an electric field of 1V and, by definition, is equal to 1.602×10^{-19} J.

developed with the fabrication of diodes, transistors and ultimately the first silicon solar cell fabricated also at Bell Labs by Pearson, Chapin and Fuller in 1954 with power conversion efficiency of 6% which demonstrated the possibility of using them for power generation by converting Sun light into electricity. The key to this impressive performance was the ability of Fuller, a chemist, to efficiently dope the silicon semiconductor and create a controlled p/n junction. The two principal ingredients of a solid-state solar cell had been developed and combined: a semiconducting material with an energy gap and an asymmetry in doping which creates an internal electric field to drive the photogenerated electrons into the external metallic electrodes.

2.1.2 Solar Radiation

The source of photovoltaic energy is the light arriving from the Sun. The total solar radiation includes photons and also several subatomic particles, such as electrons, protons, alpha particles and neutrinos, and some atomic nuclei such as carbon and nitrogen and others, comprising the solar wind plasma. Most of the solar wind particles are deflected by the magnetosphere, which protects the Earth's surface from the solar wind. When considering solar radiation with the purpose of evaluating the resource of solar energy for electricity production, only the photons are accounted for. The blackbody radiation model proposed by M. Planck (Eq. 2.1) provides a very good fit to the spectral distribution of the photons arriving at the Earth from the Sun, which is acting like a black body at temperature $T = 5960$ K. Some of the photons arriving at the outer part of the atmosphere are scattered by atoms, and others are absorbed (for example, by water in the clouds, producing dips in the wavelength range of 900, 1000, 1400 and 1900 nm or by carbon dioxide, producing dips in the wavelength range of 1800 and 2600 nm); finally, part of the radiation arriving at the surface is reflected. The spectral irradiance of the Sun's light is the power density (in units Watts per square meter) and within wavelength λ and $\lambda + \delta\lambda$ which arrives at the Earth's surface; it is shown in Fig. 2.2. For an average distance between Sun and Earth of 1AU,[2] the power density integrating all wavelengths is 1367 W/m^2, which is called the *solar constant*. Depending on the atmosphere thickness that the solar light has to cross before arriving at the surface, the spectral irradiance is slightly different. Air Mass (AM) zero is defined for the outer part of the atmosphere, while for any point on Earth's surface, Air Mass is defined as the ratio between the optical path length to the Sun and the optical path length if the Sun were in the zenith, which is the inverse of the cosine of the angular height of the Sun on the horizon as seen from this point of Earth's surface. For example, AM1.5 corresponds to the Sun elevated at an angle of 42°.

The total (also called "global") solar radiation includes direct (or beam), diffuse and albedo (or reflected) components. Several models for its calculation and empirical measurements have been presented in the past decades and important databases

---

[2] Astronomical Unit (AU) is 149, 597, 870, 700 m, i. e. roughly 150 million kilometers.

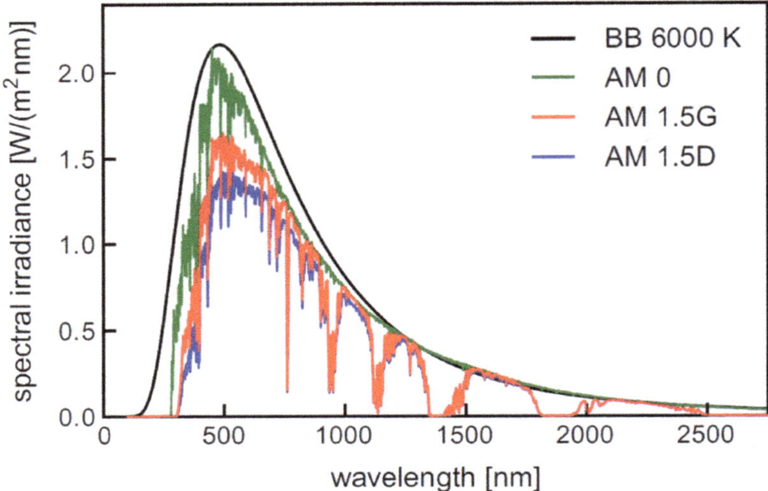

Fig. 2.2 Spectral irradiance outside the Earth's atmosphere (AM 0), on the Earth's surface for direct sunlight (AM 1.5D) and the direct sunlight together with the scattered contribution from atmosphere integrated over a hemisphere (AM 1.5 G) (according to ASTM G173-03 and in comparison to the spectrum used by Shockley and Queisser of a blackbody with a surface temperature of 6000 K (BB 6000 K). Reproduced with permission from reference [23]

have been constructed and are available. Diffuse radiation is calculated by using isotropic and anisotropic models, where one circumsolar anisotropy component is considered, or an additional horizon-dependant second anisotropy is also included [9, 17, 19]. Albedo contributions are strongly dependant on geographical location and surrounding topography or structures; therefore, the best estimations are provided by empirical databases, like the Copernicus Global Land Service[3] of the European Union, which is based on satellite observations. The most important solar radiation database is PVGIS,[4] the Photovoltaic Geographical Information System of the Joint Research Centre of the European Commision, which provides free and open access to its irradiation and meteorological database including the following, among other data:

- Solar radiation and temperature, as monthly averages or daily profiles (database and maps).
- Typical Meteorological Year data for nine climatic variables.
- Full-time series of hourly values of solar irradiance.

Other databases, such as those of the National Renewable Energy Laboratory (NREL) and the National Aeronautics and Space Administration (NASA) (in the United States of America) or other national meteorological organizations are also available. With

[3] Copernicus Global Land Service, https://land.copernicus.eu/global/products/sa.

[4] PVGIS-JRC(EU), https://ec.europa.eu/jrc/en/pvgis.

these online tools, either solar radiation data or electricity production data by using different PV technologies and system configurations are easily available and a very accurate calculation of the potential of solar electricity production at any geographical location is within the reach of anyone with Internet access.

2.1.3 *Metals and Semiconductors*

To classify materials according to their electrical properties, the best property to choose as the main criteria for the classification is resistance. First, resistance in metals is low, while in semiconductors, it is very high (and in insulators much higher), and secondly, resistance in metals increases when the material is heated, while in semiconductors, it is reduced when the material is heated. Therefore, this criteria is useful and easy to measure. Resistance is a parameter that is defined by Ohm's law: a current passing through a conductor between two points is proportional to the applied voltage across those two points

$$I = \frac{V}{R}, \tag{2.6}$$

where R is the resistance and it is measured in Ohms (Ω). Since the resistance of a piece of material depends on the shape and size of this material, it is better to define the resistivity:

$$R = \rho\frac{L}{A}, \tag{2.7}$$

where ρ is the resistivity and L, A are, respectively, the length of the conductor and the area of its cross section. The units of resistivity are Ω m. The inverse of the resistivity is the conductivity, σ, with units of Ω^{-1} m^{-1} also called "Siemens per meter" (S m^{-1}). In Fig. 2.3, a summary of resistance values is included for some metals, semiconductors and insulators; note the huge span of values for the materials. For thin films, it is convenient to define a "sheet" resistance, R_s, when the thickness of the sample is small and uniform and the area of cross section can be considered as the product of a width (W) and a thickness (t). Then, the resistance can be rewritten as follows:

$$R = \frac{\rho}{t}\frac{L}{W} = R_s\frac{L}{W}, \tag{2.8}$$

where $R_s = \rho/t$ is the sheet resistance and has units of Ω, but in order to emphasize that it refers to a thin film, it is often indicated as $\Omega/\square$. In photovoltaic technology, since many materials are used in thin films (specially for the emerging organic and hybrid technologies), the sheet resistance is commonly used to characterize the materials used in those layers.

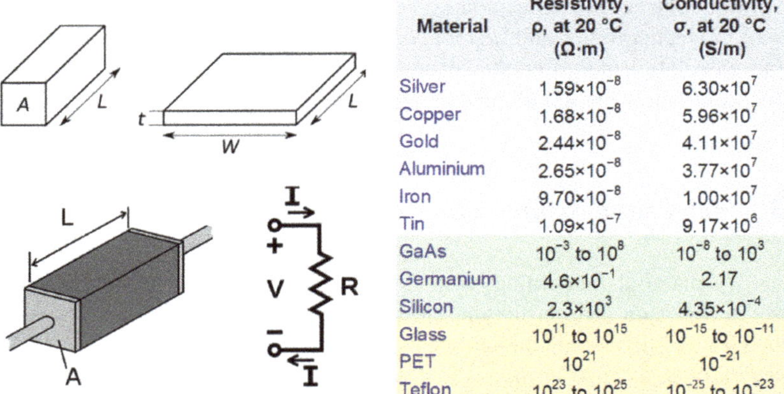

Material	Resistivity, ρ, at 20 °C (Ω·m)	Conductivity, σ, at 20 °C (S/m)
Silver	1.59×10^{-8}	6.30×10^{7}
Copper	1.68×10^{-8}	5.96×10^{7}
Gold	2.44×10^{-8}	4.11×10^{7}
Aluminium	2.65×10^{-8}	3.77×10^{7}
Iron	9.70×10^{-8}	1.00×10^{7}
Tin	1.09×10^{-7}	9.17×10^{6}
GaAs	10^{-3} to 10^{8}	10^{-8} to 10^{3}
Germanium	4.6×10^{-1}	2.17
Silicon	2.3×10^{3}	4.35×10^{-4}
Glass	10^{11} to 10^{15}	10^{-15} to 10^{-11}
PET	10^{21}	10^{-21}
Teflon	10^{23} to 10^{25}	10^{-25} to 10^{-23}

Fig. 2.3 Draft of bulk and thin film materials for whom resistance and sheet resistance are defined in the main text, and the table includes resistivity and conductivity values for representative metals (grey), semiconductors (green) and insulators (yellow)

The resistivities listed in the table of Fig. 2.3 have been measured at $T = 20$ °C and present a very large span of values, with an extremely broad range of more than thirty orders of magnitude that could be enough to classify the materials. But most importantly, the temperature dependence is very different in metals and semiconductors. For metals, the resistivity behaviour with temperature is well described by the model of Bloch and Grünesein, given by

$$\rho(T) = \rho(0) + A\left(\frac{T}{\theta_R}\right)^n \int_0^{\frac{\theta_R}{T}} \frac{x^n}{(e^x - 1)(1 - e^{-x})} \mathrm{d}x, \tag{2.9}$$

where θ_R is the Debye temperature and n depends on the kind of scattering interaction of the electrons within the material. This equation produces a constant growth of resistivity when the material is heated. For intrinsic semiconductors, an empirical model explains the behaviour of most materials:

$$\rho(T) = \rho(0)e^{-\alpha T}, \tag{2.10}$$

where α is an empirical coefficient. The exponential behaviour indicates that intrinsic semiconductors have a very broad range of resistivity, which can be strongly modified by using doping. When the values of resistivity are very high at room temperature ($\rho > 10^{10}\Omega$ m), the material can be considered as an insulator.

The resistivity versus temperature behaviour provides a good empirical classification, but an understanding of the electronic transport mechanisms requires another classification based on the structure of the energy levels of the material. It was only

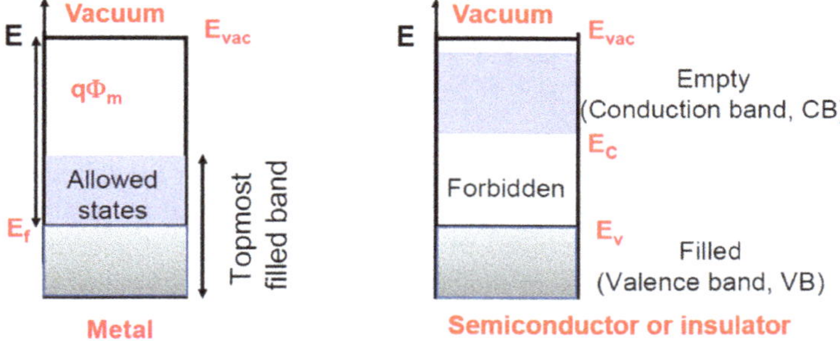

Fig. 2.4 Schematic representation of the energy levels of a metal (left) and a semiconductor (right), with the Fermi energy position within a partially filled band or within an energy gap, respectively

with the onset of the quantum theory of solids that such an explanation was provided. The materials have energy bands with states that can be filled with electrons (following Pauli's exclusion principle and the Fermi–Dirac statistics), and the ultimate electron of a given material that is accommodated in a state within an energy band (ideally at absolute temperature $T = 0$ K) establishes the Fermi energy level of this material (energies are measured with respect to a "vacuum level" which corresponds to the energy of the electron immediately out of the material). In Fig. 2.4, a simplified draft of the energetic structure of bands is presented. In a metal, the Fermi energy lies within a band, and thus, at absolute temperature $T = 0$ K, all levels below the Fermi energy are full and levels above it are empty but there is no energy gap between the filled and the empty states; therefore, statistically speaking, the Fermi energy lies within a band of allowed states which are partially filled. In a semiconductor, at absolute temperature $T = 0$ K, one band is completely filled and the next one is completely empty, both bands being separated by an **energy gap**, with energy levels in which there is no available state to accommodate electrons (there is no solution of the wavefunction at the energies within the gap); statistically speaking, in a semiconductor, the Fermi energy lies within the energy gap, the band below the Fermi energy is the **valence band** and the band above the Fermi energy is the **conduction band**. Only the energy levels at the top and bottom of those bands are useful for calculations and for measurements and are labelled, respectively, E_V and E_C in Fig. 2.4.

With the help of Fig. 2.4, some definitions can be made which will be useful to characterize the materials within the different parts of a solar cell:

- $E_{vac} - E_F$ is the **Work Function**, $q\Phi_m$ where q is the electron charge.
- $E_{vac} - E_C$ is the **Electron Affinity**, χ and does not depend on E_F.
- $E_{vac} - E_V$ is the **Ionization Energy** (first, second, third, …*binding* energy.)

- $E_C - E_V$ is the **Energy Gap**, E_G. The difference between a semiconductor and an insulator is the size of the energy gap. Roughly speaking, if $E_G < 4eV$, the material is a semiconductor, and if $E_G > 4eV$, it is an insulator, the frontier between them being a diffuse one.

The selection of metallic or semiconducting materials for the fabrication of the different parts of a solar cell will depend upon the relative values of all these magnitudes and how they are combined to optimize the process of generating excited carriers within the active layer of the cell and extracting them out of the cell. The energy band structure and the resistivity of the materials are enough to provide a link between the nanoscale quantum properties of the solid (the energy gap is a purely quantum phenomena) and the macroscopic classical characterization of an operating solar cell whose main parameters are described in the following paragraphs.

2.1.4 Equivalent Circuit and Parameters of the Solar Cell

A solar cell requires two main ingredients, the energy gap of the material which enables the possibility of absorbing photons and excite electrons, and an internal electric field to drive the photogenerated electrons out of the device and deliver an electric current (at some voltage) to an external load. Semiconductors are required to provide the energy gap, and the combination of a metal/semiconductor interface or a semiconductor with two differently doped regions (homojunction) or two different semiconductors (heterojunction) is required to provide the internal electric field. All these ingredients are included in the diode, and if this diode is capable of absorbing photons, it will behave as a solar photovoltaic device when illuminated by light. The most simple electronic circuit to represent this combination is the parallel connection of a current source and a diode as shown in Fig. 2.5 (top).

In this schematic view, the photogenerated current can be driven through an external load (R_L) or "lost" through the diode, which in this case would represent a loss of power which is not available to make work at the load (this loss is mainly due to recombination). The sign of the currents in Eqs. 2.11 and 2.12 representing this equivalent circuit is arbitrary: in conventional electronic circuits, the current is considered positive when it flows through the diode from p-type material to n-type material (from anode to cathode within the diode), but in photovoltaics, the positive sign is applied to the photogenerated current and to the delivered current to the load; then, the current through the diode, also called dark current, is subtracted from the photogenerated current. Using the Shockley equation to describe the diode, the equation for the ideal solar cell is given by

$$J(V) = J_{sc} - J_{dark}(V), \tag{2.11}$$

$$J(V) = J_{sc} - J_0 \left(e^{\frac{qV}{k_B T}} - 1 \right), \tag{2.12}$$

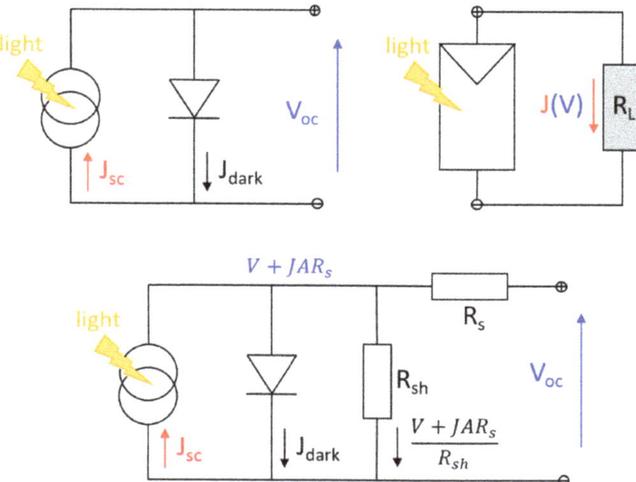

Fig. 2.5 Top: Equivalent circuit of the ideal solar cell, summarized in the right-hand side by the symbol of the solar cell, which is connected to an external load R_L. Bottom: Equivalent circuit of the real solar cell, with parasitic resistances, R_s and R_{sh}

which is the equation of the **ideal solar cell** and current densities ($J(V)$, J_{sc} and J_{dark}) are used with most common units mA/cm^2. The diode is described by the saturation current J_0 and an ideality parameter, β, which in this case of "ideal diode" is equal to one and not included in the equation for the ideal solar cell. J_{sc} is the short circuit current and $J(V)$ is the current delivered to the load and it is also called J-V or I-V characteristic curve of the solar cell. The shape of the J-V characteristic is shown in Fig. 2.6 where the dark current is shown in grey, with initially positive sign in the left-hand side of the figure (standard convention for electronics) but then it is flipped downwards (changing the sign of current) and a shift of the whole curve is applied upwards when the solar cell is illuminated and a photocurrent is created (positive sign). The photocurrent through the solar cell and through an ideal load of $R = 0\Omega$ (voltage across this load would be $V = 0$) is called **short circuit current,** I_{sc}, or J_{sc} when referred to current density. When the circuit is open ($R_L = \infty$), there is no current flowing through the load and the voltage between the terminals of the solar cell is called **open circuit voltage,** V_{oc}. If the solar cell is illuminated, all other intermediate cases with $0 < R_L < \infty$ produce an electromotive force on the load with power density $P(V) = J(V) \times V$.

The open circuit voltage (V_{oc}) does not appear explicitly in Eq. 2.12 because it refers to a single point, when the J-V characteristic curve crosses the voltage axis. It is the case when the current delivered to the load is zero, that is, $J(V) = 0$ and $V = V_{\text{oc}}$. Considering this particular point, an equation for the open circuit voltage is easily obtained:

$$V_{oc} = \frac{k_B T}{q} ln\left(\frac{J_{sc}}{J_0} + 1\right). \tag{2.13}$$

The photocurrent generated by the solar cell at short circuit conditions, J_{sc}, is a quantum phenomenon which depends on the ability of the solar cell to absorb photons and use its energy to promote electrons from the valence to the conduction band to generate charge carriers that are delivered to the external load. J_{sc} is independent of the voltage between the solar cell electrodes. Using the incident spectral photon flux ($b_s(E)$) which is the amount of photons with energy between E and $E + dE$ per unit area and unit time arriving from the Sun, and the **quantum efficiency** ($QE(E)$) of the solar cell, which is the probability that a photon arriving into the solar cell with energy between E and $E + dE$ generates an electron in the active layer that is collected by the negative electrode of the solar cell and delivered to the external load, then the short circuit current can be calculated as follows:

$$J_{sc} = q \int_0^{\infty} b_s(E) QE(E) \mathrm{d}E, \tag{2.14}$$

where q is the electron charge and the integral is calculated for all photon energies. The shape of $QE(E)$ depends on the materials in the solar cell and the cell architecture for photons with energy larger than the energy gap of the semiconducting material, while for photons with energy below the energy gap of the cell is zero. Therefore, the lower limit of the integral in Eq. 2.14 can be replaced by E_G, the energy gap of the material.

A more detailed description of the real solar cell includes two parasitic resistances in the equivalent circuit of the ideal solar cell as shown in Fig. 2.5 bottom. They are the shunt resistance (R_{sh}) and the series resistance (R_s) which accounts for different losses thus reducing the delivered power to the load, mainly recombination losses and transport losses (voltage drop due to resistance of materials and of mismatch of energy levels from the active layer to the transporting layers and the electrodes). If the parasitic resistances are considered, the equation of the ideal solar cell must be modified; the voltage drop across R_s, which is JAR_s, indicates that the voltage in the circuit branch to the left of R_s is higher than the one across the load, then $V + JAR_s$ is the voltage now biasing the diode and must be included in the Shockley equation of the diode instead of just V; the diode is no longer considered ideal and therefore, the ideality factor β must also be included in the equation (typical values for β range from 1 to 2), and finally, there is a current loss through the shunt resistance given by $(V + JAR_s)/R_{sh}$ which reduces the current delivered to the load. With all these modifications, the equation of the real solar cell becomes

$$J(V) = J_{sc} - J_0\left(e^{\frac{q(V+JAR_S)}{\beta k_B T}} - 1\right) - \frac{V + JAR_s}{R_{sh}}, \tag{2.15}$$

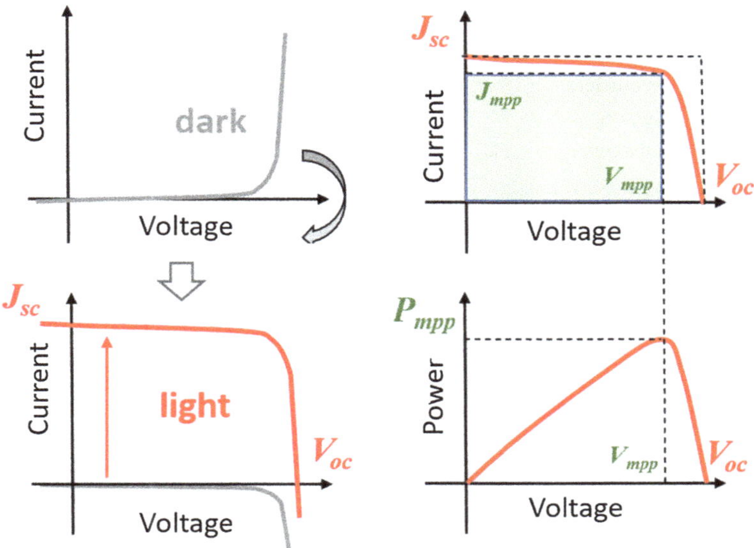

Fig. 2.6 J-V characteristic curve of a solar cell, and P-V curve which defines the *maximum power point*, the area of the green square in the J-V plot, divided by the area defined by the $J_{\mathrm{sc}} \times V_{\mathrm{oc}}$ square illustrates the graphical ratio of areas which provides the value of the filling factor, FF

where A is the active area of the solar cell. The effect of the parasitic resistance is to reduce the "squareness" of the J-V characteristic curve of the solar cell. From Eq. 2.15, it is clear that R_s must be small and R_{sh} must be large to recover the ideal case. Typical values for the parasitic resistances depend on the photovoltaic technology under consideration, but a general rule is that R_s must be lower than a few Ohms, and R_{sh} must be larger than a few hundred thousand Ohms. A good measurement of the quality of the solar cell is the "squareness" of its J-V characteristic curve which can be quantified by the **filling factor, FF**, which is defined as

$$FF = \frac{J_{\mathrm{mpp}} V_{\mathrm{mpp}}}{J_{\mathrm{sc}} V_{\mathrm{oc}}}, \tag{2.16}$$

where J_{mpp} and V_{mpp} are the current density and voltage at which the maximum power is delivered to the load, and define a special point in the J-V characteristic curve called the *maximum power point, mpp*. The current and voltage at this *mpp* point are neither the maximum current (which is J_{sc}) nor the maximum voltage (which is V_{oc}) that can be delivered to the load, but the combination in which $P(V) = J(V)V$, the delivered power, is maximum (P_{mpp}). The relationship between the J-V characteristic and the P-V curve is graphically shown in Fig. 2.6, and the maximum point of the P-V curve defines the special mpp point at which the filling factor is defined.

Table 2.1 Summary of electrical parameters used to characterize solar cells

Parameter	Symbol	Units
Power conversion efficiency	η or PCE	No units, %
Peak or nominal power	P	W_p
Power density at maximum power point	P_{mpp}	mW/cm^2
Short circuit current	I_{sc}	mA
Short circuit current density	J_{sc}	mA/cm^2
Current density at maximum power point	J_{mpp}	mA/cm^2
Open circuit voltage	V_{oc}	V
Voltage at maximum power point	V_{mpp}	V
Filling factor	FF	No units: between 0 and 1 or %
Saturation current	J_0	μA or nA
Diode ideality factor	β	No units, usually between 1 and 2

The **power conversion efficiency**, PCE (or η) of the solar cell is the ratio between the electrical power density delivered by the solar cell operating at the *maximum power point* and the power density arriving from the Sun on the active area of the cell:

$$PCE = \eta = \frac{P_{mpp}}{P_s} = \frac{J_{mpp} V_{mpp}}{P_s} = \frac{J_{sc} V_{oc} FF}{P_s}, \tag{2.17}$$

which is given in %. Since the solar cell efficiency depends on the irradiance and the temperature of the cell, all solar cells from different photovoltaic technologies must be characterized at the same ambient conditions for a fair comparison. The Standard Test Conditions (STC) have been set by the international standard IEC-60904-1 to provide the values of the solar cell parameters for any technology: solar irradiance 1 kW/m^2 with spectrum AM1.5G (defined by the international standard IEC 60904-3), cell temperature T = 25 °C and wind speed lower than 1 m/s. The parameters measured at STC are often called *peak* parameters and indicated with a p subindex in the units: for example, a module delivering 300 W at STC is said to have a *peak* power or *nominal* capacity of 300 W_p.

Table 2.1 summarizes the main parameters which are used by manufacturers to characterize the solar modules. Another group of parameters widely used are the thermal coefficients of the solar modules, which are needed to calculate thermal losses; they are empirical parameters measured in operating conditions different of the STC; for example, the *nominal operating cell temperature*, NOCT, which is the temperature of the cell measured when operating with irradiance 800 W/m^2 and at ambient temperature T = 20 °C. This NOCT parameter is used to calculate

Table 2.2 Empirically determined coefficients used to predict module back surface temperature as a function of irradiance, ambient temperature and wind speed with the SANDIA model [12]

Module type	Mount	a	b
Glass/cell/glass	Open rack	–3.47	–0.0594
Glass/cell/glass	Close roof mount	–2.98	–0.0471
Glass/cell/polymer sheet	Open rack	–3.56	–0.0750
Glass/cell/polymer sheet	Insulated back	–2.81	–0.0455
Polymer/thin film/steel	Open rack	–3.58	–0.113
22× Linear Concentrator	Tracker	–3.23	–0.130

the operating temperature of the cell at any other ambient conditions by using the linear Ross model given in Eq. 2.18, and the thermal coefficients of losses are fitted experimentally [22]

$$T_m = T_a + \frac{NOCT - 20}{800}G = T_a + KG, \tag{2.18}$$

where K is the *Ross coefficient*, it is expressed in units $^\circ$ Cm^2/W and can be defined as $K = (NOCT - 20)/800$ when G is expressed in units W/m^2. The first value reported by Ross was 0.03 °Cm^2/W for crystalline silicon and wind speed lower than 1 m/s^2 (which delivers a NOCT around 47 °C) [22]. For other thin film PV technologies such as a-Si, CIGS, CdTe in different orientations and even in BIPV applications, the obtained Ross coefficient is always around 0.03 °Cm^2/W with small deviations; only organic technologies deliver lower values (around 0.02 °Cm^2/W) but in this case, the value seems to be more dependant on the encapsulation and framing material than the organic photovoltaic cell material [26]. Manufacturers always provide the empirical NOCT for the PV modules as the main thermal parameter.

A more sophisticated thermal model for the solar cell includes the influence of wind and an exponential behaviour was proposed by researchers from Sandia National Laboratory (USA) in reference [12] and it is presented in Eq. 2.19; the two parameters a and b to be used are obtained empirically for different combinations of materials in cell, encapsulants, cover, backsheet and frames; they are found in scientific references, but very rarely reported by the manufacturers of modules; a summary is presented in Table 2.2.

$$T_m = T_a + e^{(a+bW_s)}G. \tag{2.19}$$

Outdoor tests carried out in different climatic regions have lead to more detailed models for NOCT in real operating conditions according to the international standards IEC 61215 and IEC 61646, showing that natural convection can be neglected

Table 2.3 Empirically determined coefficients used to predict cell temperature [13]

	U_0	U_1
c-Si	30.02	6.28
c-Si[a]	26.9	6.2
a-Si	25.73	6.24
CIS	22.64	3.61
CdTe	23.37	5.44

[a]Used in reference [11] and calculated as an average of values reported in [13]

Table 2.4 Summary of thermal parameters used to characterize solar cells

Parameter	Symbol	Units
Nominal operating cell temperature	NOCT	°C
Power (P_{mpp}) thermal coefficient	γ	%/°C (negative)
Current (I_{sc}) thermal coefficient	α_I	%/°C (positive)
Voltage (V_{oc}) thermal coefficient	β_V	%/°C (negative)

for wind speeds above 2 m/s, that the main effect of radiation cooling can be found during night time which is not relevant for the solar energy gain and that the effect of wind gusts and fast temperature changes is low [13]. Therefore, yet another method was proposed to calculate module temperature in different ambient conditions; it is used by the PVGIS model (European Commission Joint Research Centre, [6]):

$$T_m = T_a + \frac{G}{U_0 + U_1 W}, \tag{2.20}$$

where T_a is the ambient temperature and W is the wind speed. The coefficients U_0 and U_1 used in PVGIS have been obtained by fitting experimental data and are summarized in Table 2.3 by providing the average value for each PV technology [13].

Once the module temperature is calculated by using any of those simple models or others which include additional environmental variables such as wind direction and relative humidity, the temperature losses present a linear dependence such as *thermal coefficient* × ΔT where the *thermal coefficient* is given as a relative loss in % (with respect to nominal STC values) per temperature degree (older PV module datasheets used to provide absolute thermal losses, but it is no longer the case). Typical values are around -0.3%/°C for power and voltage losses and +0.05%/°C for current gains when the temperature of the operating module is above 25°C (opposite effect when temperature is below 25 °C) (Table 2.4).

2.2 The Basic Structure of a Solar Cell

The basic structure of a solar cell contains one active layer and two electrodes. In the active layer, fabricated with a semiconducting material, the charge carriers (electrons and holes) are photogenerated. The electrodes, fabricated with metals in most cases, selectively collect the carriers and deliver them to the external load. The inclusion of transporting layers is not compulsory but in certain technologies is required to optimize the carrier transfer from the active layer to the electrodes. The final architecture of the whole solar cell, comprising the active layer, transporting layers for electrons and holes and electrodes at both sides of the cell, must take into account a good matching of the different energy levels in the materials that guarantee the most efficient carrier extraction from the cell to the load. In the following subsections, the properties of these layers are presented, with a simplified explanation of the physical principles of their operating procedures.

2.2.1 Active Layers

The active layer of a solar cell of any technology must be fabricated with a semiconducting material. The Fermi level of this material in thermal equilibrium must be located within the energy gap. The energy gap of the semiconducting material of the solar cell will define the absorption edge for the incoming photons: as mentioned in Sect. 2.1 any photon with energy lower than the energy gap cannot excite electrons from the valence band to the conduction band (the *quantum efficiency* value at energies $E < E_G$ is zero). Any photon with energy larger than the energy gap will have a certain probability to excite an electron from the valence band to the conduction band. This initial excitation will create a bound pair electron-hole, called *exciton* whose lifetime strongly depends on the material (and therefore, the PV technology) under consideration. The minimum energy that the incoming photons need to generate an exciton is equal to the energy gap, any excess energy will promote electrons to states with higher energy than the conduction band edge (they can also be excited from deeper states below the valence band edge), and this excess energy will be thermally lost in a very fast photon–phonon scattering process (of the order of a few femtoseconds). In other words, the excess energy will heat the solar cell. It is still a technological challenge how to extract these *hot carriers* from the active layer before they lose this extra energy. For any given energy gap, there is a maximum power conversion efficiency of light into electric power (detailed balance limit) that can be ideally achieved with the corresponding active layer; this is called the "Schockley–Queisser limit", which for a single junction solar cell has a maximum of 33.16% at T = 298.15 K and AM 1.5 G for the ideal energy gap (1.34 eV, λ = 985 nm) [23, 24]. In Fig. 2.7, a curve representing the maximum power conversion efficiency is shown (with some bumps which arise from the different dips in the AM1.5G solar spectrum due to atmospheric absorption). The maximum achieved

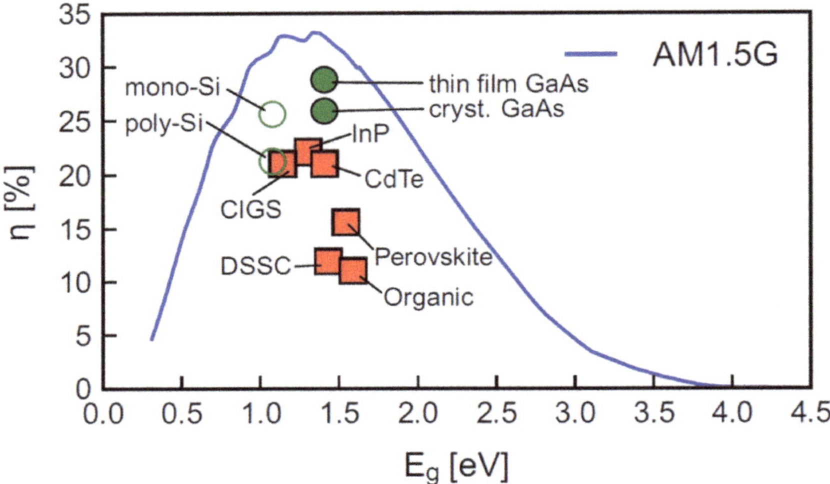

Fig. 2.7 The maximum light to electric power conversion efficiency (detailed balance limit) for a solar cell operated at 298.15 K and illuminated with the AM 1.5 G spectral irradiance (ASTM 173-03, STC) as a function of the band gap energy of the semiconducting active layer of a single junction solar cell. Efficiencies of laboratory cells of different technologies are superimposed differentiating between homojunction (circles) and heterojunction devices (squares) with an indirect band gap (empty symbols) and absorbers with direct optical transitions (full symbols). Reproduced with permission from reference [23]

efficiencies for single junction solar cells with different semiconducting active layers are superimposed in the graph, showing the relative distance with respect to their maximum achievable efficiency.

The absorption of a photon with energy larger than the energy gap can generate an exciton (with certain probability, given by the quantum efficiency as explained above). The lifetime of the photogenerated excitons strongly depends on the material of the active layer and ranges from a few femtoseconds to several microseconds; the shorter lifetimes are measured in inorganics crystalline active layers, and the longer in organic and hybrid active layers. The dissociation of the exciton into separated charge carriers (electrons and holes) requires the overcoming of its binding Coulomb energy and the subsequent separation of electron and hole to avoid geminate recombination (which is produced when the excited electron in the conduction band losses its energy and goes back to the original hole which had been left behind in the valence band) or non-geminate recombination (when the excited electron recombines with another hole whose origin was a different exciton).

The internal electrical field which separates the photogenerated carriers is created within the active layer by selective doping. This doping creates an asymmetry in the energy level distribution within the material and ultimately builds an internal electrical field in a zone of the cell around the boundary between the layers which have been doped differently; this zone is called the "depletion layer". The doping of semicon-

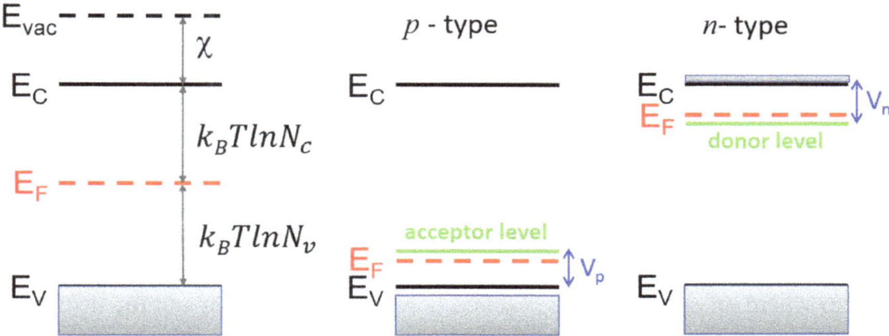

Fig. 2.8 The Fermi level of an intrinsic semiconductor lies in the middle of the band gap; N_c and N_v are the effective density of states in the conduction and valence bands, respectively. When the material is doped, the Fermi level moves towards the valence band (p-type material) and the conduction band (n-type material). If both p-type and n-type materials are in contact, a p-n junction has been created

ductors is also required to increase the conductivity of the intrinsic semiconducting material, which is usually low at room temperature (see the table in Fig. 2.3). The Fermi level of a semiconducting material lays in the energy gap because statistically, when the electrons have filled all available energy states within a band, the ultimate electron occupies the last available state in the valence band and the conduction band is completely empty (at $T = 0$ according to the Fermi–Dirac statistics and Pauli's exclusion principle, which applies to fermions since electrons are fermions with spin = 1/2). In this "intrinsic condition", the Fermi level is equally separated from the valence and conduction bands edge. If the semiconducting material is doped, the position of the Fermi level within the energy gap can be modified.

There are two kinds of doping: *n-type* and *p-type*. When the material is doped with donor impurities, it becomes a *n-type* semiconductor and the Fermi level gets closer to the conduction band; the donor impurity is a material which has one more electron in the outer shell in comparison to the host semiconductor. This extra electron has an energy (the donor level) below the conduction band edge and it is loosely bound to the donor impurity; if this electron overcomes a small energy called the ionization energy of the donor (qV_n), it will occupy an empty state in the conduction band of the material. Statistically, the Fermi level is pushed upwards in energy, just above the donor impurity level and close to the conduction band edge as shown in Fig. 2.8 (close but still below the edge, if the doping is so strong that the Femi level lies within the conduction band, then the semiconductor is considered "degenerate"). When the electron from the donor is promoted to the conduction band, the impurity is left behind positively charged, although the material remains globally neutral.

Similarly, if the material is doped with acceptor impurities, it becomes a *p-type* semiconductor and the Fermi level gets closer to the valence band; the acceptor impurity is a material that has one less electron in the outer shell in comparison to the host semiconductor. An electron from the valence band of the host can be trapped by the impurity atom, and a hole is created in the valence band. This hole will move within the valence band and statistically the Fermi level is now below the acceptor level and close to the valence band (again if the p-doping is so strong that the Fermi level lies within the valence band, the material is "degenerate"). The impurity with a trapped electron is negatively charged, but the material remains globally neutral.

If a material is selectively doped with a p-type layer and a n-type layer which are in contact, then a p-n junction is created. If the material is the same in both layers, the junction is a *homojunction*, and if the material is different for each doped layer, the junction is a *heterojunction*. In both cases, a p-n junction is a complex electronic system where an equilibrium is reached between the diffusion and the drift of the carriers within the doped material. The carriers within a material are moved by two different driving forces, leading to a diffusion motion (due to the gradient of the concentration of carriers) and to a drift motion (due to an internal electrical field created by the charges along the junction) which are opposed and reach an equilibrium depending on the doping level of both sides (the zone of the material where the internal electrical field in equilibrium is felt by the carriers is called the "depletion zone"). The equilibrium created throughout the p-n junction is altered by two mechanisms: the application of an external electrical field which modifies the drift motion of carriers (a biased diode) or the incidence of light on the material which creates additional carriers which modifies the diffusion motion of the carriers (a photodiode, which is a solar cell, under illumination). A new equilibrium is reached very quickly and the availability of power delivered to a load will depend on the impedance value of this load. The most simple load is a resistance, a passive element with values between two extreme cases $0 < R_L < \infty$. In Fig. 2.9, the relationship between the J-V curve and the energy diagrams are graphically shown for an illuminated solar cell. When the load is $R_L = 0$, the solar cell is in short circuit condition, then the maximum current (I_{sc}) is delivered to the load, but the voltage is $V = 0$, and therefore, no external power is generated; this point corresponds to a flat Fermi level throughout the solar cell, which puts the two sides (p-type and n-type) at the same voltage and the photocurrent generation is maximum (electrons move on both bands from higher to lower energy levels, but in the valence band, this motion can be represented by holes moving in the opposite direction and going upwards in the energy diagram). The other extreme case is when $R_L = \infty$ and the solar cell is in open circuit condition, the Fermi level is maximally bent, thus generating a maximum voltage between both sides of the solar cell (V_{oc}) but no current flows through the external load as indicated by the flat bands, and again no power is delivered to the load. All intermediate cases with both Fermi level and band edges bent will deliver power to the load, with both current flowing through it (I_L) and voltage built across it (V_L). For points along the J-V curve, J_L and V_L must fulfil simultaneously the equation of the solar cell (either (2.12) for the ideal solar cell or (2.15) for the real solar cell) and Ohm's law applied to the load.

When organic and hybrid photovoltaic technologies are considered, the doping effect is created in a different way: molecular or polymeric materials with different electron affinities are used in a layered or in a blend structure; the molecules with higher relative electron affinity are called "acceptors" and those with lower relative electron affinity are colled "donors" (relative to the other material within the cell).

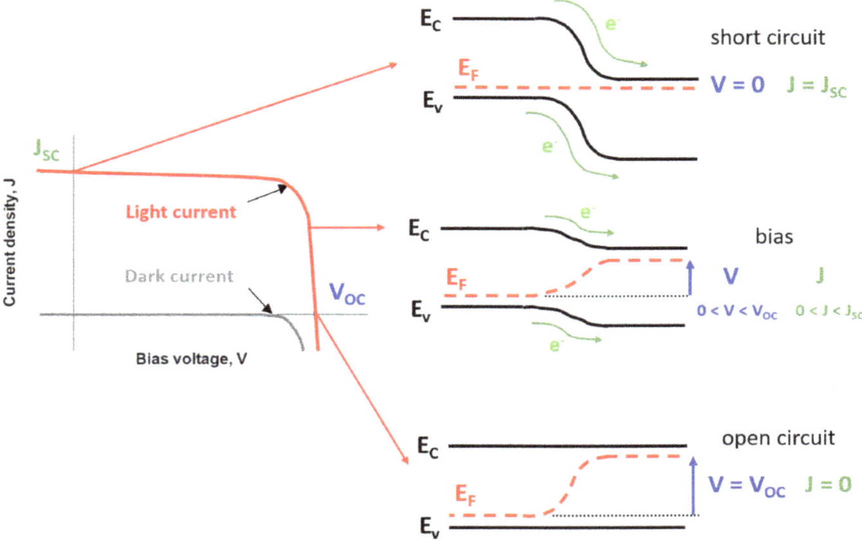

Fig. 2.9 The different points of the J-V curve of the solar cell correspond to different alignments of the band edges and the Fermi level: short circuit with flat Fermi level, open circuit with the most bent Fermi level and flat band edges and all intermediate cases with solar cell power delivered to the load

In this case, the name applies to the material and not to the doping atom, that is, the donor is a p-type material and the acceptor is a n-type material.

2.2.2 *Electrodes*

The electrodes are the outer layers of an electronic device which are connected to the external circuits. In a solar cell, which is a two-terminal device, two electrodes are required to deliver the photogenerated current to an external load. Considering the equivalent circuit of Fig. 2.5, the arrows indicate that the current through the load is positive while the dark current through the diode is negative (applying the Kirchhoff rules to the right-hand side loop of the circuit). Then the positive sign is used for the external connector to the upper branch of the circuit, and the negative sign is used for the external connector to the lower branch. The current flowing through the external load will flow from positive to negative connectors (electrons move in the opposite direction), in J-V curves where J_{sc} is considered positive, the sign for this current flowing through the load is positive. In many scientific articles, it is common to consider J_{sc} negative. Throughout this book, the applied convention for current is J_{sc} positive, J-V curves with positive current flowing through the external load as indicated by arrows in Fig. 2.5.

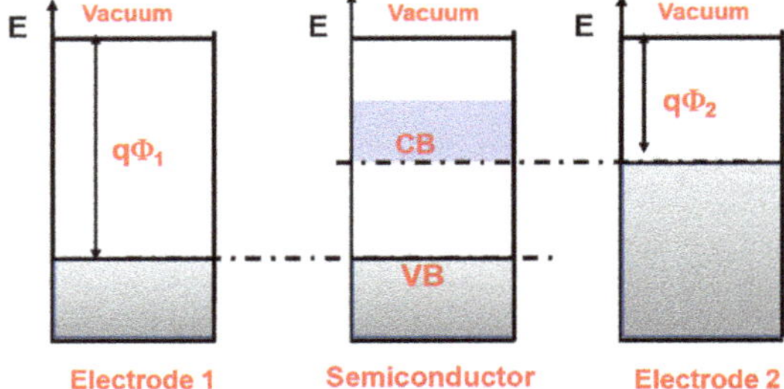

Fig. 2.10 When a semiconducting material is connected to electrodes, the optimization of charge extraction (or injection) is obtained by matching the work function of the metal in the electrode to the band edge whose carriers are to be collected

A good electrode is the one that is able to transport carriers from the device to the load and vice versa with a minimal loss of energy. Its conductivity should be high (metals or highly doped semiconductors), but the most important property of a good electrode is that its work function should match the band edges of the material to which it is connected: an electrode to extract or inject electrons into the conduction band should have a work function matched to the conduction band edge of the semiconducting material and an electrode to extract or inject electrons into the valence band should have a work function matched to the valence band edge of the semiconducting material.

Electrodes in a solar cell will extract electrons from the conduction band (current flowing into it) and inject electrons into the valence band (holes extraction, current flowing out of it). The optimal matching of the work function of the electrodes to the band edges of the active layer is shown in Fig. 2.10. The selection of electrodes with different work functions will enable the possibility to selectively block the extraction of carriers through the wrong electrode by creating potential barriers; but those barriers have an energetic cost which is a loss of voltage between both electrodes and the subsequent reduction of generated power.

Anode or cathode?

When the p-n junction is working as a diode, the convention is to consider the electrical current flowing from p-type material to n-type material when the diode is in forward bias, while no current flows in reverse bias unless the rupture voltage is reached and avalanche current is achieved (in opposite direction). In this scheme, the electrode connected to the p-type part of the diode is called **anode** and the electrode connected to the n-type part of the diode is called **cathode**. In other words, the anode is the electrode through which the current enters into the device and the cathode is the electrode through which the current leaves the device (electrons flowing in the opposite direction, that is, leaving the device through the anode and entering the device through the cathode). It is important to emphasize that the conventional names for the electrodes depend on the current flow and not on the voltage polarity. The sign of the voltage polarity will depend on the operating conditions of the device: if it is a device that provides power to a load, then the anode is negative (and cathode positive), but if the device consumes power, the anode is positive (and the cathode negative).

For solar cells in operation or for batteries during the discharging process (both devices supplying power to the load), the anode is the "negative" electrode through which the current flows into the solar cell or the battery and leaves the external load (the electrons are leaving the cell or battery and entering into the load), while the cathode is the "positive" electrode through which the current flows out of the solar cell or the battery and into the load (the electrodes are entering into the cell or battery and leaving the load).[5] This convention for signs is used by manufacturers to include labels in the connectors of the photovoltaic modules.

Following the convention of p-n junction diodes for solar cells and light emitting diodes, the electrode with lower work function is called cathode, and the electrode with higher work function is called anode. This convention is applied to all PV technologies with one exception: dye sensitized solar cells (DSSC), which usually follows the electrochemical convention of oxidation/reduction criteria. In the DSSC, the electrolyte (acting as hole conductor) is reduced in the cathode (usually a metal) and oxidized in the anode (typically TiO_2). In the DSSC technology, the anode has lower work function and extracts electrons, while the cathode has higher work function and extracts holes (injects electrons), opposite to p-n junction diode convention.

Regardless of the convention applied for electrode name, always, in all PV technologies, the higher work function electrode extracts holes from the active layer and should be matched to the valence band edge, while the lower work function electrode extracts electrons from the active layer and should be matched to the conduction band edge.

Normal or inverted geometry?

The general rule of the high/low work function criteria is used to define the best electrodes for a specific active layer, but some metals may be used as anodes or cathodes depending on the photovoltaic technology and the solar cell geometry. The concept of solar cell geometry or architecture refers to the detailed design of the stacking of several layers, comprising electrodes, transporting and active layers, for single junction cells or for tandem cells (with several junctions monolithically connected in the same device). The use of transporting layers have improved the collection efficiency of the electrodes and provided a large amount of design options for the internal architecture of solar cells.

Normal or inverted geometries are defined according to the side of incoming light and the flow of electrons through the device. These concepts are most often used in organic and hybrid emerging technologies, where one of the electrodes is

[5] When batteries are considered, another useful definition is often applied in electrochemistry: the anode is where oxidation of the electrolyte occurs (loss of electrons) and the cathode is where reduction of the electrolyte occurs (gain of electrons).

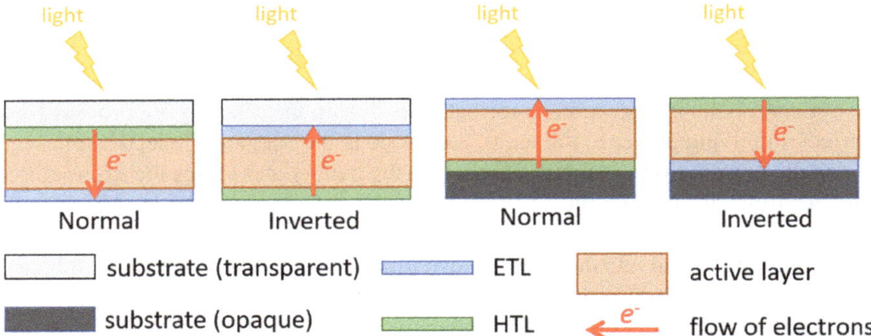

Fig. 2.11 Different configurations of solar cell geometry are defined according to the illuminated side of the solar cell and the flow of electrons through the solar cell. The classification was proposed originally by F. Krebs for organic solar cells depending on the use of transparent or opaque substrates (ETL: electron transporting layer; HTL: hole transporting layer) [14]

transparent and has been deposited on a transparent substrate, but it could also be used in commercial technologies, including bifacial solar cells. In normal geometry, the light enters the cell through a transparent electrode which is extracting holes. In inverted geometry, the light enters the cell through a transparent electrode which is extracting electrons. The materials used for the transparent electrode in organic and hybrid technologies are the same in both geometries (ITO or FTO) but they act as anode (in normal architecture) or cathode (in inverted architecture) usually with the addition of transporting layers which help to extract carriers selectively [27].

In all other PV technologies (crystalline silicon or thin film), there is no transparent electrode. The light enters the cell on the side where the electrode is designed as a grid to minimize the reduction of incoming light (a central bush-bar and fingers). Many technological advancements have focussed on the design and manufacture of electrodes that reduce the loss of light. In bifacial solar cells, the grid is used on both sides of the solar cell. According to the classification shown in Fig. 2.11 [14], that was originally proposed for organic solar cells, the most common silicon solar cell has a normal geometry with backside illumination (since it is considered a cell with non-transparent substrate); the cell is usually a p-type wafer acting as the “substrate” and a thin n-type layer on top of it, the light enters the cell through the grid electrode and into the n-type doping side of the active layer which is collecting electrons. The problem with this classification is that illumination through the electrode grid is considered backside, which is counter-intuitive for inorganic technologies. If the wafer used to manufacture the cell is n-type, and a thin layer of p-type doping layer is on the top part of the cell (the one with a grid electrode), then the geometry should be considered inverted, again with backside illumination.

2.2.3 Transporting Layers

The transporting layers are intermediate layers between the active layer and the electrodes which help to extract the photogenerated carriers from the active layer and inject them into the corresponding electrode. Ideally a transporting layer should also block the carrier of opposite sign to the one to be collected and transported.

A hole transporting layer (HTL) helps to extract holes from the active layer and inject them into the high work function electrode (usually called anode) while at the same time blocks electrons. An electron transporting layer (ETL) helps to extract electrons from the active layer and inject them into the low work function electrode (usually called cathode) while at the same time blocks holes. The word "transporting" is used to indicate the "extracting" capability of the layers.

In some technologies, the function of the transporting layers is achieved by a doping gradient from the bulk active layer to the electrode, thus creating the same effect as the layer of a different material. In emerging technologies, specific materials with tailored band edges and band gaps are introduced between the active layer and the respective electrodes as will be explained in detail in Chap. 4 devoted to cell manufacture.

2.3 Classification of PV Technologies

The practical implementation of the working principles of solar photovoltaic energy conversion described in this chapter can be achieved by the use of a broad range of materials and device architectures. A specific combination of materials and architectures used to manufacture a photovoltaic module defines the **photovoltaic technology** of this module. The materials and processes used to manufacture photovoltaic cells of a given technology are described in Chap. 4. In this final section, the main photovoltaic technologies are briefly presented and classified according to the efficiency chart regularly published by the National Renewable Energy Laboratory (USA) [8]. This classification (names and grouping) is used throughout the whole book.

Crystalline silicon technologies. The main group, currently more than 90% of the photovoltaic market, comprises solar cells that are fabricated with monocrystalline, multi-crystalline or thin film crystalline silicon. Silicon heterostructures (HIT) are a new type of cells at a research stage that may become important in the coming years.

Thin film technologies. The second most important group, with a market share of around 10% in the past years (slightly lower since 2020), comprises a large variety of materials with the common characteristic that the solar cells are thin films of materials of a few micrometers thick (by contrast with the 150 to 300 micrometers thick crystalline silicon cells). Three technologies are classified within this group: amorphous silicon (a-Si), cadmium telluride (CdTe) and copper indium gallium diselenide (CIGS).

III-V technologies. They are cells that are crystalline but also very thin heterostructures of alloys composed by elements of the groups III and V of the periodic table. They are cells with high power conversion efficiency used in space applications or on Earth with concentrating systems. The devices are tandems of monolithically connected cells, with two terminals and a stack of two, three or four junctions (and more in some cases); single junction gallium arsenide (GaAs) cells can be considered as a subgroup of this technology. They are usually fabricated on a germanium substrate, and other germanium containing cells are also included in this group.

Emerging technologies. This group contains a variety of technologies manufactured with different materials that are still in the research and development stage, and although some of them are starting to be commercialized, they still do not have a significant share of the market. This group include: organic solar cells (polymeric, small molecules or combinations of both), dye sensitized solar cells, perovskite solar cells, perovskite–silicon tandem monolithic solar cells, quantum dot solar cells and inorganic CZTS cells (sometimes included in the thin film group); there are single junctions or tandem cells within each group or in combinations with cells of different groups.

The parameter used to rank PV technologies is the power conversion efficiency (PCE). Other electrical parameters, such as short circuit current, open circuit voltage, or fill factor, are also used for the electrical characterization of the modules and they

Table 2.5 Examples of power conversion efficiency of best research cells and best commercial modules for photovoltaic technologies. Summary of data from the National Renewable Energy Laboratory (NREL) efficiency chart and data published 1st July 2021 [8]

PV technology	Best research cell PCE(%)	Best module PCE(%)
mono-Si (concentrator)	27.6	
mono-Si (non-conc.)	26.1	24.4
multi-Si	23.3	20.4
HIT-Si	26.7	
a-Si	14.0	9.8
CdTe	22.1	19.0
CIGS	23.4	19.6
III-V (4 junction, conc.)	47.1	38.9
III-V (4 junction, non-conc.)	39.2	31.2[a]
Organic	18.2	11.7
Dye-sensitized	13.0	
Perovskite	25	17.9
Perovskite-Si tandem	29.5	
Quantum dot	18.1	
CZTS	12.6	

[a]For non-concentrator triple junction III-V modules

are useful for module configuration and PV system design; they will be described in more detail in Chap. 4. Lifetime is also important, but all commercial technologies guarantee lifetimes that go beyond 25 years, and this parameter is only used to compare emerging technologies where shorter lifetimes are still a problem for commercialization. The power conversion efficiency tables are constantly updated, the most referenced efficiency chart is the National Renewable Energy Laboratory (NREL), which is updated every six months and since 1993 is published as an article in the journal Progress in Photovoltaic: Research and Applications. In Table 2.5, a summary of the most recent PCE values at the time of writing is presented for the best research cell and for the best commercial module [8].

References

1. Adams W, Day R (1877) The action of light on selenium. Proc R Soc Lond A25
2. Ashcroft N, Mermin N (1976) Solid state physics. HRW International Editions, Holt, Rinehart and Winston
3. Becquerel E (1839) Mémoire sur les effets électriques produits sous l'influence des rayons solaires. Comptes Rendus 9:561–567
4. Becquerel E (1839) Recherches sur les effets de la radiation chimique de la lumiére solaire, au moyen des courants électriques. Comptes Rendus 9:145–148
5. Einstein A (1905) Über einem die Erzeugung und Verwandlung des Lichtes betreffenden heuristischen Gesichtspunkt (Generation and conversion of light with regard to a heuristic point of view). Annalen der Physik 17:132–148
6. Faiman D (2008) Assessing the outdoor operating temperature of photovoltaic modules. Prog PhotovoltS: Res Appl 16(4):307–315. https://doi.org/10.1002/pip.813
7. Fritts CE (1883) On a new form of selenium cell, and some electrical discoveries made by its use. Am J Sci s3-26(156):465. https://doi.org/10.2475/ajs.s3-26.156.465, http://www.ajsonline.org/content/s3-26/156/465.abstract
8. Green MA, Dunlop ED, Hohl-Ebinger J, Yoshita M, Kopidakis N, Hao X (2021) Solar cell efficiency tables (Version 58). Prog PhotovoltS: Res Appl 29(7):657–667
9. Hay JE (1993) Calculating solar radiation for inclined surfaces: practical approaches. Sol Radiat, Environ Clim Chang 3(4):373–380. https://doi.org/10.1016/0960-1481(93)90104-O
10. Hertz H (1887) Ueber einen Einfluss des ultravioletten Lichtes auf die electrische Entladung. Annalen der Physik 267(8):983–1000. https://doi.org/10.1002/andp.18872670827
11. Huld T, Amillo AMG (2015) Estimating PV module performance over large geographical regions: the role of irradiance, air temperature. wind speed and solar spectrum. Energies 8(6):5159–5181. https://doi.org/10.3390/en8065159
12. King DL, Boyson WA, Kratochvil JA (2004) Photovoltaic Array Performance Model. Tech. Rep. SAND2004-3535, Sandia National Laboratories, Albuquerque, NM, and Livermore, CA (USA). https://doi.org/10.2172/919131, https://www.osti.gov/biblio/919131, volume: 8
13. Koehl M, Heck M, Wiesmeier S, Wirth J (2011) Modeling of the nominal operating cell temperature based on outdoor weathering. Sol Energy Mater Sol Cells 95(7):1638–1646. https://doi.org/10.1016/j.solmat.2011.01.020
14. Krebs FC, Gevorgyan SA, Alstrup J (2009) A roll-to-roll process to flexible polymer solar cells: model studies, manufacture and operational stability studies. J Mater Chem 19(30):5442–5451. https://doi.org/10.1039/B823001C
15. Mott NF (1968) Metal-insulator transition. Rev Modern Phys 40(4):677–683. https://doi.org/10.1103/RevModPhys.40.677

16. Mott NF (1990) Metal-insulator transitions, 2nd edn. CRC Press (Francis & Taylor), London. https://doi.org/10.1201/b12795
17. Muneer T (1990) Solar radiation model for Europe. Build Serv Eng Res Technol 11(4):153–163. https://doi.org/10.1177/014362449001100405
18. Nelson J (2003) The physics of solar cells. Imperial College Press and distributed by World Scientific Publishing Co. https://doi.org/10.1142/p276, https://www.worldscientific.com/doi/abs/10.1142/p276, _eprint: https://www.worldscientific.com/doi/pdf/10.1142/p276
19. Perez R, Ineichen P, Seals R, Michalsky J, Stewart R (1990) Modeling daylight availability and irradiance components from direct and global irradiance. Solar Energy 44(5):271–289
20. Planck M (1900) Über eine Verbesserung der Wien'schen Spectralgleichung (On an Improvement of Wien's Equation for the Spectrum). Verhandlungen der Deutschen Physikalischen Gesellschaft 2:202–204
21. Planck M (1900) Zur Theorie des Gesetzes der Energieverteilung im Normalspectrum (On the Theory of the Energy Distribution Law of the Normal Spectrum). Verhandlungen der Deutschen Physikalischen Gesellschaft 2:237–245
22. Ross JRG (1976) Interface design considerations for terrestrial solar cell modules. In: Conference record (A78-10902 01-44). New York, Institute of Electrical and Electronics Engineers, Baton Rouge, La. USA, pp 801–806. https://ui.adsabs.harvard.edu/abs/1976pvsp.conf.801R
23. Rühle S (2016) Tabulated values of the Shockley-Queisser limit for single junction solar cells. Solar Energy 130:139–147. https://doi.org/10.1016/j.solener.2016.02.015
24. Shockley W, Queisser HJ (1961) Detailed balance limit of efficiency of p-n junction solar cells. J Appl Phys 32(3):510–519. https://doi.org/10.1063/1.1736034
25. Thomson JJ (1897) XL. Cathode Rays. Lond Edinb Dublin Philos Mag J Sci 44(269):293–316. https://doi.org/10.1080/14786449708621070, https://doi.org/10.1080/14786449708621070. Publisher: Taylor & Francis
26. Toledo C, López-Vicente R, Abad J, Urbina A (2020) Thermal performance of PV modules as building elements: analysis under real operating conditions of different technologies. Energy Build 110087. https://doi.org/10.1016/j.enbuild.2020.110087, http://www.sciencedirect.com/science/article/pii/S0378778820300955
27. Würfel P, Würfel U (2016) Physics of solar cells: from basic principles to advanced concepts, 3rd edn. Wiley. https://www.wiley.com/en-us/Physics+of+Solar+Cells%3A+From+Basic+Principles+to+Advanced+Concepts%2C+3rd+Edition-p-9783527413126

Chapter 3
Assessment of Sustainability

Concepts such as conservation or sustainability are today widely accepted and used to qualify almost any social and economic activity. The more widely used, the more diffuse its meaning becomes, and sometimes this indefinition is intentional. It is a socially accepted term with a massive positive consensus, but the indefinition surrounding the concept risks weakening its meaning and losing its political effects. As Bryan G. Norton indicates in [60] (p. 168):

> It is no doubt useful, in policy discussions, to have a term like 'sustainability', which like 'conservation' in days of old, can stand as a label for the many activities of environmentalist. The danger is that the term, like 'conservation' before it, will become a cliché. Nobody oposses it because nobody nows exactly what it entails.

Sustainability is, therefore, a widely used concept whose meaning is open to different interpretations. Let's analyse briefly its meaning with the objective to propose specific methodologies and tools that can be used to evaluate qualitatively, but also quantitatively, if any good or service is sustainable. According to the Cambridge English Dictionary, sustainability is:

> the quality of being able to continue over a period of time,

or more specifically, when the environment is considered, it is

> the quality of causing little or no damage to the environment and therefore able to continue for a long time.

This short definition is too general and diffuse and therefore requires a clarification of the concepts used in the definition. An initial scope for the sustainability definition can be limited to a specific good or service (the word product is used to encompass both meanings). Then according to the dictionary definition, sustainability is the quality of any product to extend its operational lifetime for a certain period of time (ideally a long period of time, although how long is ill defined). Alternatively,

A. Urbina, *Sustainable Solar Electricity*, Green Energy and Technology,
https://doi.org/10.1007/978-3-030-91771-5_3

and focussing on the second definition provided by the dictionary, the required maintenance and replacements of the product should guarantee that this period of time is long enough to repair the negative environmental impacts (damage) created during the manufacture, operation and end-of-life stages of the initial and subsequent products.

The adjective sustainable can be applied to specific goods or services which are often linked in more complex structures or systems. The scale of the systems to be qualified as sustainable ranges from single objects to whole cities, regions or countries (if only considering a geographical scale), and philosophical and moral questions also arise when the concept is applied to human groups or to the conservation of biodiversity [60]. Furthermore, social organization and economical development in the past have consequences in the present which affect different geographical regions. The best example is the problem of climate change and how greenhouse gas emissions from a small set of countries in the past 150 years are now affecting all countries in the world, thus generating the concept of climate justice. Sustainability can, therefore, lead to political concepts related to the so-called *globalization* that go well beyond the analysis of environmental impacts of a single object manufacture. When a specific technology, such as photovoltaic technology, is evaluated in order to measure its degree of sustainability, the scale of application of the concept spans from the well-defined industrial manufacture processes to the more diffuse geostrategical implications.

The methods and tools to evaluate the degree of sustainability depend on the scale of the systems to be evaluated. These methods and tools are under constant development since those issues were raised in two important United Nations documents that contributed to a more detailed definition of the concept of sustainability.

Sustainability is linked to the concept of human development. The World Commission on Environment and Development defined sustainability in the famous Brundtland's **Our common future** report [9].

> Humanity has the ability to make development sustainable - to ensure that it meets the needs of the present without compromising the ability of future generations to meet their own needs. The concept of sustainable development does imply limits - not absolute limits but limitations imposed by the present state of technoloqy and social organization on environmental resources and by the ability of the biosphere to absorb the effects of human activities.

This definition includes a much broader scope that overcomes a single product. Sustainability is linked to the *present and future needs* of the people which establishes a temporal link with a time scale of human generations; it also extends to the whole set of products that a generation requires to meet his needs. Sustainability is linked to present and future human development, but the preservation of environment was secondary at that stage. The evolution of technology and socioeconomic organization could guarantee human welfare in the future, but this quality is considered independent of the preservation of the environment as a task by itself, environment is included in terms of supplier of resources and sink of effects of the human activities; nature is in a subsidiary position which depends on the technological and socioeconomical development of each generation.

A different link between preservation of the environment and human development was declared in the "Earth's Summit", the United Nations Conference on Environment and Development, held in Rio de Janeiro (Brazil) in 1992 [84]. The Report of the conference proclaimed 27 principles which created a new framework for the work of the United Nations Organization on human development and environment. Some of the principles are inspired in the previous Brundtland's declaration of 1987, but five years later, in 1992, the definitions of sustainability and conservation go into deeper details, thus avoiding empty meanings. It is worth to reproduce here a few of the principles proclaimed in the **Rio Declaration on Environment and Development** [84].

> Principle 1: Human beings are at the centre of concerns for sustainable development. They are entitled to a healthy and productive life in harmony with nature.

Brundtlands's concept of *needs* is extended to include a healthy and productive life, that should be in harmony with nature, therefore linking human wellbeing with environment conservation. Harmony is not linked to the stage of technology and the ability to consume natural resources.

> Principle 3: The right to development must be fulfilled so as to equitably meet developmental and environmental needs of present and future generations.

This principle builds upon Brundtland's definition, with the word *needs* and the temporal extension linked to human generations, but goes beyond it because it includes a new concept, the *environmental* needs. Human beings have environmental needs, although they are not clearly defined. This link is reinforced in

> Principle 4: In order to achieve sustainable development, environmental protection shall constitute an integral part of the development process and cannot be considered in isolation from it.

Sustainability has also a spatial, geographical meaning, either in present time or in future time. The basic needs, the healthy and productive life, must be guaranteed to all human beings at any time. It is a geographical extension of the concept of sustainability, entailing an implicit justice demand. The practical requirements to guarantee this geographical part of sustainability are extremely demanding, specially with regard to the socioeconomic organization. Already Brundtland's report pointed out this challenge [9].

> Poverty is not only an evil in itself, but sustainable development requires meeting the basic needs of all and extending to all the opportunity to fulfil their aspirations for a better life. A world in which poverty is endemic will always be prone to ecological and other catastrophes.

It is a very powerful paragraph indeed, linking endemic poverty to ecological catastrophe. This link was emphasized in the Rio Declaration [84].

> Principle 5: All States and all people shall cooperate in the essential task of eradicating poverty as an indispensable requirement for sustainable development, in order to decrease the disparities in standards of living and better meet the needs of the majority of the people of the world.

> Principle 6: The special situation and needs of developing countries, particularly the least developed and those most environmentally vulnerable, shall be given special priority. International actions in the field of environment and development should also address the interests and needs of all countries.

Sustainability is, therefore, linked to reducing socioeconomic disparities, and specially to eradicate poverty. Nevertheless, it seems to focus only on the environmental damage caused by endemic poverty, and not on the impacts caused by developed countries. This asymmetry is only partially smoothed by

> Principle 8: To achieve sustainable development and a higher quality of life for all people, States should reduce and eliminate unsustainable patterns of production and consumption and promote appropriate demographic policies.

Not only poverty may cause environmental catastrophe, but also unsustainable patterns of production and consumption should be reduced (there is no specific mention to developed countries with higher levels of consumption). The unsustainable pattern may also refer to the size of the consumption, in developed countries the problem arises because there is a huge consumption rate *per cápita* (not mentioned in the Rio declaration) and in developing countries because there is a large population growth (mentioned in the Rio declaration when it points to *appropriate demographic policies*).

It is clear that a difficult equilibrium must be obtained when an official declaration is intended to be approved by all countries. Nevertheless, Brundtland's report and the Rio declaration 30 years ago already contained good definitions and all the concepts required to establish a clear meaning for sustainability. The time dimension of sustainability means intergenerational solidarity, and the spatial dimension of sustainability means solidarity between peoples living now on the planet. And in both cases, environmental protection of nature is a compulsory requirement to fulfil both dimensions of solidarity.

From an economical point of view, two paradigms have been discussed since earlier definitions of sustainability were proposed: *weak sustainability* and *strong sustainability*. Weak Sustainability (WS) can be considered an extension of the neoclassical welfare economics developed by Robert Solow and John Hartwick [26, 29, 73]. WS considers that the important concept is the aggregated stock of natural capital (the environment), human capital (skills) and human-made capital (infrastructure) and how it evolves in the future; since the aggregation is the quantity that must be kept constant, it does not matter if the environment is contaminated or if nonrenewable resources are consumed if they are used to increase the aggregated stock of capital, for example, built infrastructures, (roads, factories, machinery, etc…including schools and universities) which can be used to produce goods and services and to increase human skills. In the WS paradigm, natural capital can be depleted if the means to increase other forms of capital in the future are provided, hence Eric Neumayer calls it the *substitutability paradigm*. Opposed to it, Strong Sustainability (SS) regards natural capital as non-substitutable when considered for the production of goods and services, as a sink to negative externalities and as a producer of environmental services and therefore it is also called the *non-substitutability paradigm*

[58]. Norton argues that strict comparison of WS and SS approaches to sustainability cannot be carried out since the methodology used for their evaluation is very different and there is no agreement on a common approach and on what capital should be considered as the true subject of analysis [60]; valuation methods for monetization of externalities caused by economic growth are still under discussion; also, the cost of opportunity of not preserving the natural capital for future generations requires a methodological consensus that is far from reached. But despite these uncertainties, there is an increasing consensus that at least certain forms of natural capital must be preserved as life-supporting functions required for the survival of mankind, despite the uncertainties that future technical developments may bring in order to artificially restore the lost functionalities of natural capital; in the end, it is a discussion of a balance between the economical evaluation of future risks and the cost of opportunity of using the required capital to reduce the environmental impacts of economic growth. The best example that following this path a very broad consensus can be achieved is the publication in October 2006 of the landmark study called the Stern Review on "The Economics of Climate Change", which concluded with this clear statement [76].

> This Review has assessed a wide range of evidence on the impacts of climate change and on the economic costs, and has used a number of different techniques to assess costs and risks. From all of these perspectives, the evidence gathered by the Review leads to a simple conclusion: the benefits of strong and early action far outweigh the economic costs of not acting.

All these different perspectives and paradigms require a scientific or technical consensus about quantification of inputs that contribute to whatever definition of sustainability is considered and how it is achieved in different categories. This quantification is built upon the measurement of well-defined parameters and on calculations of how these parameters contribute to qualify a certain activity as sustainable.

In this book, the objective is to evaluate the sustainability of photovoltaic technology. This objective narrows the broad scope of the principles cited above and focuses on a specific human activity: production and consumption of energy. There is a strong link between access to energy and human development, and there is also a strong link between the production of energy and environmental destruction.

Solar electricity produced by photovoltaic technology is considered a clean (or green) technology, but this cleanliness (or greenness) is a relative concept. A detailed quantification of the sustainability of solar electricity in comparison with other means of energy production is the main objective of this book. It is very difficult to analyse all dimensions of sustainability of solar electricity and all the broad options of photovoltaic technologies. In the following sections, the methodological procedure applied in the book for this task is presented. Two main approaches will be applied: on the one hand, a well-established and regulated procedure, **Life Cycle Assessment (LCA)**, is explained in Sect. 3.1, and on the other hand, a more general socioeconomic analysis including policy regulations whose methods and scenarios are explained in Sect. 3.2.

3.1 Environmental Sustainability: Life Cycle Assessment Applied to Energy Systems

The quantitative evaluation of the sustainability of any good or service (the *products*[1]) is a difficult task. One of the techniques used to understand and address the environmental impact of producing and consuming the products is **Life Cycle Assessmet (LCA)**. LCA is a methodological tool which is defined and regulated by two standards of the International Organization for Standardization: ISO14040 and ISO14044 were prepared by the Technical Committee ISO/TC 207 (Environmental management, Subcommittee SC 5, Life Cycle Assessment), both were approved in 2006 and later revised and confirmed in 2016.[2] The life cycle assessment is the compilation and evaluation of all inputs and outputs of a product throughout its full life cycle, comprising all consecutive and interlinked stages from the raw material acquisition or generation from natural resources, use, end-of-life treatment, recycling and final disposal. According to these standards, any LCA should include at least the following phases (Fig. 3.1).

- the goal and scope definition of the LCA,
- the life cycle inventory analysis (LCI) phase,
- the life cycle impact assessment (LCIA) phase and
- the life cycle interpretation phase.

The results of the different phases for the LCA study of any product can be compared with results for similar or substitutional products, but this comparison will only be useful (and fair) if the context, scope and methods of data gathering for the inventory and impact assessment calculation are transparent at every step of the LCI and LCIA phases. The ISO standards recommendations are good practice to improve the methodology and enable the possibility of comparison of results of different LCA studies. Since LCA can be applied to very different fields, the definitions contained in the standard are open to different interpretations, thus leading to methodological discussions in the scientific community. This is specially the case for the LCIA phase, where several methods of impact assessment have been proposed and evaluated, thus leading to a constant evolution in the methodology. Furthermore, depending on the field of application, some LCIA methods are more adequate than others.

The Institute for Environmental Protection of the Joint Research Centre (European Commission) has launched the **The International Reference Life Cycle Data System (ILCD)** which is a set of documents, databases and software tools that provides recommendations on models and characterization factors that should be used for impact assessment in a broad range of applications [45, 46]. Other organizations have also provided methodological tools to implement LCA studies: the Society of

---

[1] In the international standards described in this section, the term product includes goods and services.

[2] The previous versions ISO 14040:1997, ISO 14041:1998, ISO 14042:2000 and ISO 14043:2000 were cancelled and replaced by the new standards.

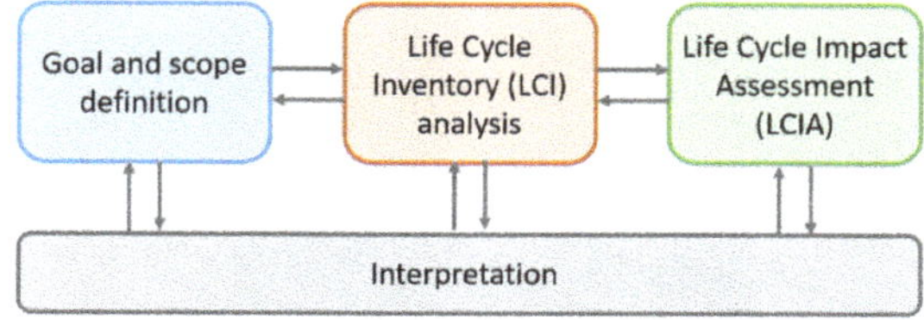

Fig. 3.1 Life cycle assessment framework showing the four phases of a LCA study

Environmental Toxicology and Chemistry (SETAC), a not-for-profit, worldwide professional organization; and private companies such as PRé Sustainability B. V. (developers of SimaPro, LCA software); Sphera Solutions GmbH (developers of GaBi, LCA software) and GreenDelta GmbH (developers of the open-source OpenLCA software). Also a public–private multi-stakeholder partnership has been launched in the framework of the United Nations Environment Programme (UNEP): the **Life Cycle Initiative**, aiming to provide education and training in "life cycle thinking" and access to LCA databases as well as a library of recommended impact assessment factors through the "Resource Efficiency through Application of Life Cycle Thinking" (REAL) project and the Global LCA Data Access (GLAD) network.

A summary of the scientific discussion on these issues regarding the four phases of an LCA is presented in the following subsections, with a special focus on energy related products.

3.1.1 Goal and Scope of the LCA

LCA is a method focussed on environmental evaluation of well-defined products; it is a relative approach in which the product under assessment may have global impacts, but those impacts arise from the product specification and its particular relationship to the environment. It is not a classical "dynamical systems" methodology approach, although well-documented LCA studies can be included as parts of a global dynamical model if the time variable is correctly addressed.

The functions and performance requisites of the product under study should be clearly specified in the scope of a LCA. More precisely, a LCA is structured around a well-defined **functional unit (FU)**. The FU definition is a crucial step in any LCA and the results will strongly depend on this initial selection; the FU can be considered as *the quantified performance of a product system for use as a reference unit* (as it is literally defined in the ISO14040 standard). This "circular" definition provided by the standard is very open and requires a careful analysis which will depend on the product under evaluation. The FU provides a reference that can be used for mathematical normalization of results and to which other LCA results of similar products performing the same function can be compared.

The selection of the functional unit and the definition of the goal and scope of the LCA study determine the complexity and the methodology to be applied in the

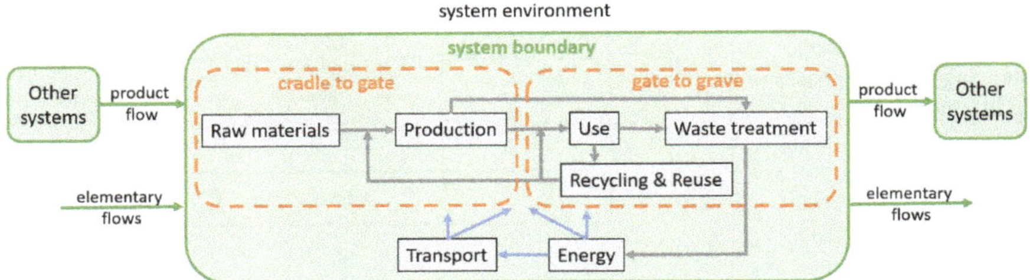

Fig. 3.2 Product system of a LCA study, the system boundary includes all stages of the product lifetime. Other systems and the environment are out of the system boundary

subsequent phases. The ISO standards provide a set of definitions which helps to determine the scope of the LCA. Once the FU has been defined, the next step is to build a hierarchy of processes and to define which phases of the lifetime of the product are included in the scope of the LCA. The processes that contribute to the manufacture or delivery of the product (good or service) constitute an initial *cradle-to-gate* stage, while all other later processes such as use, maintenance and end of life constitute the *gate to grave* stage. A draft of the main life cycle stages of a product system is presented in Fig. 3.2. The LCA may comprise only one or both stages (in this latter case, it is called *cradle to grave*) in which all contributing processes are allocated.

The product system is determined by the system boundary. All processes within a system boundary are linked by flows of intermediate products, including recycled products and waste from production or end of life; also transport and energy contributions may be included within the system boundary. Flows to and from the environment through this boundary exist and they are called elementary flows (from or to the environment) or product flows if they are linked to other product systems as shown in Fig. 3.2. The elementary flows include natural resources (as an input) and releases to air, water and land (as an output), and therefore, their accountability is important for the quantification of environmental impact of the process under study. Other releases may be considered such as odour, noise, vibration, radiation, etc.... The scope of the LCA should include information about the functions of the product system, the system boundary, the allocation procedures to be used in the following LCA phases and a list with all data requirements, applied assumptions and its limitations.

Depending on the scope of the LCA study, the modelling of all processes and flows within the system boundary may require a complex network connecting the different processes, which in turn are a network of activities that transform the inputs into outputs. Once the functional unit, scope and system boundary has been established in the first phase of the LCA, an inventory of all interconnected processes and activities leading to a detailed list of inputs and outputs has to be carefully carried out.

3.1.2 Life Cycle Inventory Analysis (LCI) Phase

The LCI is the phase of life cycle assessment involving the compilation and quantification of inputs and outputs for a product throughout its life cycle. At the end of this phase, the LCI result is an outcome that catalogues the flows crossing the system boundary and provides the starting point for life cycle impact assessment. An LCI phase can also be considered as a study by itself when an interpretation phase follows the compilation of the inventory and results are analysed and presented without including an impact analysis. These are called *LCI studies*.

In a bottom-up approach, the LCI starts with the identification of all unit processes. The unit process is the smallest element considered in the life cycle inventory analysis for which input and output data are quantified (the *flows*); they are organized in networks that build more complex processes and ultimately lead to the product system. In this view, the system boundary is defined as the set of criteria specifying which processes are part of a product system. A top-down approach can also be used, but ideally, both directions (bottom-up or top-down) should lead to the same network that can be represented using a process flow diagram. The compilation of the LCI is an iterative process and it can be modified as a result of a deeper analysis of the processes (or a better understanding of the underlying physical and chemical mechanisms) during the work. It may even require a modification of the goal and scope of the LCA study.

The common practice of developing the product system network of processes is called **attributional** approach and depicts the reality of the analysed system's processes and life cycle stages in close analogy to the supply chain, use stage and end of life. On the other hand, when the LCI modelling framework aims at identifying the consequences of a decision in the foreground system on other processes and systems, it is called **consequential** approach; the actual processes are not depicted (for example, the suppliers of a specific product supply chain as an attributional model does) but it models the forecasted consequences of decisions that have to be included as decision trees in the LCI phase of the LCA study [45, 46]. The consequential approach is strongly affected by market analysis and economic modelling of the unit processes, since the decision-makers usually choose the most cost-effective approach (independently of the environmental impacts unless they are monetized).

The more tedious part of the LCI is the careful compilation and listing of all inputs and outputs of the product system. The input is any product, material[3] or energy flow that enters a unit process. Similarly, the output is a product, material or energy flow that leaves a unit process. Any of two or more products coming from the same unit process or product system is a co-product. A special kind of inputs and outputs are those which are exchanged with the environment and cross the system boundary, it is called *elementary flow*, which is a material or energy entering the system being studied that has been drawn from the environment without previous human transformation, or material or energy leaving the system that is

[3] Products and materials include raw materials, intermediate products and co-products. Secondary raw materials may include recycled materials.

released into the environment without subsequent human transformation (emissions to air, discharges to water and soil and radiation). Data may be obtained by in-situ experimental work in the location (laboratory or factory) where the unit processes are carried out, but they may also be taken from scientific publications. Usually a LCA study will have a mixture of different sources of data. The LCI compilation requires an intensive search for data collection but also a calculation to quantify the amount of inputs and outputs required for the functional unit. Data and calculations for each unit process and for the whole product system should be clearly presented and explained, either if they are the result of an experimental measurement or of a model simulation. All unit processes must obey the laws of mass and energy conservation, and therefore, mass and energy balances provide a useful check on the validity of a unit process description. In most cases, a cut-off criteria are applied to specify the amount of materials or energy flows that are excluded from the study to avoid an exaggerated level of hierarchy in the product system network; the cut-off is associated to the lower threshold of environmental impacts that will be considered (low environmental significance). The decision to include waste treatment within the LCA system boundary is linked to the cut-off effect on the final LCA result and may result in a refining of the system boundary itself; a balance must be achieved and explained in the final report.

Part of the materials entering into the processes are finally embedded in the manufactured product, but others are not. The *ancillary* input is a material input that is used by the unit process producing the product, but which does not constitute part of the product, while the waste substances or objects are those which the holder intends or is required to dispose of (may include hazardous waste which is strongly regulated). Once the LCI has been completed, the calculation of impacts may proceed in the next phase.

The energy input into the processes is said to be embedded in the manufactured product. When the output of the system process under study is more than one product (independently of the selection of one single functional unit), all inputs and outputs must be allocated to the final product. Some outputs are waste, but the allocation procedure should include only products and not waste; care must be taken to avoid allocation of outputs to waste. Sometimes this allocation is also applied to parts of the final product; criteria linked to the physical and chemical processes for allocation to parts must be taken into account for this purpose. If reuse and recycling are included within the system boundary, they are treated as processes of the system (including energy recovery), but this will imply that some unit processes linked to extraction and processing of raw materials are shared by more than one product (either final or intermediate); furthermore, some materials from recycling may not have the original properties of raw material and, therefore, modify the functional parameters of the final product, which have implications in the functional unit of the LCA. Some methodological discussions are still going on about allocation procedures with open and closed loops when reuse and recycling processes are included within a system boundary [20].

In a LCA study applied to energy technologies, some of the flows crossing the system boundary need to be considered with special attention in any of the life cycle phases: production and use of fuels, electricity and heat; transportation (strongly linked to energy consumption); emissions released to the air (specially greenhouse gases because of its implications in climate change). The ISO standard mentions that [47].

> the calculation of energy flows should take into account the different fuels and electricity sources used, the efficiency of conversion and distribution of energy flow, as well as the inputs and outputs associated with the generation and use of that energy flow.

Care is required not to count twice the energy contents of raw materials, and part of this energy is released to the environment and not included in the final product as part of its embedded energy; this is called feedstock energy, mostly arising from heat of combustion of a raw material. It is important to emphasize that the process energy is the energy input required for operating the process or equipment within a unit process, *excluding energy inputs for production and delivery of the energy itself*; again, care is required not to count this energy twice.

No compulsory distinction between renewable and non-renewable sources of energy is included in the ISO standards. In a LCA related to energy processes and sustainability, this distinction should be considered and therefore go beyond the established ISO14040 or ISO14044 basic recommendations.

3.1.3 Life Cycle Impact Assessment (LCIA) Phase

The LCIA is the phase of life cycle assessment aimed at understanding and evaluating the magnitude and significance of the potential environmental impacts for a product system throughout the life cycle of the product. The LCIA associates the data compiled in the inventory to selected environmental impact categories (classification) and calculates the value of all the indicators associated with each category (characterization) to provide a unique aggregated category indicator result per functional unit. One LCI result can be assigned to one or more categories. The compilation of the LCIA category indicator results for the different impact categories is referred to as an **LCIA profile**. There are several characterization models to carry out this task, and each model involves the selection of equations and characterization factors (parameters) to calculate the value of the category indicators with its corresponding magnitude and units; therefore, the results of the LCIA phase depend on the method selected for the calculation and a strong commitment to transparency must be adopted throughout all the procedure. The LCIA should provide all information about assumptions and other decisions in order to enable critical review and reporting. Roughly, all LCIA methods can be grouped into two families: methods determining impact category indicators at an intermediate position of the impact pathways (midpoint) and damage-oriented methods aiming at more easily interpretable results in the form of damage indicators at the level of the ultimate social concerns (endpoint) [52].

The LCIA calculation is applied to each individual **midpoint impact category**, and one or more midpoint categories will end in a **category endpoint**, which is an attribute or aspect of the natural environment (ecosystems), human health or resources, identifying an environmental issue giving cause for concern; the category endpoints are clustered in **endpoint areas of protection**, now called **damage categories** [42]. The impact categories should represent the aggregated impacts of inputs and outputs of the product system on the category endpoint(s) through the category indicators. The most common categories used by different LCIA methods (sometimes with slightly different names) are as follows: climate change (sometimes called global warming), ozone depletion, human toxicity (sometimes separating cancer and non-cancer effects), particulate matter (respiratory inorganics), ionizing radiation, photochemical ozone formation, acidification, eutrophication, ecotoxicity, land use, resource depletion, etc.... Examples of impact categories calculated for some case studies are provided by the standard ISO/TR 14047 [49].

Depending on the method used for the LCIA phase, between ten and thirty categories at midpoint will deliver a smaller set of categories at the endpoint. Both sets of categories and the **damage pathways** used to move from midpoint to endpoint impact categories differ from one method to another. The quantification of the indicators and characterization factors associated with each category are in constant evolution and depend on the knowledge of the physical, chemical and biological processes occurring in the ecosystems. The best methodology to be applied for each LCA study will depend on the LCA functional unit, goal and scope, and both midpoints and endpoints have advantages and disadvantages that must be clearly stated in the interpretation phase of the LCA [4, 21].

The characterization models deliver values for each indicator and it is a first result of the LCIA phase. But further calculations can be carried out (not mandatory, they are optional elements in the ISO standards); they are the normalization, grouping and weighting procedures. Normalization is applied when the magnitude of a category indicator is divided by a well-known reference value (it depends on the selection of the functional unit) or to provide results on a *per cápita* basis, or for a specific geographical area. Grouping is sorting and ranking the impact categories into value-choice sets, hierarchies or priorities. Weighting is used to convert and aggregate indicators across impact categories, this task requires some decision-making about parameters used in the calculation and therefore introduce some subjectivity in the LCIA results, which is being discussed by the scientific community in a constant effort to develop and improve the LCA methodology; when weighting is applied, data and indicator values prior to weighting should be always available. In the scope of the LCA study, a clear description of the model used in the LCIA phase and the parameters applied in the weighting operation must be included, and in the final interpretation phase, it is strongly recommended to include sensitivity and uncertainty analysis of the LCA results. It should be emphasized that although often used in publications, there is no scientific basis for reducing LCA results to a single overall score or number, since weighting is subjective and requires value choices.

LCIA includes mandatory and optional components (according to ISO14040 standard). The selection of impact categories, category indicators and characterization

Table 3.1 Some examples of Life Cycle Impact Assessment (LCIA) methodologies currently available

Methodology	Developed by	Country of origin	References
CML 2002	CML[a]	Netherlands	[28]
Eco-indicator 99	PRé[b]	Netherlands	[25]
EDIP 2003	DTU[c]	Denmark	[30, 31, 67–69, 86]
EPS 2000	IVL[d]	Sweden	[74, 75]
Impact 2002+	EPFL[e]	Switzerland	[12, 51, 52, 56, 63–65, 70]
LIME	AIST[f]	Japan	[32–34, 50]
LUCAS	CIRAIG[g]	Canada	[80]
MEEuP	VHK[h]	Netherlands	[54]
ReCiPe	CML[a], PRé[b], RIVM[i]	Netherlands	[14, 39–42, 72, 78, 88–91]
Swiss Ecoscarcity 07	E2[j], ESU[k]	Switzerland	[7, 22, 23, 57]
TRACI 2.0	US EPA[l]	USA	[2, 3, 5, 35–38, 59]

[a]Centrum voor Milieuwetenschappen (Institute of Environmental Sciences) Universiteit Leiden
[b]PRé Sustainability B. V.
[c]Danmarks Tekniske Universitet
[d]Svenska Miljöinstitutet (Swedish Environmental Research Institute)
[e]École Polytechnique Fédérale de Laussane
[f]Research Institute of Science for Safety and Sustainability (National Institute of Advanced Industrial Science and Technology)
[g]Centre international de référence sur le cycle de vie des produits, procédés et services at Polytechnique Montreal
[h]Van Holsteijn en Kemna B. V., Delft
[i]Rijksinstituut voor Volksgezondheid en Milieu (National Institute for Public Health and the Environment)
[j]E2 Management Consulting A. G.
[k]ESU-services Ltd.
[l]United States Environmental Protection Agency

models; the assignment of LCI results to the selected impact categories (classification) and the calculation of category indicator results (characterization) as defined above are all mandatory components of the LCIA phase (Table 3.1).

Some methods are more focussed on midpoint categories (for example, CML 2002, TRACI, EDIP 2003), while others include midpoint and endpoint categories (ReCiPe, Impact 2002+, LIME, Swiss Ecoscarcity) or only endpoint categories (Eco-indicator 99, EPS2000). A detailed analysis of existing environmental LCIA methodologies for use in LCA has been carried out by the Institute for Environment and Sustainability of the Joint Research Centre (European Commission) [45, 46]. In these reports, a detailed analysis of the impact categories and pathways covered by each LCIA method is provided, including how normalization and weighting are performed. The Life Cycle Initiative, which is a joint project between UNEP and SETAC, has been working to propose a comprehensive LCA framework in order to combine midpoint and endpoint-oriented methods and recommend best practices to carry out LCIA studies [56]. Furthermore, the new framework takes a worldwide perspective, tackling the problem of spatial differentiation of impacts so that LCA will progress towards a tool meeting the needs of both developing and developed countries [30, 52, 69]. In particular, special attention has been devoted to differentiate between continents regarding the human intake of toxic substances, and although some significant continental variations have been found, they are lower than the differences between substances and much even lower if cumulative risks are considered [55]. In the framework of the United Nations Environment Programme, the Life Cycle Initiative published a report to evaluate current LCIA methods and proposed a roadmap for the harmonization of indicators and categories, which should not be seen as static, but rather in constant evolution and capturing the cross-cutting scientific knowledge of physical and chemical processes and best practice to quantify the impacts by evaluating current indicators and developing new ones in each damage category [53, 81, 82].

A critical assessment of how each LCIA method quantifies impact indicators and elaborates damage pathways from midpoint categories to endpoint areas of protection (damage categories) has been carried out by the International Reference Life Cycle Initiative (ILCD) of the European Commission Joint Research Center [45, 46]. As an example, in Fig. 3.3, a draft of the damage pathways of the ReCiPe LCIA method is presented; a group of 17 midpoint impact categories is quantified and end up in three broad areas of protection (human health, ecosystems and resource depletion) [42].

The different LCIA methods evaluate several impact categories whose definition varies from one method to another despite referring to the same impact. Many of them are very similar in their definition and the units used to quantify them. A small set of the most commonly used impacts is briefly described below:

Climate change quantifies the global warming potential due to greenhouse gases emissions that increase radiative forcing in the atmosphere. It may be defined over various time horizons. The reference unit for this indicator is kg CO_{2eq}, a unit which aggregates several gases normalized by their respective radiative force: carbon dioxide, methane and dinitrogen oxide.

Freshwater eutrophication quantifies the spilling of nutrients such as phosphates and nitrates resulting from human activities into freshwater environments; the breaking of equilibrium in complex ecological systems such as rivers, lagoons or coastal reservoirs triggers the growth of plankton and the decrease of available light and oxygen (hypoxia) in aquatic environments which ultimately leads to the loss of algae and animal life. The reference unit is kg PO_4^{3-} (phosphate ion) equivalents.

Human toxicity quantifies the toxic potential of substances in the human body. It is commonly divided into two categories: non-cancer and cancer effects. The calculation considers a large set of substances and the exposure or dose that the human body sustains. In some cases, specific categories for radioactive substances and ionizing radiation are also defined. The reference unit for this indicator is kg of 1,4-dichlorobenzene equivalents emitted to urban air.

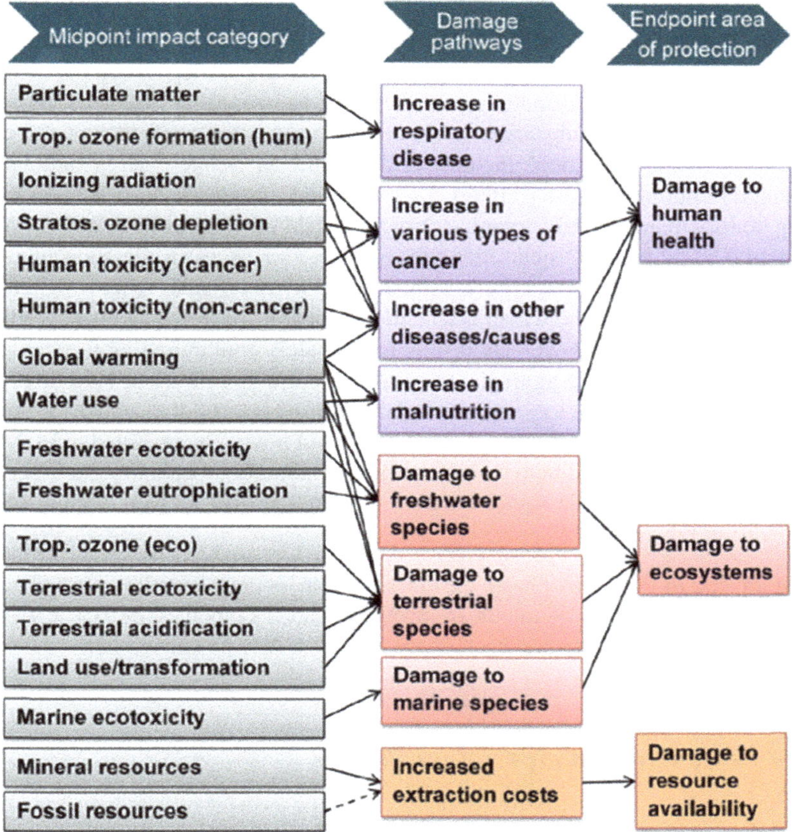

Fig. 3.3 Overview of the impact categories that are covered in the ReCiPe 2016 method and their relation to the areas of protection. The dotted line means that there is no constant mid-to-endpoint factor for fossil resources (Reproduced with permission from reference [42])

Freshwater ecotoxicity quantifies the toxicity of released substances to living organisms other than humans; it is measured in 1,4-dichlorobenzene equivalents emitted to freshwater.

Mineral depletion quantifies the global reduction of available mineral resources, based on the estimation of current available reserves. Since the reserves quantification is constantly updated, this category requires a careful description of the database used (usually the United States Geological Survey annual reports). Some methods also include fossil fuel depletion in this category. The characterization is based on economical considerations and includes a modelling of cost damage of the mineral depletion, which is calculated as the marginal cost increase per kg extracted and it is measured in kg iron (Fe) equivalents.

Particulate matter quantifies all particulate matter emissions. The reference unit is kg PM10 (particles with up to 10 μm diameter) emitted to air. Some LCIA methods consider different particulate matter categories depending on its size.

Photochemical oxidant formation is an impact category that evaluates the contribution of individual substances to ozone formation, and therefore, it is also called simply "ozone formation" in some LCIA methods. Ozone is an hazardous substance to humans causing inflamed airways and lung damage. It is measured in kg non-methane volatile organic compound (NMVOC) emitted to air.

Terrestrial acidification quantifies the release to air of inorganic substances that ultimately increase soil acidity, mostly due to acid rain (oxides react with humidity in the air to form acids) which is hazardous to plant species. The reference unit is kg of sulphur dioxide (SO_2) equivalents ultimately contaminating the soil.

Land occupation is the aggregation of all agricultural and urban land directly or indirectly occupied by a system throughout its life cycle. Global landscape impacts are still under methodological discussions about how to be included in LCIA categories. It is measured in m^2-a (square meters annualy), a quantity that represents how land (measured in square meters) is occupied over a given amount of time (measured in years).

At the end of the LCIA phase, the classification, characterization, normalization, grouping and weighting elements are provided and documented in such a way that enables the life cycle interpretation phase of the LCA to be carried out.

3.1.4 Life Cycle Interpretation Phase

The Life cycle interpretation is the phase of the life cycle assessment in which the findings of either the inventory analysis or the impact assessment, or both, are evaluated in relation to the defined goal and scope in order to reach conclusions and recommendations. A critical evaluation of results should be accomplished in order to ensure completeness and consistency between the obtained LCA study results and the recommendations of the ISO14040 and ISO14044 standards. In the interpretation phase, both *sensitivity analysis*, which is the estimation of the effects of the choices and assumptions made regarding uncertainties in the data, allocation methods or calculation of category indicator results, and a *uncertainty analysis*, which is the procedure to quantify the uncertainty introduced in the results due to the cumulative effects of model imprecision and input uncertainties. A critical assessment of data quality and variability must be included in the final report. Other significant issues that may affect the LCI compilation and LCIA calculations should be identified and addressed in this phase.

Depending on the goal and scope of the LCA, the interpretation phase may offer information structured in different ways as recommended by the ISO14044, p. 46 [48].

a differentiation of individual life cycle stages; e.g. production of materials, manufacturing of the studied product, use, recycling and waste treatment;
b differentiation between groups of processes; e.g. transportation, energy supply;
c differentiation between processes under different degrees of management influence; e.g. own processes where changes and improvements can be controlled, and processes that are determined by external responsibility, such as national energy policy, supplier specific boundary conditions;
d differentiation between the individual unit processes; this is the highest resolution possible.

Once the structure of results presentation has been selected, further analysis can be carried out to evaluate the relevance of individual or grouped data, such as *contribution* to each life cycle stage or to each category indicator, *dominance* of significant contributions after a ranking has been established, *influence*, which is the possibility to influencing the environmental issues and *anomaly identification*, when unusual or surprising deviations from expected or normal results are observed and require a double check of data and calculations.

Finally, a report summarizes all phases of the LCA study. The style and extension of the report vary depending on the audience to which it is intended. Conclusions, limitations and recommendations are presented to the commissioner of the LCA study, to stakeholders or to any other interested party which may use the LCA results for different purposes ranging from very specific product manufacture improvement to very general policymaking recommendations. When comparison of different products or technologies is carried out, the publication of results must be specially careful to guarantee a fair comparison; a good selection of the functional unit is critical for this fairness in comparative studies and an interpretation of a detailed sensitivity analysis must be provided.

Sometimes the LCA report is combined with the use of multicriteria decision methods to generate prioritized recommendations for a specific application, specially if the LCA is used to compare different technological alternatives, which is common practice when energy systems are compared [10].

3.2 Socioeconomic Sustainability: Energy and Sustainable Development

LCA as regulated by ISO14040 and ISO14044 standards are strongly focussed on the environmental impacts of a product assessment. The social and economic dimension of a product manufacture, use and disposal are out of the scope of a standard LCA, but a global sustainability approach should include a socioeconomic analysis and therefore go beyond what is regulated in the standard. Several international groups are working with this purpose in developing a **Life Cycle Sustainability Assessment (LCSA)** methodology.

As Jeroen Guinée *et al* indicated more than ten years ago, the challenge for a broad sustainable focus should go beyond a new LCA extended methodology, it is more about the integration of multidisciplinary approaches in a common framework [27].

> Unlike LCA, LCSA rather is a framework of models than a model in itself: a transdisciplinary integration framework for disciplinary models and methods, selected and interlinked for addressing and answering a specific life cycle sustainability question. LCSA is a framework for looking from one viewpoint, i.e. the life cycle viewpoint, to sustainability questions and only providing life cycle answers and no other; risk assessment (RA) is, for example, not part of this framework. However, RA is very relevant for certain sustainability questions and should then be added to or performed instead of LCSA-tools.

The scope of a LCSA framework includes and integrates environmental impacts (Life Cycle Assessment), economic impacts (Life Cycle Costing) and also social impacts (Social Life Cycle Assessment). In many countries, sustainability is becoming a requisite for public policies and this requirement is being rapidly transferred to the private sector of the economy, ranging from small stakeholders to big multinational corporations. The task for a comprehensive approach is difficult since many trade-offs exist between socioeconomic and environmental impacts and a common base for quantitative assessment is still under development [87]. Two main issues remain open: to standardize the **Life Cycle Costing (LCC)** methodology and to develop tools to monetise the environmental externalities of products throughout its life cycle in order to include these impacts in any LCC method.

The economical evaluation of a product (considered as a good or service in the broad LCA interpretation) is a mandatory task to address the sustainability. The monetary calculations of the cost of ownership of the product and the cost of the service provided should be considered to evaluate the socioeconomic impacts. The economic parameters can be included in a LCC assessment method, which is often merged with conventional LCA studies in a joint approach which is very useful for market analysis and environmental economics [50].

3.2.1 Life Cycle Costing and Total Cost of Ownership

Life Cycle Costing (LCC) is a methodology that aims at calculating all the costs incurred during the life cycle of a product. The main categories of costs that are accounted are as follows: research and development, design and prototyping, production, use, maintenance, recycling and disposal. Depending on the application of the product, the cost boundaries, cost categories and cost bearers may change; similarly, different quantification, aggregation and interpretation of results can be used in a LCC study and recommended good practice codes have been developed by SETAC, although no standardization has been carried out so far [16, 79]. Similar to LCC is the methodology to calculate the Total Cost of Ownership (TCO) of a product, although in this case, the analysis is more business and market oriented, with special attention to the acquisition cost (either purchase price or lease cost) of the products and the final calculation also includes net profit margins for companies in all stages of the product lifetime. Since TCO is strongly dependant on purchase decisions of buyers, the value-choice for parameters in the TCO calculation is broad and difficult to quantify since it strongly depends on the short-term retail market fluctuations. Advertising campaigns can influence public decisions and buyers willingness to pay when purchasing any product; its impact is difficult to evaluate in a LCC of the product. Furthermore, when considering environmental impacts, a new wave of "green" marketing is now pervading almost all marketing campaigns of any product. The real economic impact of this green marketing is still under debate and it is strongly influenced by geographical and cultural differences [11, 13, 66].

Some legal directives are becoming mandatory in several countries. They provide a simplified list of issues to be included in a LCC study. For example, the European Union directive 2014/24/EU Subsection 3 *Award of the contract*, Article 68. Life-cycle costing (point 1) [62].

> Life-cycle costing shall, to the extent relevant cover parts or all of the following costs over the life cycle of a product, service or works:
>
> (a) costs, borne by the contracting authority or other users, such as:
>   (i) costs relating to acquisition,
>   (ii) costs of use, such as consumption of energy and other resources,
>   (iii) maintenance costs,
>   (iv) end of life costs, such as collection and recycling costs.
>
> (b) costs imputed to environmental externalities linked to the product, service or works during its life cycle, provided their monetary value can be determined and verified; such costs may include the cost of emissions of greenhouse gases and of other pollutant emissions and other climate change mitigation costs.

Importantly, the EU directive includes an explicit mention to externalities, but it does not make compulsory its inclusion in the LCC since it depends on the provision that *their monetary value can be determined and verified*, this is the key question: the determination and verification of externalities [6, 15, 71]. Externality evaluation involves a monetization of impacts (environmental and social) that are excluded from the LCC scope used by companies to calculate their production costs. The impacts can be quantified by the LCA methodology in the environmental impact categories, and this LCA quantification has to be translated into economic units and be included in the study, that is, "internalized" in the LCC methodology. The social impacts are multidimensional and a social LCA methodology is still not clearly defined, making the internalization of the social externalities a more challenging task. Only public bodies are starting to include sustainability considerations in their purchase orders, with some policies towards sustainable public procurement (SPP) slowly advancing; according to United Nations Environment Programme, only around 30% of public bodies worldwide include some kind of enforcement of SPP or ecolabelling policies [83].

When energy systems are considered, the parameters to evaluate the monetary cost of the means of production and the energy produced are well defined. With a focus on energy socioeconomical evaluation, two parameters are broadly used: the TCO of the energy production system and the levelized cost of energy that this system is producing. The International Energy Agency provides good definitions for these parameters that are used in his World Energy Model [44].

3.2.2 Levelized Cost of Energy (LCOE)

The levelized cost of energy is the total cost of an energy system divided by the produced energy. Both the cost and the production are calculated for the whole lifetime of the energy system. The total cost is the sum of the building cost of the energy plant, including financing the initial capital cost, its operation and maintenance (O&M) cost, fuel cost and (eventually) its decommissioning cost. When a fossil fuel or nuclear system is considered, the fuel consumption during the production stage represents a large part of the expenditures during the operational lifetime of the plant; for most of the renewable energy systems, there is no fuel cost (in particular photovoltaic or wind electricity production). The LCOE is used as a comparative parameter between different energy production systems and it is widely used by governments and international agencies to take or recommend energy policy decisions. Depending on some assumptions, slightly different equations are used to calculate the LCOE of the produced energy; the main approach is to consider all costs and investments as net present values (NPV, that is, discounted costs) and also includes a discounted energy generation assuming a lifetime for the energy system; with this assumption, a simple equation can be used for the LCOE calculation:

$$\mathrm{LCOE} = \frac{\sum_{t=1}^{n} \frac{I_t + (O\&M)_t + F_t}{(1+r)^t}}{\sum_{t=1}^{n} \frac{E_t}{(1+r)^t n}}, \tag{3.1}$$

where

- I_t is the investment in year t, including financing and first-year lump payments,
- $(O\&M)_t$ is the operations and maintenance expenditures in year t,
- F_t is the fuel expenditure in year t (it may include a CO_2 emission cost),
- E_t is the energy generation in year t,
- r is the discount rate and
- n is the lifetime of the energy system (in years).

A different formulation is used by the Department of Energy (DOE) of the United States, and in particular by the National Renewable Energy Laboratory (NREL). It has the advantage that some of the value-choice options implicitly included in Eq. 3.1 are made explicit, and therefore, a more detailed calculation of different energy generation conditions for each technology can be accomplished. The DOE formulation is:

$$\mathrm{LCOE_{DOE}} = \frac{CC \times CRF + (O\&M)_{\mathrm{fixed}}}{8760 \times CF} + F \times HR + (O\&M)_{\mathrm{variable}}, \quad (3.2)$$

where

- CC is the capital cost,
- CRF is the capital recovery factor,
- $O\&M$ are the operation and maintenance expenditures (fixed and variable),
- CF is the capacity factor (a fraction between 0 and 1, it is sometimes expressed in hours/year),
- F is the fuel expenditure and
- HR is the heat rate, which is the efficiency of the power plant to convert fuel into electricity.

In this equation, the capital expenditure has been turned into annual payments by using the capital recovery factor (CRF), which is defined as

$$\mathrm{CRF} = \frac{r(1+r)^t}{(1+r)^t - 1}, \quad (3.3)$$

where r is the discount rate (the assumed effective rate at which future income streams are discounted) and t is time (in years). If $r = 0$, then $CRF = 1/t$ is used instead of Eq. 3.3. In both Eqs. 3.1 and 3.1, the discount rate can be different for each technology. Some formulations of $\mathrm{LCOE_{DOE}}$ include the present value of a depreciation factor in the first term of Eq. 3.2, which depends on the tax rate paid; it has not been included here (for more information, check NREL model documentation in [8] and Open Energy Data Initiative[4]). In both cases, the LCOE delivers a price at which the energy has to be sold to recover the initial capital investment and the interest paid if all or part of the capital has been financed with a loan. It is the minimum price to recover all investments and O&M costs (including fuel) during all the life of the project.

Alternative methods have been proposed in which some of the values are not discounted, for example, the *undiscounted cost of energy (UCOE)*, or the *discounted costs of cost energy (DCCOE)*, or the *total cost of energy (TCOE)* where the capital costs are financed on an annuity basis over a defined lifetime. The most widely used Eq. 3.1 depends on some assumptions for the calculations such as the lifetime of the energy system or the capacity factor (that may present strong seasonal and annual variations), and although this value-choice dependability of the LCOE calculation can be slightly improved by the other metrics, only more sophisticated methods such as those based on the Monte Carlo simulations may capture all subtleties of the LCOE calculations [1].

---

[4] OpenEI database: https://openei.org/.

3.2.3 Value-Adjusted Levelized Cost of Electricity (VALCOE)

When renewable energy systems are considered, the best practice is to calculate the LCOE for the electricity produced by the energy system. Both the installed power capacity and the produced energy are calculated for an electrical system by using Eq. 3.1. Then, comparison with any other fossil fuel or nuclear resource is based on electricity production and electricity end-user or wholesale prices. This option reduces the uncertainty in the LCOE calculations although it keeps a strong dependance on the different socioeconomic scenarios that are projected during the lifetime of the energy system and which differ from one geographical region to other. In order to deal with these uncertainties, the International Energy Agency has proposed a new metric to calculate the cost of electricity: the regional *value-adjusted levelized cost of electricity (VALCOE)* that was presented for the first time in the World Energy Outlook 2018 [43].

The VALCOE is calculated as a correction to the LCOE value by adding three elements of value: energy, capacity and flexibility. Each of these elements is calculated as the difference between an average value for the whole energy system and the average value of the specific energy technology under consideration. It can be calculated for all technologies and provides a measure of technology competitiveness in a given geographical location which may be affected by market and policy conditions. The dependance on intermittency of the energy source (renewables), capacity factors or flexibility required by electrical grid managers are some of the causes of uncertainty in the standard LCOE calculation that are addressed by the VALCOE calculation. The value differences between dispatchable and intermittent electricity generation technologies are better captured in the VALCOE definition, since different operational patterns are accounted for. Nevertheless, the impact of policies regarding special tax provisions, subsidies or other support measures are not included in the VALCOE, and therefore, the vision of investors is not fully captured in the new approach (neither it was in the LCOE).

The VALCOE is composed of LCOE and energy, capacity as well as flexibility value. Its calculation is graphically represented in Fig. 3.4 and goes as follows [44].

$$\text{VALCOE}_x = \text{LCOE}_x + [\bar{E} - E_x] + [\bar{C} - C_x] + [\bar{F} - F_x]. \tag{3.4}$$

The energy value (for the average energy system of a given region or country and for the specific energy technology (x) in the same location) is calculated with statistical information about the wholesale electricity prices and output volumes for each technology. The wholesale price is based on the marginal cost of generation and does not include scarcity pricing or other monetized policies specific to each country. Data are calculated on an hourly basis (USA, European Union, China and India) or in four load segments, which depend on the capacity installed and the electricity demand: baseload (demand higher than 5944 hrs/year), low-midload (between 5944 and 3128), high-midload (between 3128 and 782) and peakload (lower than 782). Local policies may have a strong impact on the duration of energy generation by

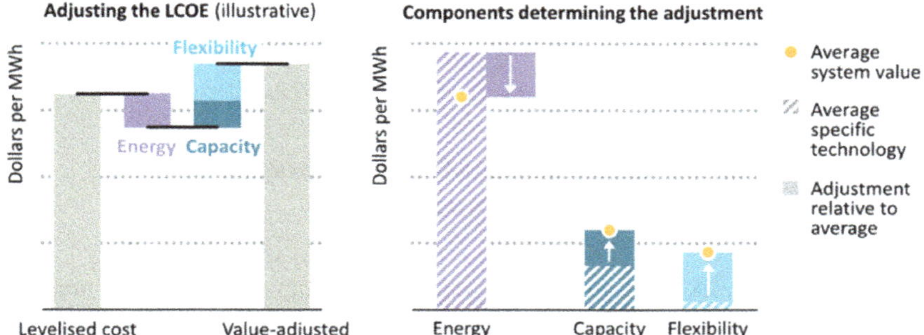

Fig. 3.4 Moving beyond the LCOE to the value-adjusted LCOE. Combining costs and value provides a more robust basis for evaluating competitiveness across technologies than costs alone (International Energy Agency—World Energy Model Documentation 2020 detailed in [44])

each technology and the price of electricity paid to companies depending on the load segment in which that specific energy system is mostly working. The adjustment for the capacity value mostly depends on the intermittency of the energy generation, affecting more to renewable technologies; for dispatchable technologies, the capacity value is near unity and only reduced if the unplanned outage occurs, while for renewables, it is calculated on an hourly basis by using International Renewable Energy Agency (IRENA) regional data. Finally, the flexibility value depends on the share of renewables in the electricity mix of a given country and it is corrected by a multiplier which reflects available market data about the monetized flexibility of generation capacity (possible rewards or tax credits). All details about the calculations and the database are included in the documentation of the World Energy Model of the IEA [44].

Nevertheless, either with LCOE or with VALCOE modelling, the capacity factor and lifetime of the energy systems are still open to a large uncertainty, and even the robust model used by the International Energy Agency to produce their World Energy Outlook reports cannot avoid them. The capacity factors are based on statistical data of their massive database of power plants built and operated worldwide, which can be affected by market and policy factors with strong geographical variations. The other source of uncertainty is the assumption of lifetimes for the energy systems: roughly, the IEA assumes a range between 45 and 60 years for existing fossil fuel plants and nuclear plants, for hydropower plants 50 years, for wind and photovoltaic plants a distribution centred around 25 years and for bioenergy power plants 25 years. These lifetimes are too optimistic for fossil fuels and too pessimistic for renewable sources: for example, a PV plant is built with PV modules whose guarantee is today between 25 and 30 years for T_{80}, that is, the PV modules will still produce electricity with a power conversion efficiency which is 80% of its initial efficiency in year one. Therefore, the expected lifetime of PV plants can be much longer than the average 25 years considered by the IEA model, although it is a good starting point for LCOE and VALCOE calculations, since in this case real lifetime production will in most cases outperform the initial assumptions of the IEA [44].

3.2.4 Circular Economy, Environmental Footprints and Sustainable Development

Circular economy, like sustainability, is a general concept with different meanings, but it is becoming a new paradigm for economic growth. Only a few actions are clearly perceived as steps towards a circular economy: extending the lifetime of products and increasing reuse and recyclability of its components, either biological or technical. And yet these minimal steps towards circular economy will require big changes in the behaviour of society, mainly in rich countries. A population constantly buying new products is one of the main drivers of the actual economic model. The transport of raw materials and manufactured goods through large distances is also an inherent component of the model. Both characteristics are opposed to the implementation of circular economy and will require big behavioural changes in society: accept reused objects and favour recycled products in the retail shops, keep bought objects in use for longer times and contribute to collection and recycling of waste products. Similarly, companies must accomplish important changes to move from planned obsolescence to extended lifetimes of manufactured products and include the design for recycling and better end-of-life treatment in the manufacturing process. Both end-users and producers need access to information that allows them to make decisions towards circular economy. The role of interpretation phase of Life Cycle Assessment and the diffusion of results reaching broader public beyond academic or technical groups is key for the contribution of LCA to the circular economy.

Material recycling and reducing energy use are two important components of a future circular economy, but they are not enough to guarantee sustainability. The always increasing demand of resources of a growing population requires a circular economy approach to redefine the global economic growth and reach all aspects of production and consumption cycles. Although the first steps are slowly being implemented, there is still a long road to circular economy [61].

Recently, the European Union has moved forward with the proposal of tools that are claimed to integrate LCA methodology and extend it in such a way that may have a direct impact on personal, organizational and institutional behaviour. They are the Product Environmental Footprint (PEF) and the Organization Environmental Footprint (OEF) methodologies, proposed by the European Commission in the following reports: [17, 18]. These methods have also been applied to the evaluation of sustainability of electricity production from photovoltaics [77, 85]. Sometimes, the calculated footprints are limited only to carbon or water, which is a serious limitation with respect to previous LCA approaches.

The new proposals have been criticized on several points: on the one hand, they are redundant to LCA methods in many of the new proposed categories and on the other hand, the introduction of new routes to include recycling of substances and products may lead to inaccurate accountability of impacts. This is because some characterization factors in inventory flows are missing and others are difficult to apply at regional level and have been replaced by European averages leading to possible inaccuracies. Other weaknesses have been also pointed out and recommendations for improving the PEF and OEF have been proposed by Professor Matthias Finkbeiner; he considered that PEF and OEF instead of being a breakthrough in sustainability evaluation may become a breakdown, reducing the implementation of LCA approaches in European Union environmental policies and he claims that *"May the sadly missed sense of (LCA) reality finally find its way into PEF!"* [19]. The reply to this criticism from Directorate-General for the Environment of the European Commission was based mainly on the aim to avoid the proliferation of methods for LCA implementation that can be used within the scope of the ISO14040 and ISO14044 standards, thus leading to different results for the same processes, and to narrow the scope of calculations and LCIA methods to be applied; this was identified as a need by several companies that have requested to join the European Commission initiative, of which seventeen out of ninety were selected for the first pilot plans for implementation of the PEF and OEF methods. Also some of the technical concerns are being addressed to update the methodology and to include the recommendations derived from the pilot plans [24].

The risk of moving from well-established LCA methods to new concepts is high. LCA has been developed throughout many years thanks to the expert work of several research groups; it has led to standards that regulate the methodology and allow researchers, stakeholders and end-users to work with clear concepts and provide clear and measurable recommendations to companies and institutions. May LCA results be easily communicated to policymakers? Will LCA methods be finally useful to implement sustainability policies? Perhaps the only remaining challenge for LCA methodology is how to communicate results to the general public.

Other concepts under development may deviate from this well-regulated methods: carbon, water or more generally environmental footprints are attractive concepts, but they are still not well defined in scope and methodology for its calculation, either referred to as product footprint or to organization footprint. Furthermore, when the economical considerations move beyond cost calculations (ownership of systems or price of products) and aims to broader "circular economy" outreach, it risks again that not well-defined concepts ends up dominating the debate about sustainability; this fact may push environmental considerations to a backstage discussion of experts and focussing the interest of society into ill-defined environmental impacts or reductionist economical considerations.

The challenge remains: How to make Life Cycle Assessment a tool useful for society that influences future sustainability policies? How to combine LCA with other emerging tools used to evaluate sustainability without losing its strict scientific methodology and well-established standardization? How to define and quantify sustainability in all its multiple dimensions, interactions and complexity?

References

1. Aldersey-Williams J, Rubert T (2019) Levelised cost of energy—a theoretical justification and critical assessment. Energy Policy 124:169–179. https://doi.org/10.1016/j.enpol.2018.10.004
2. Bare J (2002) Developing a Consistent Decision-Making Framework by Using the U.S. EPA's TRACI. Tech. Rep. Record ID: 62551, Systems Analysis Branch, Sustainable Technology Division, National Risk Management Research Laboratory, Environmental Protection Agency, United States
3. Bare J (2011) TRACI 2.0: the tool for the reduction and assessment of chemical and other environmental impacts 2.0. Clean Technol Environ Policy 13(5):687–696. https://doi.org/10.1007/s10098-010-0338-9
4. Bare JC, Hofstetter P, Pennington DW, de Haes HAU (2000) Midpoints versus endpoints: the sacrifices and benefits. Int J Life Cycle Assess 5(6):319. https://doi.org/10.1007/BF02978665
5. Bare JC, Norris GA, Pennington DW, McKone TE (2003) TRACI-the tool for the reduction and assessment of chemical and other environmental impacts. J Ind Ecol 6(3–4):49–78. https://doi.org/10.1162/108819802766269539. Publisher: John Wiley & Sons Ltd
6. Bielecki A, Ernst S, Skrodzka W, Wojnicki I (2020) The externalities of energy production in the context of development of clean energy generation. Environ Sci Pollut Res 27(11):11506–11530. https://doi.org/10.1007/s11356-020-07625-7
7. Braunschweig A, Brand G, Scheidegger A, Schwank O (1998) Weighting in Ecobalances with the Ecoscarcity Method. Ecofactors 1997. Tech. Rep. Environment Series N° 297, Swiss Federal Agency for the Environment, Forests and Landscape (SAEFL)—BUWAL Agency (now BAFU/FOEN), Bern. https://www.researchgate.net/publication/301765987_Ecobalances_Weighting_in_Ecobalances_Ecofactors_1997, https://ghgprotocol.org/Third-Party-Databases/BUWAL
8. Brown M et al (2020) Regional Energy Deployment System (ReEDS) Model Documentation: Version 2019. Tech. Rep. NREL/TP-6A20-74111, National Renewable Energy Laboratory, Golden, Colorado, USA. https://www.nrel.gov/docs/fy20osti/74111.pdf
9. Brundtland GH, Khalid M (1987) Report of the world commission on environment and development. "Our common future". A/42/427, Oxford University Press. https://digitallibrary.un.org/record/139811?ln=en
10. Campos-Guzman V, Socorro García-Cascales M, Espinosa N, Urbina A (2019) Life cycle analysis with multi-criteria decision making: a review of approaches for the sustainability evaluation of renewable energy technologies. Renew Sustain Energy Rev 104:343–366. https://doi.org/10.1016/j.rser.2019.01.031. Go to ISI://WOS:000458294500026, type: Journal Article
11. Chwialkowska A, Bhatti WA, Glowik M (2020) The influence of cultural values on pro-environmental behavior. J Clean Prod 268:122305
12. Crettaz P, Pennington D, Rhomberg L, Brand K, Jolliet O (2002) Assessing human health response in life cycle assessment using ED10s and DALYs: Part 1-cancer effects. Risk Anal 22(5):931–946. https://doi.org/10.1111/1539-6924.00262. Publisher: John Wiley & Sons Ltd.
13. Dangelico RM, Vocalelli D (2017) "Green Marketing": an analysis of definitions, strategy steps, and tools through a systematic review of the literature. J Clean Prod 165:1263–1279. https://doi.org/10.1016/j.jclepro.2017.07.184
14. De Schryver AM, Brakkee KW, Goedkoop MJ, Huijbregts MAJ (2009) Characterization factors for global warming in life cycle assessment based on damages to humans and ecosystems. Environ Sci Technol 43(6):1689–1695. https://doi.org/10.1021/es800456m. Publisher: American Chemical Society
15. Di Maio F, Rem PC, Baldé K, Polder M (2017) Measuring resource efficiency and circular economy: A market value approach. Resour Conserv Recycl 122:163–171. https://doi.org/10.1016/j.resconrec.2017.02.009
16. Estevan H, Schaefer B (2017) Life Cycle Costing. State of the art report. Tech. Rep. SPP Regions (Sustainable Public Procurement Regions) Project Consortium, 2017, ICLEI - Local Governments for Sustainability, European Secretariat. https://sppregions.eu/fileadmin/user_upload/Life_Cycle_Costing_SoA_Report.pdf

17. EU (2013a) Building the Single Market for Green Products Facilitating better information on the environmental performance of products and organisations. Tech. Rep. COM/2013/0196, European Union
18. EU (2013b) Commission Recommendation of 9 April 2013 on the use of common methods to measure and communicate the life cycle environmental performance of products and organisations. Tech. Rep. 2013/179/EU, European Union
19. Finkbeiner M (2014) Product environmental footprint-breakthrough or breakdown for policy implementation of life cycle assessment? Int J Life Cycle Assess 19(2):266–271. https://doi.org/10.1007/s11367-013-0678-x
20. Frischknecht R (2010) LCI modelling approaches applied on recycling of materials in view of environmental sustainability, risk perception and eco-efficiency. Int J Life Cycle Assess 15(7):666–671. https://doi.org/10.1007/s11367-010-0201-6
21. Frischknecht R, Stucki M (2010) Scope-dependent modelling of electricity supply in life cycle assessments. Int J Life Cycle Assess 15(8):806–816. https://doi.org/10.1007/s11367-010-0200-7
22. Frischknecht R, Steiner R, Braunschweig A, Egli N, Hildesheimer G (2006) Swiss Ecological Scarcity Method: the new version 2006. In: Proceedings of the 7th international conference on EcoBalance. Tsukuba, Japan, p 5. https://www.researchgate.net/publication/237790160_Swiss_Ecological_Scarcity_Method_The_New_Version_2006
23. Frischknecht R, Steiner R, Jungbluth N (2009) The Ecological Scarcity Method Eco-Factors 2006. Tech. rep., Federal Office for the Environment (FOEN), Bern, Switzerland. https://www.bafu.admin.ch/dam/bafu/en/dokumente/wirtschaft-konsum/uw-umwelt-wissen/methode_der_oekologischenknappheitoekofaktoren2006.pdf.download.pdf/the_ecological_scarcitymethodeco-factors2006.pdf
24. Galatola M, Pant R (2014) Reply to the editorial "Product environmental footprint-breakthrough or breakdown for policy implementation of life cycle assessment?" written by Prof. Finkbeiner (Int J Life Cycle Assess 19(2):266-271). Int J Life Cycle Assess 19(6):1356–1360. https://doi.org/10.1007/s11367-014-0740-3
25. Goedkoop M, Spriensma R (2001) The eco-indicator 99. A damage oriented method for life cycle impact assessment, methodology report, 3rd edn. Pré Sustainability. https://pre-sustainability.com/legacy/download/EI99_annexe_v3.pdf
26. Gowdy J (2005) Toward a new welfare economics for sustainability. Ecol Econ 53(2):211–222. https://doi.org/10.1016/j.ecolecon.2004.08.007
27. Guinée JB, Heijungs R, Huppes G, Zamagni A, Masoni P, Buonamici R, Ekvall T, Rydberg T (2011) Life cycle assessment: past, present, and future. Environ Sci Technol 45(1):90–96. https://doi.org/10.1021/es101316v
28. Guinée (Ed) JB (2002) Handbook on life cycle assessment, eco-efficiency in industry and science, 1st edn, vol 7. Springer, Netherlands. https://doi.org/10.1007/0-306-48055-7
29. Hartwick JM (1977) Intergenerational equity and the investing of rents from exhaustible resources. Am Econ Rev 67(5):972–974. http://www.jstor.org/stable/1828079. Publisher: American Economic Association
30. Hauschild MZ, Potting J (2005) Spatial differentiation in life cycle impact assessment—the EDIP2003 methodology, environmental news, vol 80. Danish Ministry of the Environment. Environmental Protection Agency, Copenhagen. https://www2.mst.dk/udgiv/publications/2005/87-7614-579-4/pdf/87-7614-580-8.pdf
31. Hauschild MZ, Wenzel H (1998) Environmental assessment of products. volume 2: scientific background, 1st edn. Springer, United States. https://www.springer.com/gp/book/9780412808104
32. Hayashi K, Itsubo N, Inaba A (2000) Development of damage function for stratospheric ozone layer depletion. Int J Life Cycle Assess 5(5):265. https://doi.org/10.1007/BF02977578
33. Hayashi K, Okazaki M, Itsubo N, Inaba A (2004) Development of damage function of acidification for terrestrial ecosystems based on the effect of aluminum toxicity on net primary production. Int J Life Cycle Assess 9(1):13–22. https://doi.org/10.1007/BF02978532

34. Hayashi K, Nakagawa A, Itsubo N, Inaba A (2006) Expanded damage function of stratospheric ozone depletion to cover major endpoints regarding life cycle impact assessment (12 pp). Int J Life Cycle Assess 11(3):150–161. https://doi.org/10.1065/lca2004.11.189
35. Hertwich EG, Pease WS, Koshland CP (1997) Evaluating the environmental impact of products and production processes: a comparison of six methods. Sci Total Environ 196(1):13–29. https://doi.org/10.1016/S0048-9697(96)05344-2
36. Hertwich EG, Pease WS, McKone TE (1998) Environmental policy analysis: evaluating toxic impact assessment methods: what works best. Environ Sci Technol 32(5):138A-144A. https://doi.org/10.1021/es9840403. Publisher: American Chemical Society
37. Hertwich EG, McKone TE, Pease WS (1999) Parameter uncertainty and variability in evaluative fate and exposure models. Risk Anal 19(6):1193–1204. https://doi.org/10.1023/A:1007094930671
38. Hertwich EG, Mateles SF, Pease WS, McKone TE (2001) Human toxicity potentials for life-cycle assessment and toxics release inventory risk screening. Environ Toxicol Chem 20(4):928–939. https://doi.org/10.1002/etc.5620200431. Publisher: John Wiley & Sons Ltd
39. Huijbregts MA, Rombouts LJA, Ragas AMJ, van de Meent D (2005) Human-toxicological effect and damage factors of carcinogenic and noncarcinogenic chemicals for life cycle impact assessment. Integr Environ Assess Manag 1(3):181–244. https://doi.org/10.1897/2004-007r.1
40. Huijbregts MA, Struijs J, Goedkoop M, Heijungs R, Jan Hendriks A, van de Meent D (2005) Human population intake fractions and environmental fate factors of toxic pollutants in life cycle impact assessment. Chemosphere 61(10):1495–1504. https://doi.org/10.1016/j.chemosphere.2005.04.046
41. Huijbregts MAJ, Steinmann ZJN, Elshout PMF, Stam G, Verones F, Vieira M, Hollander A, Zijp M, van Zelm R (2016) ReCiPe 2016 v1.1 : A harmonized life cycle impact assessment method at midpoint and endpoint level Report I: Characterization. Tech. Rep. RIVM report 2016-0104a, Rijksinstituut voor Volksgezondheid en Milieu RIVM. https://www.rivm.nl/bibliotheek/rapporten/2016-0104.html
42. Huijbregts MAJ, Steinmann ZJN, Elshout PMF, Stam G, Verones F, Vieira M, Zijp M, Hollander A, van Zelm R (2017) ReCiPe 2016: a harmonised life cycle impact assessment method at midpoint and endpoint level. Int J Life Cycle Assess 22(2):138–147. https://doi.org/10.1007/s11367-016-1246-y
43. IEA (2019) World Energy Outlook 2018. Tech. rep., International Energy Agency. https://www.iea.org/reports/world-energy-outlook-2018
44. IEA (2020) World Energy Model. Documentation. Tech. rep., International Energy Agency. https://www.iea.org/reports/world-energy-model
45. ILCD (2010) ILCD Handbook: Analysis of existing Environmental Impact Assessment methodologies for use in Life Cycle Assessment. Tech. rep., Institute for Environment and Sustainability, Joint Research Centre, European Commission, European Union. https://eplca.jrc.ec.europa.eu/uploads/ILCD-Handbook-LCIA-Background-analysis-online-12March2010.pdf
46. ILCD (2011) ILCD Handbook: Recommendations for Life Cycle Impact Assessment in the European context. Tech. Rep. JRC 61049/EUR 24571 EN, Institute for Environment and Sustainability, Joint Research Centre, European Commission, European Union. https://eplca.jrc.ec.europa.eu/uploads/ILCD-Handbook-Recommendations-for-Life-Cycle-Impact-Assessment-in-the-European-context.pdf, iSBN 978-92-79-17451-3 ISSN 1018-5593 10.278/33030
47. International Organization for Standardization (2006a) ISO 14040:2006 Environmental management - Life cycle assessment - Principles and framework. https://www.iso.org/standard/37456.html. Technical Committee : ISO/TC 207/SC 5 Life cycle assessment ICS : 13.020.10 Environmental management 13.020.60 Product life-cycles
48. International Organization for Standardization (2006b) ISO 14044:2006 Environmental management - Life cycle assessment - Requirements and guidelines. https://www.iso.org/standard/38498.html. Technical Committee : ISO/TC 207/SC 5 Life cycle assessment ICS : 13.020.10 Environmental management 13.020.60 Product life-cycles

49. International Organization for Standardization (2012) ISO/TR 14047:2012 Environmental management - Life cycle assessment - Illustrative examples on how to apply ISO 14044 to impact assessment situations. https://www.iso.org/standard/37456.html. Technical Committee : ISO/TC 207/SC 5 Life cycle assessment
50. Itsubo N, Sakagami M, Washida T, Kokubu K, Inaba A (2004) Weighting across safeguard subjects for LCIA through the application of conjoint analysis. Int J Life Cycle Assess 9(3):196–205. https://doi.org/10.1007/BF02994194
51. Jolliet O, Margni M, Charles R, Humbert S, Payet J, Rebitzer G, Rosenbaum R (2003) IMPACT 2002+: a new life cycle impact assessment methodology. Int J Life Cycle Assess 8(6):324. https://doi.org/10.1007/BF02978505
52. Jolliet O, Müller-Wenk R, Bare J, Brent A, Goedkoop M, Heijungs R, Itsubo N, Peña C, Pennington D, Potting J, Rebitzer G, Stewart M, de Haes HU, Weidema B (2004) The LCIA midpoint-damage framework of the UNEP/SETAC life cycle initiative. Int J Life Cycle Assess 9(6):394. https://doi.org/10.1007/BF02979083
53. Jolliet O, Antón A, Boulay AM, Cherubini F, Fantke P, Levasseur A, McKone TE, Michelsen O, Milà i Canals L, Motoshita M, Pfister S, Verones F, Vigon B, Frischknecht R, (2018) Global guidance on environmental life cycle impact assessment indicators: impacts of climate change, fine particulate matter formation, water consumption and land use. Int J Life Cycle Assess 23(11):2189–2207. https://doi.org/10.1007/s11367-018-1443-y
54. Kemna R, van Elburg M, Li W, van Holsteijn R (2005) Methodology Study Eco-design of Energy-using Products (MEEUP). Final Report. Tech. rep., Van Holsteijn en Kemna BV, commisioned by DG ENTR, Unit ENTR/G/3 European Commision, Delft. https://ec.europa.eu/docsroom/documents/11846/attachments/3/translations/en/renditions/native
55. Margni M, Jolliet O (2006) Continent-specific intake fractions and characterization factors for toxic emissions: does it make a difference? Int J Life Cycle Assess 11(1):55–63. https://doi.org/10.1065/lca2006.04.012
56. Margni M, Gloria T, Bare J, Sepälä J, Steen B, Struijs J, Toffoletto L, Jolliet O (2008) Guidance on how to move from current practice to recommended practice in Life Cycle Impact Assessment. Tech. rep., UNEP-SETAC Life Cycle Initiative. https://www.researchgate.net/publication/235678766_Guidance_on_how_to_move_from_current_practice_to_recommended_practice_in_Life_Cycle_Impact_Assessment
57. Müller-Wenk R (1994) The ecoscarcity method as a valuation instrument within the SETAC-framework. In: Udo of. Haes, Jensen HA, Klöpffer AA, Lindfors W, Brussels L-G (eds) Integrating impact assessment into LCA. Proceedings of the LCA symposium held at the fourth SETAC-Europe congress. Proceedings of the LCA symposium held at the fourth SETAC-Europe congress, pp 115–120
58. Neumayer E (2013) Weak versus strong sustainability, 4th edn. Edward Elgar Publishing, Cheltenham, UK. https://doi.org/10.4337/9781781007082, URLhttps://www.elgaronline.com/view/9781781007075.xml
59. Norris GA (2002) Impact characterization in the tool for the reduction and assessment of chemical and other environmental impacts. J Indus Ecol6(3-4):79–101. https://doi.org/10.1162/108819802766269548. Publisher: John Wiley & Sons, Ltd
60. Norton BG (2003) Searching for sustainability. Cambridge University Press, Cambridge
61. Olsen SI (2019) The long road to a circular economy. Integr Environ Assess Manag 15(4):492–493. https://doi.org/10.1002/ieam.4170. Publisher: John Wiley & Sons Ltd
62. Parlament E, Council E (2014) Directive 2014/24/EU of the European Parliament and of the Council of 26 February 2014 on public procurement and repealing Directive 2004/18/EC. https://eur-lex.europa.eu/legal-content/EN/TXT/PDF/?uri=CELEX:32014L0024. Document 32014L0024
63. Payet J (2005) Assessing toxic impacts on aquatic ecosystems in LCA. Int J Life Cycle Assess 10(5):373. https://doi.org/10.1065/lca2005.09.003
64. Pennington DW, Margni M, Ammann C, Jolliet O (2005) Multimedia fate and human intake modeling: spatial versus nonspatial insights for chemical emissions in Western Europe. Environ Sci Technol 39(4):1119–1128. https://doi.org/10.1021/es034598x. Publisher: American Chemical Society

65. Pennington DW, Margni M, Payet J, Jolliet O (2006) Risk and regulatory hazard-based toxicological effect indicators in life-cycle assessment (LCA). Human Ecol Risk Assess Int J 12(3):450–475. https://doi.org/10.1080/10807030600561667. Publisher: Taylor & Francis
66. Polonsky MJ (2011) Transformative green marketing: impediments and opportunities. J Bus Res 64(12):1311–1319. https://doi.org/10.1016/j.jbusres.2011.01.016
67. Potting J, Hauschild M (1997) Part II: spatial differentiation in life-cycle assessment via the site-dependent characterisation of environmental impact from emissions. Int J Life Cycle Assess 2(4):209. https://doi.org/10.1007/BF02978417
68. Potting J, Hauschild M (1997) Predicted environmental impact and expected occurrence of: Actual environmental impact part 1: the linear nature of environmental impact from emissions in life-cycle assessment. Int J Life Cycle Assess 2(3):171. https://doi.org/10.1007/BF02978815
69. Potting J, Hauschild M (2005) Background for spatial differentiation in LCA impact assessment - The EDIP2003 methodology., Environmental News, vol 80. Danish Ministry of the Environment. Environmental Protection Agency, Copenhagen. https://www2.mst.dk/Udgiv/publications/2005/87-7614-581-6/pdf/87-7614-582-4.pdf. Environmental Project No. 996 2005
70. Rosenbaum RK, Margni M, Jolliet O (2007) A flexible matrix algebra framework for the multimedia multipathway modeling of emission to impacts. Environ Int 33(5):624–634. https://doi.org/10.1016/j.envint.2007.01.004
71. Roth IF, Ambs LL (2004) Incorporating externalities into a full cost approach to electric power generation life-cycle costing. Effic Costs Optim Simul Environ Impact Energy Syst 29(12):2125–2144. https://doi.org/10.1016/j.energy.2004.03.016
72. Sleeswijk AW, van Oers LF, Guinée JB, Struijs J, Huijbregts MA (2008) Normalisation in product life cycle assessment: an LCA of the global and European economic systems in the year 2000. Sci Total Environ 390(1):227–240. https://doi.org/10.1016/j.scitotenv.2007.09.040
73. Solow RM (1974) Intergenerational equity and exhaustible resources12. Rev Econ Stud 41(5):29–45. https://doi.org/10.2307/2296370
74. Steen B (1999a) A systematic approach to environmental priority strategies in product development (EPS). Version 2000 - General system characteristics. Tech. Rep. CPM report 1999:4, Centre for Environmental Assessment of Products and Material Systems, Chalmers University of Technology, Sweden. http://citeseerx.ist.psu.edu/viewdoc/download?doi=10.1.1.552.1248&rep=rep1&type=pdf
75. Steen B (1999b) A systematic approach to environmental priority strategies in product development (EPS). Version 2000 - Models and data of the default method. Tech. Rep. CPM report 1999:5, Centre for Environmental Assessment of Products and Material Systems, Chalmers University of Technology, Sweden. http://citeseerx.ist.psu.edu/viewdoc/download?doi=10.1.1.368.6714&rep=rep1&type=pdf
76. Stern N (2007) The economics of climate change. Cambridge University Press, United Kingdom
77. Stolz P, Frischknecht R, Wyss F, de Wild-Scholten MJ (2016) PEF screening report of electricity from photovoltaic panels in the context of the EU Product Environmental Footprint Category Rules (PEFCR) Pilots. Tech. rep., Treeze Ltd., Switzerland and SmartGreenScans, Netherlands. http://pvthin.org/wp-content/uploads/2020/05/174_PEFCR_PV_LCA-screening-report_v2.0.pdf
78. Struijs J, van Dijk A, Slaper H, van Wijnen HJ, Velders GJM, Chaplin G, Huijbregts MAJ (2010) Spatial- and time-explicit human damage modeling of ozone depleting substances in life cycle impact assessment. Environ Sci Technol 44(1):204–209. https://doi.org/10.1021/es9017865. Publisher: American Chemical Society
79. Swarr TE, Hunkeler D, Klöpffer W, Pesonen HL, Ciroth A, Brent AC, Pagan R (2011) Environmental life-cycle costing: a code of practice. Int J Life Cycle Assess 16(5):389–391. https://doi.org/10.1007/s11367-011-0287-5
80. Toffoletto L, Bulle C, Godin J, Reid C, Deschênes L (2007) LUCAS—A New LCIA method used for a canadian-specific context. Int J Life Cycle Assess 12(2):93–102. https://doi.org/10.1065/lca2005.12.242

81. UNEP, Setac, Initiative LC, (2016) Global Guidance for Life Cycle Impact Assessment Indicators (Volume 1). Tech. rep, United Nations Environment Programme
82. UNEP, Setac, Initiative LC, (2019) Global Guidance for Life Cycle Impact Assessment Indicators (Volume 2). Tech. rep, United Nations Environment Programme
83. United Nations Environment Programme (2017) Global review of sustainable public procurement 2017. Tech. Rep. Job No: DTI/2113/PA, United Nations Environment Programme. https://wedocs.unep.org/bitstream/handle/20.500.11822/20919/GlobalReview_Sust_Procurement.pdf?sequence=1&isAllowed=y, iSBN 978-92-807-3658-8
84. United Nations Organization (1992) Report of the united nations conference on environment and development (Earth's Summit), vol A/CONF.151/26/Rev.l (Vol. I), united nations publication, sales no. e.73.ii.a.14 edn. Rio de Janeiro (Brazil). https://documents-dds-ny.un.org/doc/UNDOC/GEN/N92/836/55/PDF/N9283655.pdf?OpenElement
85. Wade A, Stolz P, Frischknecht R, Heath G, Sinha P (2018) The Product Environmental Footprint (PEF) of photovoltaic modules-Lessons learned from the environmental footprint pilot phase on the way to a single market for green products in the European Union. Progress Photovoltaics: Res Appl 26(8):553–564. https://doi.org/10.1002/pip.2956. Publisher: John Wiley & Sons Ltd
86. Wenzel H, Hauschild MZ, Alting L (1997) Environmental assessment of products. Vol 1—Methodology, tools and case studies in product development, 1st edn. Springer US. https://www.springer.com/gp/book/9780412808005
87. Zamagni A, Pesonen HL, Swarr T (2013) From LCA to life cycle sustainability assessment: concept, practice and future directions. Int J Life Cycle Assess 18(9):1637–1641. https://doi.org/10.1007/s11367-013-0648-3
88. van Zelm R, Huijbregts MA, Harbers JV, Wintersen A, Struijs J, Posthuma L, van de Meent D (2007) Uncertainty in msPAF-based ecotoxicological effect factors for freshwater ecosystems in life cycle impact assessment. Integr Environ Assess Manag 3(2):203–210. https://doi.org/10.1897/IEAM_2006-013.1. Publisher: John Wiley & Sons Ltd
89. van Zelm R, Huijbregts MAJ, van Jaarsveld HA, Reinds GJ, de Zwart D, Struijs J, van de Meent D (2007) Time horizon dependent characterization factors for acidification in life-cycle assessment based on forest plant species occurrence in Europe. Environ Sci Technol 41(3):922–927. https://doi.org/10.1021/es061433q. Publisher: American Chemical Society
90. van Zelm R, Huijbregts MA, den Hollander HA, van Jaarsveld HA, Sauter FJ, Struijs J, van Wijnen HJ, van de Meent D (2008) European characterization factors for human health damage of PM10 and ozone in life cycle impact assessment. Atmos Environ 42(3):441–453. https://doi.org/10.1016/j.atmosenv.2007.09.072
91. van Zelm R, Huijbregts MAJ, Posthuma L, Wintersen A, van de Meent D (2009) Pesticide ecotoxicological effect factors and their uncertainties for freshwater ecosystems. Int J Life Cycle Assess 14(1):43–51. https://doi.org/10.1007/s11367-008-0037-5

Part II
Life Cycle Assessment of Solar Electricity

The Life Cycle Assessment (LCA) of the photovoltaic systems (the *product*) and the electricity produced by them (the *service*) requires a very clear statement of the scope and the functional unit (FU) used for the LCA study. The main part of the book is devoted to the two stages of the whole life cycle of a PV system: first, the PV system manufacture phase (from cradle to gate), starting with raw materials production and ending with the PV module delivery (at the gate of the factory); then, the second stage focuses on the use phase and the end-of-life phase (including recycling and landfilling) and requires additional tools to calculate the electricity produced during the operational phase.

The organization of Part II is the following: in Chap. 4, the production steps of each main kind of photovoltaic technology are presented including a comparison of the different industrial production routes. In Chap. 5, the materials required for photovoltaic cells and modules manufacture (with information from the inventory part of the LCAs) are presented and discussed in detail, with a focus on impacts of mining and on the limits imposed by scarce or toxic materials. Life Cycle Impact Assessment (LCIA, a part of LCA) methodology is used in this stage, including a discussion of the slightly different results provided by the different LCIA methodological approaches. Chapter 6 is devoted to the specially important energy impacts of the PV module manufacture; the embedded energy in the modules (or cumulative energy demand for its production). Also, the energy payback time (EPBT) has been included in this chapter; it is a parameter broadly used to assess the sustainability of electricity production but which is strongly dependant on the operational phase of the PV system life, including the geographical location where it is operated, and some authors consider that it is not a reliable parameter. This is a methodological discussion that needs a detailed analysis of the most recent recommendations by the International Energy Agency Photovoltaic Power Systems Program (IEA-PVPS) Task 12 working group. Chapter 7 is devoted to a detailed presentation and analysis of the 15 impact categories (in some cases more) of most common LCA approaches; the production routes described in Chap. 4 are analysed from the LCA perspective, with a discussion of the strengths and weaknesses of each route. The focus will be shifted to end-of-life and recycling issues in Chap. 8, and the final chapter of Part

II is devoted to Balance of System components with a more detailed analysis of the use of batteries for energy storage.

Several functional units may be used in a LCA study of a PV system. All have some advantages and disadvantages, and the scope of the study is strongly dependant on the selection of the FU. The most used ones are AC electricity delivered to the grid and quantified in kWh (defined as a service), and a group of FUs that can also be considered as a "reference flow" since they are products at different stages of the life cycle of the PV system: module surface quantified in m^2 and electrical power quantified in kW_p (either nominal power of PV modules or DC-rated power of the PV system). They are briefly described below:

kWh of AC electricity delivered to the grid. It is the FU recommended by the IEA-PVPS Task 12 report on Methodology Guidelines on Life Cycle Assessment of Photovoltaic. It evaluates a delivered service by the PV system (the electricity output) and therefore evaluates a quantified performance of a product system as recommended by ISO guidelines for LCA; the main advantage of this FU is that it provides a direct comparison between PV systems of different technologies and system configurations and with any other electricity-generating technology. The disadvantage is that it requires the consideration of many parameters that are strongly dependant on the specific PV system design and the environmental conditions of the location where the system is operating (they affect the performance ratio to be used in the calculation of the electricity output). Only in a full cradle-to-grave LCA approach, this FU can be properly calculated.

m^2 of modules. It is a FU for a cradle-to-gate scope, where the life cycle of the module fabrication is assessed, sometimes also including an end-of-life phase. But it should not be considered a cradle-to-grave scope since the use phase is not analysed. Furthermore, since different balance of system components must be evaluated in a use phase, this FU is not useful to compare the performance of different PV systems. Additionally, since power conversion efficiency is not considered, a fair comparison of PV technologies is not possible: a module with poor efficiency may deliver better LCA results per m^2 (in terms of reduced impacts), while other much better modules which required additional manufacturing steps to increase their efficiency may deliver poorer LCA results per m^2. In building integrated PV systems (BIPV), where the module coverage of a façade or roof is an important function strongly dependant on the surface size, an LCA with this functional unit may be useful and should be carried out including the contribution of additional elements to attach or integrate the modules in the building.

kW_p nominal peak power at module level. The DC power delivered by the module in standard conditions can be used as a FU. It should be taken into account that systems with equal kW_p may deliver very different amounts of electricity (kWh) depending on their operational conditions and the performance of the BoS components. If cabling, support structures, regulator, etc...are included, the kW_p FU will correspond to the rated DC power of the PV system.

As recommended by the IEA-PVPS Task 12 report on Methodology Guidelines on Life Cycle Assessment of Photovoltaic, transparency in the LCA reporting is of paramount importance; it is required for a correct interpretation of results and to enable a fair comparison between PV technologies and with other energy technologies. When a LCA study is communicated, the following information should be clearly included:

1. PV technology;
2. Type of system (e.g. roof-top, ground mount, fixed tilt or tracker);
3. Module-rated efficiency and degradation rate;
4. Lifetime of photovoltaic modules and balance of system (BoS) components;
5. Location of installation;
6. Annual irradiation and expected annual electricity production with the given orientation and inclination or system's performance ratio.

The motivation behind this transparency requirement is that a LCA study needs the input of technical parameters that depend on the manufacturing process of the photovoltaic module of a given technology but also there is an important contribution of the final design details of the whole photovoltaic system including BoS components. The location where the system is built and will produce electricity (the output depends on irradiance and temperature) and the performance ratio of the system working in operating conditions during a long period of time will have a strong impact on the results. There are additional relationships between geography and technology: the factory where the silicon, cells and modules are fabricated will consume energy with a specific electricity grid mix of the country (or region) where the factory is operating, and this is also a dynamical relationship, since this mix is time-dependent, and therefore, the results of any LCA study will also depend on the time considered for the PV module manufacture.

Chapter 4
Production of PV Modules

The performance of a solar cell is measured using the same parameters for all PV technologies. Nowadays, a broad range of power conversion efficiencies can be found, either in laboratory solar cells or in commercial PV modules, as was shown in Chap. 2; the working principles of solar electricity generation may differ from one PV technology to another, but have a common basis: the photovoltaic effect and the need to extract the photogenerated carriers from the active layer of the solar cell. A small company devoted to PV systems design and installation (either small BIPV systems or large PV plants at MW scale) will not pay much attention to the manufacturing process of the PV module that is being installed. The technical specification sheets will include the electrical and thermal parameters described in Chap. 2, but no information about the manufacturing process (although sometimes the labelling codes may provide an insight of the solar cell structure). Despite this uniformity in the information to installers and general public, very large differences exist between the manufacturing processes from one technology to another.

The initial classification proposed by Professor Martin Green is strongly related to the manufacturing processes: a first generation of cells manufactured from silicon-wafers, a second generation of cells, mostly fabricated by deposition of a thin film of materials on a substrate and therefore using much less material in the active layer, and a third generation, similar to the second one, in which the manufacturing techniques are also based on a deposition of a thin film on a substrate (although more sophisticated methods are used) and aiming at very high power conversion efficiency cells. With this rough criteria in the initial classification, the generations led to three groups (one for each generation) that could also be classified based on cost, either cost per square meter of module or the cost per nominal power, both costs are linked by the power conversion efficiency as presented in [42]. The cost of manufacturing a solar cell will roughly depend on the amount of material to be included in the cell and the difficulty of the processing (material, cell and module); the first generation

A. Urbina, *Sustainable Solar Electricity*, Green Energy and Technology,
https://doi.org/10.1007/978-3-030-91771-5_4

was (and still is) the benchmark technology with the biggest share of market and the one setting the baseline cost to be reduced by the other competing technologies. The second, "thin film" generation, could potentially be cheaper because it uses much less material per solar cell, and the third generation, although it is also a "thin film" technology, it uses scarce and expensive materials and complex manufacturing processes making them an expensive technology in cost per square meter, but if the efficiency is high enough, it could become a cheap, competitive technology, if the cost per nominal power is reduced below the first-generation silicon-wafer solar cells.

But these expectations have not happened yet despite intensive research effort in the past decades. Several strategies within the so-called third generation have been implemented to boost the power conversion efficiency and go beyond the single junction Shockley–Queisser limit and fabricate high efficiency solar cells. The most successful approach is the fabrication of tandem cells by deposition of several stacks of materials in a thin film device while other advanced approaches include the creation of intermediate levels within the band gap or silicon band gap engineering by creation of quantum well structures in thin film silicon devices including intercalated layers of isolating materials (dielectrics) at nanometer scale [22]. The tandem approach based on III-V elements has achieved the record efficiencies (see efficiency charts mentioned in Chap. 2 or the constant updating in the NREL web site,[1] also published periodically by M. Green in the journal Progress in Photovoltaics, Reseach and Applications) and they are currently used in spatial applications such as communication satellites or the International Spatial Station (ISS), where they have proved to be radiation resistant and showed lifetimes in space above 20 years [52, 83, 137]. In terrestrial applications, this advanced third generation high efficiency cells are used under concentrated Sunlight and thus requiring precise tracking systems. They are still not competitive enough to gain a significant market share.

The new set of emerging technologies that have been developed in recent years (organic and hybrid technologies as mentioned in Chap. 2) does not fit in the third-generation group although they are already reaching very good efficiencies. They share with the second generation the thin film concept of low material usage and the technique of fabrication based on deposition on a substrate. The emerging technologies aim at getting a competitive cost (per square meter and if efficiencies are high enough, also per nominal power) but also targetting to very specific market niches for applications that require flexibility, low weight or highly modulable light absorption bands. Many of these new approaches are close to reach the market but it remains to be seen if they can dent into the crystalline silicon market share, or perhaps other markets, such as spatial applications thanks to its low weight and flexibility [108].

In this chapter, a classification based on production methods is considered for the organization of the sections. It is inspired in the classification proposed by Professor Martin Green, but a new group has been introduced to cope with the new organic and hybrid emerging technologies. Additionally, the rationale behind the classification

[1] Efficiency charts published and updated by the National Renewable Energy Laboratory (NREL) in Golden, Colorado, USA.

is more based on the manufacturing process of each technology: wafer-based silicon in the first group, with special attention to silicon process; amorphous silicon in a second group; a third one for other thin film technologies fabricated by different deposition methods on a substrate (specially CdTe and CIGs, but also others with some mention to new approaches); a III-V group (thin film tandems, oriented to high efficiency); and finally, organic and hybrid emerging technologies, also characterized by deposition methods on a substrate, but also including the new perovskite-silicon tandem technology, where the silicon part is wafer-based.

4.1 Crystalline Silicon Technology

The crystalline silicon technology manufacturing process is based on the fabrication of the solar cell from a crystalline or polycrystalline silicon wafer. There are three big steps: silicon processing to fabricate the wafer, cell manufacture from this wafer, and a final step of cell encapsulation towards the full module manufacture. Rarely the three steps are carried out in the same location, and therefore transport stages are important in this technology (not only in the very initial mining of the raw materials, but also within the PV technology process system).

4.1.1 Silicon Processing: From Raw Material to Solar Grade Ingots

Silicon is an atomic element Si with atomic number 14 and three main isotopes: ^{28}Si with 92.23% abundance, ^{29}Si with 4.67% abundance and ^{30}Si with 3.1% abundance. Other isotopes are radioactive with atomic weight ranging from 24 to 44 and with very short lifetimes and with only ^{32}Si traces present in the Earth's crust due to its relatively longer lifetime (152 years) [82]. Silica, or silicon dioxide (SiO_2) in the form of sandstone or quartz, together with a variety of silicates, are the most abundant materials on Earth's crust; measured by mass, Si accounts of 27.7% of Earth's crust [124].

Silica is the raw material for silicon manufacture. It is chemically reduced to remove the oxygen and convert silica into elementary silicon. There are several methods for the reduction reaction, but in the photovoltaic industry, it is the thermal reduction in the presence of coke with high content of carbon and few impurities the one that is used. The reaction proceeds as follows:

$$SiO_2 + 2C \rightarrow Si + 2CO \tag{4.1}$$

$$2SiC + SiO_2 \rightarrow 3Si + 2CO \tag{4.2}$$

An excess of SiO_2 will avoid the formation of unwanted silicon carbide (SiC). The high temperature required for the reduction is obtained in an electric arc furnace, with carbon electrodes, which rises the temperature of a large chamber (of 10 m^3 or more) to 2000 °C. Once the reaction is complete, Si with purity around 99% is obtained; the main impurities are Fe, Al, Ti, Mn, C, Ca, Mg, B and P, with iron the most abundant (being the processed material in most cases a ferrosilicon alloy). The presence of impurities during the silica reduction process leads to toxic elements that need to be removed from the furnace and filtered, thus posing some environmental potential risk.

The purity obtained in the arc furnace is not enough for solar cell manufacture, it is called metallurgical grade silicon (MG–Si), with good purity around 99% but not enough for electronic applications. A few years ago, the main market for high purity silicon was the microelectronics industry, requiring purity in excess of 9N or 10N (nine nines or ten nines) 99.99999999% (with less than 0.2 parts per billion *ppb* impurities) called semiconductor-grade or electronic-grade silicon (EG–Si). For solar cell manufacture, the purity requirement is less demanding, and purity of 6N or 99.9999% (1 part per million *ppm* of impurities) is enough for **solar grade** silicon (SOG–Si). Therefore, the arc-furnace Si product requires a further purification process.

The manufacture of solar grade silicon by purification of metallurgical grade silicon is dominated by a few industrial processes which have in common the need to produce intermediate silicon compounds which are then chemically and/or thermally processed to obtain silicon with the desired purity. The processes are energy consuming and have a low yield, thus producing a large amount of waste; furthermore, most of the intermediate products and reactants are toxic, explosive or may produce a high environmental damage. The dominant industrial process is the hydrogen reduction and/or thermal decomposition of tri-chlorosilane ($SiHCl_3$) known as the *Siemens process*. In Fig. 4.1, a schematic representation of the main stages (numbered 1 to 4 in the figure) of the Siemens process are presented, they are as follows:

1. **Production of tri-chloro silane from metal grade silicon**. The MG-Si material is crushed to small size particles (around 2 mm) and then hydrochlorinated by HCl gas in a fluidized bed reactor (FBR) at 573 K. Several reactions are produced during this process; the summary, with reaction yield around 90% depending on the conditions, is
$$Si + 3HCl \rightarrow SiHCl_3 + H_2 \tag{4.3}$$
2. **Purification of tri-chloro silane**. By using the different boiling points of the impurities and the tri-chloro silane, a repeated distillation procedure yield high purity $SiHCl_3$.
3. **Production of polycrystalline silicon**. The tri-chloro silane gas mixed with hydrogen is introduced in a metal bell jar reactor where Si cylinders acting as seeds are heated up to 1373–1423 K by an electrical current. The Si deposition occurs by hydrogen reduction and/or thermal decomposition in a complex chemical process occurring in the hydrogen gas atmosphere of the bell jar and on

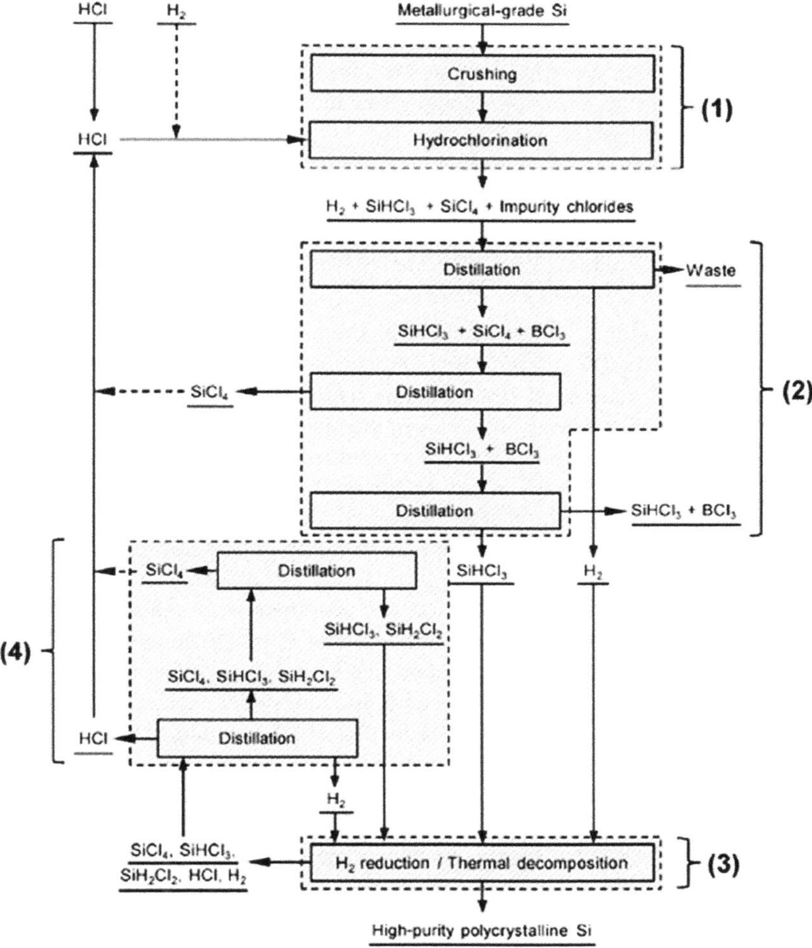

Fig. 4.1 Overview of the Siemens process for silicon purification, showing the four main stages of the production described in the text: production (1) and purification (2) of $SiHCl_3$, production of poly-Si by H_2 reduction and/or thermal decomposition (3) and recovery and reuse of by-products (4) (Reproduced with permission from [136, 139])

the surface of the rods. A summary of the process is presented in the following chemical reactions:

$$4SiHCl_3 \rightarrow Si + 3SiCl_4 + 2H_2 \tag{4.4}$$

$$SiHCl_3 + H_2 \rightarrow Si + 3HCl \tag{4.5}$$

It can be considered a chemical vapour deposition process, and the yield and speed of the process depend on the hydrogen pressure and the jar and rod temperatures. An average growth speed is 1mm per hour, the seed grows from 10mm to 250mm in a week. The concentration of impurities that are present in the final Si solid rods are included in Table 4.1.

4. **Recovery and reuse of by-products.** The different by-products produced in stages 2 and 3, mainly $SiCl_4$, $SiHCl_3$, SiH_2Cl_2 and H_2 are recovered, separated by condensation into the liquid phase and re-distilled, then they can be partially reused in stages 1 and 3 or used in external processes such as production of high purity SiO_2 (for example, the $SiCl_4$).

The final product is a cylindrical rod with diameter between 140 mm to 150 mm (in some cases up to 400 mm rods are obtained) and 2 m length. The rods are often crushed into smaller pieces of 10 to 100 mm (*chunks*) or 1 to 10 mm (*chips*) before further processing. The energy efficiency of the Siemens process is quite low. A large amount of energy has to be fed into the system mainly in stage 3 to heat up the Si seed rods, most of this energy is radiated from the rods, but the bell jar has to be cooled down to avoid deposition of Si on the wall; therefore, most of the energy is lost in a water-cooling system that recover heat from the jar. Additionally, the Siemens production speed is very slow.

Nevertheless, the Siemens process is still dominant in the production of solar grade silicon (still around 90% worldwide in 2021). Other processes are challenging this dominant position; they can be classified into three groups: modifications of the Siemens process based on hydrogen reduction and/or thermal decomposition of silane-based gases, metallothermal reduction of silicon halides by Zn or Al, and metallurgical purification methods to upgrade metallurgical grade silicon.

The second most used process so far, known as the Komatsu process is a modified Siemens process, in which silane (SiH_4) is thermally decomposed at slightly lower temperatures (1073 K), thus leading to a more energy efficient production, although the use of silane makes the overall manufacture more complex [140]. The Komatsu process accounts for all the commercial SOG-Si that is not produced by the Siemens

Table 4.1 Comparison of the amount of impurities present in solar grade silicon manufactured by different purification methods and market share for solar cell production

Impurity	Siemens (solar, bell jar) (value range)	Komatsu (FBR) (value range)	UMG-Si (value range)
P (donor)	0.3–5 ppba	0.3–20 ppba	300–1000 ppba
B (acceptor)	0.1–5 ppba	0.3–20 ppba	500–2000 ppba
Total metals	20–50 ppbw	30–1000 ppbw	10–1000 ppbw
C	0.25–1 ppma	0.5–10 ppma	50–200 ppma
O	0.5–5 ppmw	10–100 ppmw	(100 ppmw)
Market share (%)	90	10	n/a

ppm = parts per million; ppb = parts per billion (atomic or weight)
Data source [15, 139]

process [136]. Another improvement of both the Siemens and Komatsu processes has been the introduction of fluidized bed reactors (FBR) instead of the bell jar; the Si rods acting as seeds are replaced by the injection of Si particles in the reaction chamber providing a much larger reaction area than the rods and therefore increasing the reaction speed, the final product are spheroids with 0.1 mm to 1 or 2 mm diameter. By using the FBR an important energy saving is achieved although other practical problems remain to be solved for a full commercial success [30, 31].

A detailed review of these alternatives was published by Yasuda et al in 2014 and they pointed out to the improved metallurgical processes leading to a new class of silicon called *upgraded metallurgical grade silicon (UMG-Si)* as the most promising due to its relative simplicity and high yield, overcoming some of the problems posed by the Siemens and Komatsu processes [139]. Nevertheless, the purity obtained in UMG-Si is lower than other methods, leading to lower quality in the subsequent crystallization processes and performance of the manufactured crystalline silicon cells (Table 4.1).

The high purity silicon needs to be further processed into ingots that will be used by the photovoltaic industry for solar cell production. The next step is to obtain highly crystalline material from the purified silicon. Two main crystallization processes are widely used, with a market equally shared between them with small variations in recent years: the single crystal Czochralski method (CZ) and the multicrystalline directional solidification method (DS).

The CZ method was developed in 1916 for the crystallization of metals by Jan Czochralski when he was working in the german company AEG (Allgemeine Elektrizitäts Gesellschaft). He published an article in 1918 describing the method: a simple idea based on pulling a crystalline seed out of a crucible with molten metal, the crystal grows in a columnar shape while it is pulled out and the liquid material slowly solidifies [24]. This apparent simplicity when applied to molten silicon requires a high degree of technical sophistication in order to control the growth of single-crystalline silicon and to avoid its contamination by impurities from the crucible. The method was applied to grow single-crystalline semiconductors by Bell Labs since 1948 and has been technically developed to manufacture large size rods of monocrystalline material (an ingot of several kilograms which is a single crystal). A variation of the CZ method is the molten floating zone (FZ) where a moving coil surrounding a silicon rod is heated to a temperature near the melting point temperature of silicon and moved up and down until all the rod is crystallized [111]. The DS method is based on the melting of silicon loaded in a crucible which is allowed to cool down with a refined temperature control. The material solidifies by crystallization from multiple seeds that create single-crystalline domains that grow up to several cm^3 size. Both in the CZ and DS methods, the ingots (cylindrical and squared, respectively) have to be cut with special wire saws into wafers of the desired size, with sectional boule diameter between 165 mm (wafer area: 155 cm^2) and 205 mm (wafer area: 237 cm^2) and thickness, ranging from 180 μm to 300 μm; since the wire saw diameter is around 120 μm, more than 30% of the crystalline material is lost during the cutting stage and it is one of the main bottlenecks to increase the yield of crystalline (either single or multi) cells from ingot to wafer; additionally, the mechanical requirements

Table 4.2 Properties of ingots and bricks from monocrystalline (CZ) and multicrystalline (DS) silicon crystallization process

Parameter	Mono-Si	Multi-Si
Energy consumption	40 kWh/kg	10 kWh/kg
Crystallization yield	>90%	70–80%
Growth rate	100–200 mm/min	5–10 mm/h
Weight	40–150 Kg	100–300 Kg
Size	Cylindrical	Square
	D = 15 cm, L = 1–3.5 m	66 cm × 66 cm × 20 cm

of the wafers will also make very difficult any further processing of Si cells with thickness lower than 80μm below [39] (Table 4.2).

Another crystallization method is the growth of multi-crystalline ribbons from a thin layer of silicon powder by two consecutive melting steps achieved by using focused incoherent light as the heat source. Sheets of 80 × 150 mm^2 with a thickness of 350 μm were achieved in laboratory scale fabrication since the early 90s [27]. Nevertheless, due to the poorer crystallinity of the ribbons and an excess of impurities when compared to CZ or DS methods the industrial production of ribbon-Si solar cells has not been capable to gain a significant market share.

4.1.2 Crystalline Solar Cell Manufacture

The evolution of the solar cell manufacture from the first Si solar cell with less than 1% power conversion efficiency (1941) to the 25% milestone that was achieved in 2009 is strongly linked to the research carried out at the University of New South Wales (UNSW, Australia) by Professor Martin Green [43]. Now best cells are at 27.6% (IBC technology). The basic research led to technological developments that have been transferred from laboratory to industry very fast, leading to commercial modules that are today approaching the research cell milestone (best modules in 2021 reached 24.4% as shown in the NREL efficiency chart published in [47]).

A simplified route for crystalline silicon solar cell manufacture is presented in Fig. 4.2 starting from silicon single crystal wafer and ending in a fully operational solar cell. The basic processing steps are linked to the cell structure presented in Sect. 2.2. The process starts from a p-type Si monocrystalline wafer (for multicrystalline wafers the process is similar), aims at the fabrication of a vertical p-n junction with the n-type material on top of the wafer (top is used to indicate the side of incoming light) and ends up with two kinds of contact: a grid on the top to allow light penetration and a uniform back contact.

The common fundamental processing steps, that may have some variations depending on the manufacturer and the technological family, are the following:

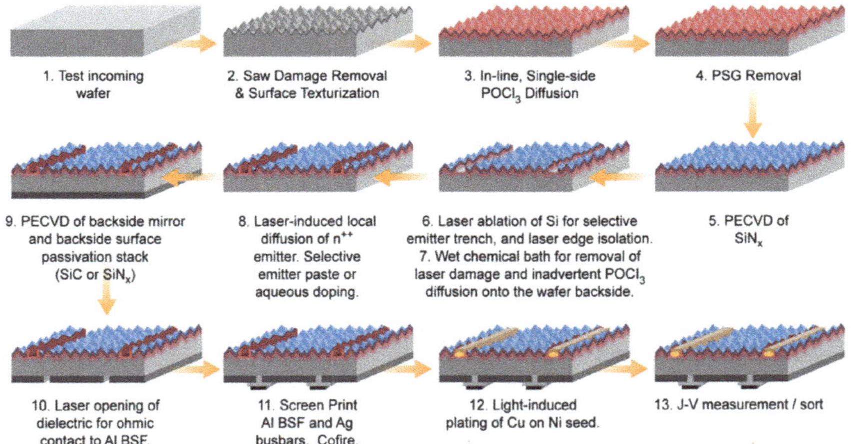

Fig. 4.2 Model process flow for fabrication of crystalline silicon cells (Reproduced with permission from [39])

Wafer testing and texturation. All wafers have been cut in a wire saw and have surface damage. After testing the wafer quality (usually with an in-line photoconductance test of minority carrier lifetime in all wafers or in a random selection of wafers within a production batch), and checking the presence of microcracks by IR absorption attenuation [116, 119], the wafers are textured by exposing the surface to an aqueous solution of NaOH or KOH with isopropyl alcohol [116, 119]. The etching of the different crystallographic orientations proceed at different speeds creating a textured surface of random micropyramids of a few microns size (between 5 μm and 15 μm) that improves the light absorption by multiple reflection/refraction events of the incoming light; more advanced structures which combine pyramids, grooves and regular patches have improved light trapping by texturation of both surfaces of the wafer [16].

Creation of p-n junction. Usually, the wafers have been cut from p-type Si ingots (doped by acceptor impurities, such as boron) and the p-n junction is created by incorporation of donor impurities (phosphorus) to the textured top side of the wafer; this process reverses the bulk p-type doping near the surface to a thin layer of n-type doping ($<1\,\mu m$), called the *emitter*. This process is achieved by heating the wafers up to 900 °C in the presence of phosphorus oxychloride ($POCl_3$) gas within a quartz furnace; the phosphorus atoms are diffused into the top surface of the wafer [138]. A thin glassy layer of silica or phosphosilicate (PSG) is also formed and has to be removed by acid attack with HF dip, a process which also eliminates the phosphorus diffused in the edges of the wafer to avoid shortcuts and is called *edge isolation* [53].

Anti-reflecting coating. The micropyramids reduce significantly the reflection of incoming light, but an additional layer with a different refractive index provides further reduction of reflectivity of the cell surface. The most common layer is hydrogenated silicon nitride (Si_3N_4) and it is created when a reaction between ammonia and silane gases is triggered by plasma-enhanced chemical vapour deposition (PECVD) in a reaction chamber at 400 °C, the advantage of silicon nitrides to be used as ARCs is that it also contributes to surface passivation by reducing surface recombination and they are now the standard in the industry; by modifying the proportion of Si and N in hydrogenated oxides (SiN_x:H), the refractive index can be modulated since Si-rich films have higher refractive index and N-rich films have lower refractive index and therefore a modulation of incoming light can be reached [102]. Other oxides such as SiO_2, SiC, TiO_2 or diamond-like carbon coatings have also been used, but they do not have an effective surface passivation additional functionality.

Fabrication of contacts. Screen-printed contacts from silver (Ag) or aluminium (Al) pastes have been the most used method for contact fabrication on both sides of the solar cell. The front side is made from Ag paste creating a grid of thin *fingers* with a compromise in section between small size to minimize shadows on the cell and large size to optimize charge transport and reduce resistance, with rectangular cross section typically ranging from 45–100 μm × 10–20 μm [10]); the fingers are connected to two or three *busbars* that run across the surface of the cell and collect the charge from the fingers. After screen printing of front contacts, the cells are dried in an oven at around 200 °C. Then, the back side is either fully covered or printed again with a *finger* grid with Al (or Ag or a mixture Ag/Al) and the corresponding busbars (this kind of cell is called Al alloyed back surface field, Al–BSF). After new drying process a final step of paste-firing is carried out at around 810 °C in order to eliminate unwanted paste additives and to create a reliable bonding of the contacts to the cell surface by lead boron-silicate glass frit ($PbO–B_2O_3–SiO_2$) which is contained within the Ag paste. This frit etches and difusses through the $SiN_x{:}H$ ARC layer to form a direct bond with good electrical contact to the underlying n-type emitter region [39]. The contact fabrication, including local diffusion of dopants near the contact points to reduce recombination and to improve charge extraction, and specially the development of buried contacts, has evolved in the past decades and in both cases involving the use of laser technology.

Testing and shorting of cells At the end of the cell production line, all manufactured cells are tested by a flash J-V characterization to measure the electrical parameters of each cell. Then, they are automatically sorted into groups depending on the value of photocurrent at maximum power point. The modules will be fabricated with strings of serially connected cells of each group to avoid current mismatches within a module. A production line will therefore manufacture batches of modules with slightly different performance according to this sorting procedure.

The simple process described above and drafted in Fig. 4.2 has evolved since the late 80s when the power conversion efficiency of best c-Si solar cells was around 20%. In 1989, a new concept was proposed by Professor Martin Green at University of

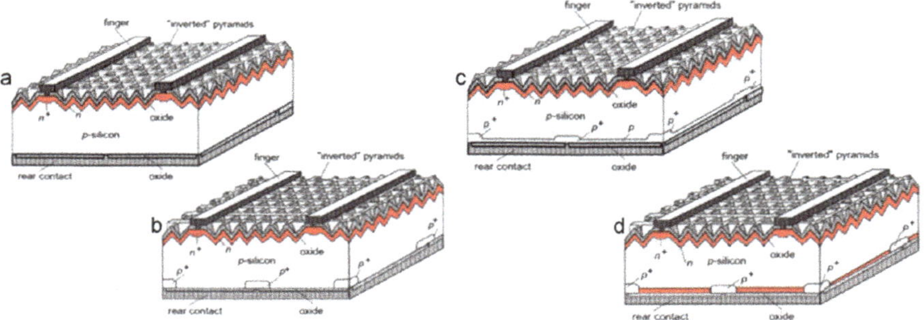

Fig. 4.3 The Passivated Emitter and Rear Cell (PERC) family: **a** Simple PERD cell (Passivated Emitter, Rear Directly-contacted); **b** PERL cell (Passivated Emitter, Rear Locally-doped); **c** PERT cell (Passivated Emitter, Rear Totally-diffused); **d** PERF cell (Passivated Emitter, Rear Floating-junction). The PERC configurations now most widely implemented are the PERL and PERT (Reproduced with permission from [44])

New South Wales: the Passivated Emitter and Rear Cell (PERC) that allowed a rapid development of several technological families until reaching the 25% milestone in 2009, 20 years later [43]. The structure and evolution of the PERC family is presented in Fig. 4.3.

The technological improvements were carried out on p-type silicon wafers, although tests with n-type were also carried out, where the results were poorer due to the difficulty of reliable doping of large surfaces by diffusion of boron. Three basic modifications were introduced in the back contact processing: reduction of rear surface recombination by a reduced metal/semiconductor contact area and the use of a dielectric layer to passivate the surface with the simultaneous increment of internal reflection by the use of a metal reflector acting as a back mirror. Aluminium oxide (Al_2O_3), was used with success as a good passivator with low refractive index for p-type surfaces. Also top surfaces were improved by local diffusion of phosphorus to create n^+ regions near the metal/semiconductor top contacts and better texturation strategies (inverted pyramids). Local diffusion of boron created p^+ regions to improve back contacts. For the local diffusion of dopants and contact fabrication, a combination of laser and photo-lithography was carried out with constant technological improvements mainly developed by University of New South Wales and Fraunhofer Institute for Solar Energy [44, 107]. The evolution of the family is indicated in Fig. 4.3: (a) Simple PERD cell (Passivated Emitter, Rear Directly-contacted); (b) PERL cell (Passivated Emitter, Rear Locally-doped); (c) PERT cell (Passivated Emitter, Rear Totally-diffused); (d) PERF cell (Passivated Emitter, Rear Floating-junction). The PERC configurations now most widely implemented are the PERL and PERT. An additional improvement is the use of advanced laser technology to create grooves in the front side of the cells to be filled with metals and create buried contacts by electroless plating, light induced electro-plating, screen printing, ink jet printing, or aerosol printing with Cu and Ni instead of the standard screen-printed Ag (left drawing of Fig. 4.4). More recent research has proposed the modification of the

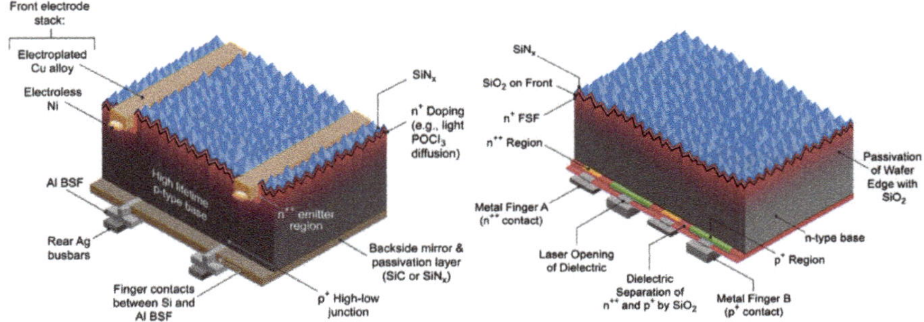

Fig. 4.4 Buried contact solar cell on a p-type wafer and interdigitated back contact cell (IBC) on a n-type wafer (Reproduced with permission from [39])

cell architecture to put both contacts in the rear side of the cell; this technological approach is called *interdigitated back–contact cell (IBC)* and it has advantages like strongly reducing shadowing in the front face and increasing the emitter to metal area coverage thus reducing series resistance (and increasing filling factor), but also poses technical challenges like the requirement to use high quality n-type Si wafers with lifetime of photogenerated carriers in excess of a few ms to allow the carriers that have been photogenerated near the front face to travel to the back of the cell where the interdigitated p-n junctions are located; front, back and edge surfaces can be passivated with SiO_2 which is a technical advantage. The IBC Si solar cell today has the power conversion efficiency record of single crystal homojunction Si cells, 26.1%, but still no large scale commercial manufacture is available [1, 39, 48, 57]. A more recent approach is based on heterojunction cells fabricated by depositing an intrinsic thin layer of hydrogenated amorphous silicon (a-Si:H) followed by a p-type a-Si:H layer on top of n-type Si wafers, and a n-type a-Si:H layer on the back of the wafer creating a back surface field; this kind of c-Si based cell is called heterojunction with an intrinsic thin layer technology (HIT), initially developed with screen-printed contacts [123], and later combined with the interdigitated back contact approach by alternating p-type and n-type a-Si:H patterned layers deposited on the back side of a n-type c-Si wafer to deliver a record breaking efficiency of 26.6% [142, 143].

The industry has invested strongly to improve the production lines for manufacturing high efficiency, cost competitive commercial modules, although power conversion efficiencies of modules (around 22%) are still far from the values achieved by the best laboratory cells; there is room for improvement by reducing rear surface recombination, specially in the emitter region, and by reducing resistive losses in charge collection contacts and finger/busbar wires. The boron-doped p-type Si wafers, wether mono or multicrystalline, and PERC familiy technological variations with buried contacts are now dominating the commercial approach although some companies are moving to the use of n-type P-doped Si wafers to avoid the problem of recombination in boron-oxygen defects found in Czochraslki p-type B-doped Si ingots [91] The c-Si-based technology world record power conversion efficiency for a medium size module is 24.4%, held by Kaneka, for an heterojunction interdigitated back contact (HJ-IBC) approach [47].

4.2 Thin Film Technologies

In this group, different technologies are included. They have in common that the amount of material required in the manufactured cell is lower than in the crystalline wafer approach. The thin film technologies are not fabricated from a wafer cut out of an ingot or brick of previously purified and crystallized material; the main fabrication procedure is the deposition of thin films from chemical precursors on top of a substrate within a reaction chamber; other procedures consider printing techniques in a less demanding environment. The thickness of thin film cells are around ten times lower than the thickness of wafer-based cells, ranging from a a few or tens of μm in comparison to the hundreds of μm required for c-Si cells. A second characteristic of the solar cells grouped under the thin film category is that their production is potentially compatible with lower cost procedures since the crystalline quality of the material is not as demanding as the c-Si wafer category and therefore the cost per square meter of module can be reduced [42, 94]. Regarding module assembly, thin film technologies share a common advantage: the deposition of the materials can be carried out on large surfaces (up to a few square meters) and the individual cells are "cut" by different techniques, mostly laser-based, before final contacts are deposited to create a serially connected string within each module, where the array of interconnected cells has been produced at once; this manufacture procedure contributes to further reducing the final cost of the modules. The technologies considered in this thin film group are amorphous silicon, cadmium telluride, chalcopyrite and kesterite solar cells.

Other technologies are also thin film technologies, like III-V solar cells or emerging organic and hybrid solar cells, with active layers as thin as a few μm, but they have properties that differentiate them from the thin films grouped here in what could be considered a "second generation" between the first generation (based on crystalline silicon cells) and the third one, with higher cost per square meter of module although they require low amount of material (the III-V), or an "emerging" generation with a potentially very low cost because they are based on abundant and low cost organic and hybrid materials that can be easily processed. Both III-V and emerging technologies will be presented in other sections.

4.2.1 Amorphous Silicon

Amorphous silicon (a-Si) has been used to produce solar cells since the early 80s, and it is the most developed thin film technology. Silicon is deposited on different substrates by plasma enhanced chemical vapour deposition (PECVD) from chemical precursors such as silane (SiH_4), which is the most widely used for amorphous silicon cells, or a mixture of SiH_4 and GeH_4 for graded alloys of amorphous silicon and germanium cells; the deposition is sequential and creates layers of doped and intrinsic material. Other methods of deposition such as sputtering or "hot wire" techniques are being developed. The substrates are mainly glass covered with a thin layer of silicon oxide (SiO_2) and a transparent conducting oxide (usually indium-doped tin oxide, ITO, or fluor-doped tin oxide, FTO) or plastic (with ITO or FTO) and sometimes on metal sheets (copper or steel). Finally, a zinc oxide (ZnO) layer is deposited followed by a metallic back contact, usually from Al paste by screen printing. Texturation strategies similar to the ones used for crystalline technology may be applied both on the substrate covered by the conducting oxide or on the final ZnO layer.

The amorphous silicon has very different properties compared to the crystalline silicon. In the amorphous material, the nearest neighbours of the silicon atoms resemble the diamond-like fcc structure of the crystalline material, but the structure is distorted and some atoms are not bonded to four neighbours, thus leaving an unused valence orbital called *dangling bond* and the long range order is lost, thus delivering an amorphous material. The dangling bond of the silicon atoms may be neutral, positively or negatively charged. In the vapour phase, hydrogen is added and incorporated in the structure to saturate the dangling bonds of the amorphous structure which are linked to a loss of efficiency in the first days or weeks of insolation of the cells: the cells suffer a light induced degradation process called "Staebler-Wronski" effect that may produce a 30% loss of the initial efficiency [134]. The addition of hydrogen to the structure stabilizes the performance of the cells and its understanding contributed to an improvement in the performance of a-Si:H cells. Passivation with 5–10% of hydrogen makes the a-Si material suitable for stable and efficient solar cells. If the gases used to deposit the silicon are very diluted in hydrogen, a microscopic domain of crystalline material is formed within a matrix of amorphous material, this is called "microcrystalline" or "nanocrystalline" silicon, with improved stability more similar to crystalline material [85].

An advantage of the distortion of the atomic structure from crystalline to amorphous material is the relaxation of the optical selection rules for the absorption of photons which gives rise to a direct band gap in the disordered material (in crystalline silicon the band gap is indirect and the light absorption is low). This increased absorption, almost one order of magnitude higher compared to c-Si at optical wavelengths, is the reason why the thickness of the amorphous silicon cell can be reduced. Disorder and the addition of hydrogen also contributes to increase the band gap up to 1.7 eV, a too high value (not optimal for solar energy conversion), but if hydrogen is not added, the solar cell has very poor transport properties due to the high defect density of the unpassivated cells [106]. The a-Si:H material can be p-type and n-type doped, but the inclusion of impurities alters the dangling bond equilibrium, reduces the lifetime of minority carriers and makes the doping less effective. Additionally, when the density of dangling bonds is very high, the Fermi level is pinned amongst the remaining defect states, which also act as charge traps and recombination centres thus dominating the carrier transport properties of the solar cell. Since diffusion

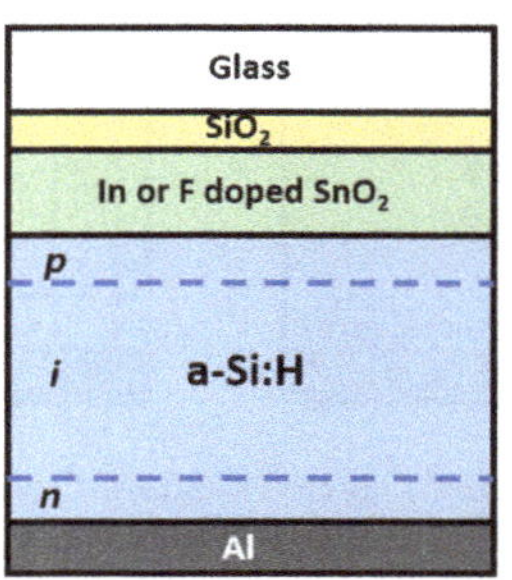

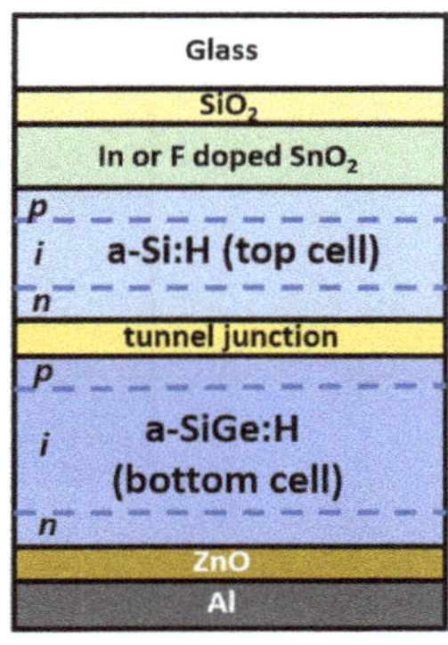

Fig. 4.5 Structure of a-Si:H single junction (left) and a-Si:H/a-SiGe:H double junction (right) solar cells (layers thickness not at scale)

lengths are short in doped a-Si:H, an undoped intrinsic region is included between the p-type and n-type regions delivering a p-i-n structure (Fig. 4.5). The intrinsic region increases the thickness of the cell thus improving its absorption properties. The built-in bias between p and n regions is dropped along the intrinsic region creating an electrical field which drives charge separation [94]. Several strategies to improve power conversion efficiency of these cells are used; despite its large band gap (1.7 eV) the open circuit voltage is low (around 0.9 V) due to the high activation energies of the amorphous material which results in a low built-in bias; this can be improved by adding a wider band gap emitter such as a-SiC:H. The inclusion of additional layers can be extended to a multijunction tandem design, where two or more p-i-n cells are monolithically connected in series, the cells use a relatively thin intrinsic layer with larger thickness in the back cell in order to match the photogenerated current. If the material for each of the cells in the tandem include alloys, the absorption profile can be tuned since each of the cells will have a different energy gap; cells with a-SiC:H alloy in the active layer have a wider gap, while those with a-SiGe:H have a narrower gap (Fig. 4.5); tandem cells with two, three or four stacked cells have been manufactured with efficiencies up to 13.6%, in some cases including a graded Ge content in the absorber layer [109]. When the tandem cells include both amorphous and microcrystalline silicon material, they are called "hybrid micromorph" devices [80, 81].

4.2.2 *Cadmium Telluride*

Cadmium telluride (CdTe) was used to produce solar cells with power conversion efficiencies above 5% since early 60s [23]. It is a semiconducting material compounded with atoms from groups II and VI of the periodic table. Many II-VI combinations are possible, but only CdTe and zinc telluride (ZnTe) can be doped both n-type and p-type. While ZnTe has a too large band gap to be a good absorber (2.26 eV), CdTe has a direct band gap of 1.54 eV at room temperature and good absorption coefficient above 10^5 cm^{-1} at a wavelength of 700 nm, and a layer as thin as a few μm can

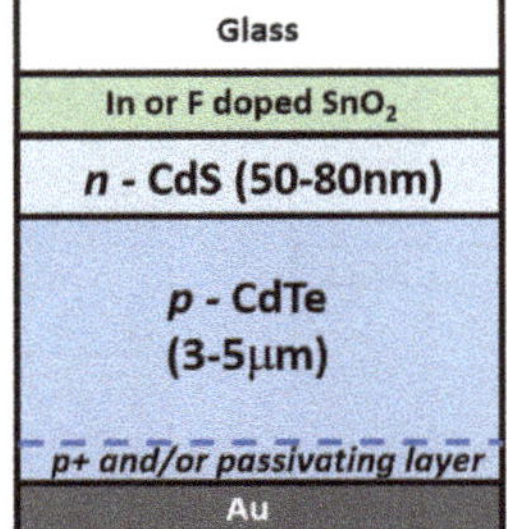

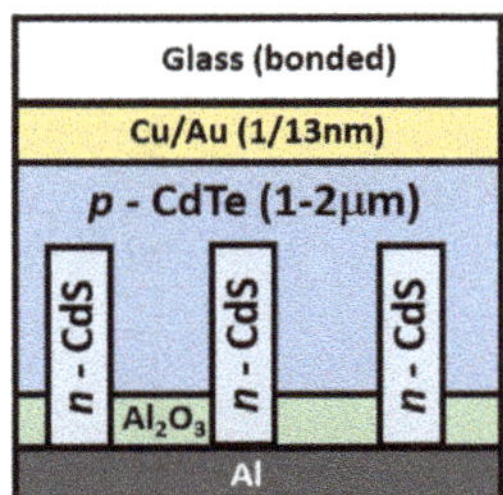

Fig. 4.6 Structure of CdTe cells. Left: layered n-CdS/p-CdTe heterojunction cell grown on a glass substrate with Au back contact. Right: n-CdS nanopillars/p-CdTe matrix heterojunction, with ultra-thin Cu/Au front contacts (layer thickness not at scale)

absorb up to 90% of incident light [61]. It was considered as one of the best alternatives to crystalline silicon to produce cheap and efficient solar cells. Nevertheless, there are important technical difficulties to manufacture high quality monocrystalline or polycrystalline films; CdTe crystallizes with a wurtzite crystal structure, but the size of the crystallites is a few μm wide and the grain boundaries (with excess Te atoms) give rise to defect states deep in the band gap, thus creating recombination centres that reduce the efficiency of the cells. Furthermore, the high intraband defect density of states reduces the p-type doping efficiency, although several treatments have been developed to saturate the intraband traps and rise power conversion efficiency. An additional difficulty arises at the CdS-CdTe heterojunction due to lattice mismatch or to the formation of compounds such as $CdTeO_3$ that increases junction recombination [94].

The typical CdTe cell structure is presented in Fig. 4.6. The p-n heterojunction cell is formed by a n-type CdS layer and a p-type CdTe layer, deposited sequentially on a glass substrate covered with a transparent conductive oxide (typically indium or fluor-doped tin oxide). An extra doping process of CdTe with lithium or copper to improve the conductivity near the rear contact is carried out before deposition of the back metal contact. This thin p^+ layer is required to improve ohmic contact since CdTe has a high work function which makes difficult to find a metal with work function higher than the hole affinity of CdTe (–5.78 eV). Different metals or alloys and compounds such as Au, Al and ZnTe/Cu are being tested to overcome this problem [61].

Surprisingly, in the first years of development of CdTe solar cells, a very simple fabrication procedure was reported by Nakayama et al.; both CdS and CdTe layers were screen printed from a paste on a glass substrate covered with In_2O_3. The process was followed by a thermal annealing in nitrogen atmosphere (800 °C) and immersion in an aqueous cuprous solution to create a Cu_2Te p-type layer that was contacted with silver paint; the resulting "ceramic" layer was around 10 μm thick and showed a power conversion efficiency of 8.1% [90]. Despite this impressive advance with a potentially very cheap printing technique, most of ulterior developments were carried out with vacuum evaporation or chemical bath techniques since the barrier of 10% power conversion efficiency was very difficult to surpass due to the poor control of the thin layer crystallinity and the high density of defects. Only in mid-90s a jump in power conversion efficiencies was achieved, reaching 16% with a high control of deposited layers; for example, CdS as thin as 50nm on ITO/glass substrate was achieved [5]. Several techniques were optimized: physical vapour deposition (PVD), chemical vapour deposition (CVD), close space sublimation (CSS), electrodeposition (ED), metal-organic CVD (MOCVD), radio frequency sputtering (RFS) and molecular beam epitaxy (MBE) [61]. Activation of the solar cells by heating in $CdCl_2$ is an essential step to improve material quality, but this step involves the use of toxic chemical compounds. Alternative routes for the activation have been proposed such as annealing the CdS/CdTe structure in a mixture of argon and a non-toxic gas containing Cl_2 such as chlorodifluoromethane ($CHClF_2$) at 400 °C [103]. Also, sputtering in argon atmosphere with a low content of CHF_3 incorporates fluorine to the CdS layer, which grows with higher quality (more dense and defect free), with the additional benefit of forming CdF_2 at the grain boundaries, which are passivated and thus delivering a better CdS/CdTe interface [100]. Good contacts to electrodes have been a challenge in this technology and buffer layers between the transparent conduction oxide and the CdS layer were included, such as SnO_2 and ZnO_2. The ternary compound cadmium zinc telluride has also been used as active layer of solar cells since its band gap can be tuned from 1.45 eV to 2.26 eV depending on the proportion of zinc added to the alloy ($Cd_{1-x}Zn_xTe$) thanks to the band engineering of the II-VI alloys (see Fig. 4.8); post-treatment of deposited films with $ZnCl_2$ vapour at 400 °C improved the crystallinity of the alloy, reduced the density of defects and delivered a high $V_{oc} = 0.78$ V [61]. A promising approach still with low power conversion efficiencies but with a high potential to reduce the use of materials in ultra-thin cells was also reported [28, 29]: the cells are fabricated by growing single-crystalline nanopillars of n-type CdS directly on an aluminium substrate by vapour–liquid–solid (VLS) process using an anodic alumina (Al_2O_3) template and gold nanoparticles as seeds, the template is then partially etched and replaced by a 1 μm thick polycrystalline matrix of p-type CdTe by chemical vapour deposition (CVD); the cell is completed with the top electrical contact, fabricated by the thermal evaporation of Cu/Au (1 nm/13 nm) and delivering good ohmic contact with the CdTe layer thanks to the high work function of gold (Fig. 4.6). Due to the low transparency of either the aluminium substrate or the Cu/Au top contact, the power conversion efficiency of this cell was low (6%), but it could be improved and it opens the door to nanostructured CdTe solar cells. This approach has inspired the fabrication of core/shell nanowires for nanostructured solar cells looking for low-cost solution-based processing, such as CdS/CdTe, or CdS/Cu_2S [2, 121].

Also module manufacture was improved by deposition of materials on very large surfaces (>1 m^2) and using laser scribing to create a serially connected string of cells. This progress allowed manufacturing companies to be optimistic, since the predicted requirements for competitivity against crystalline silicon had been surpassed by the end of the 90s [76]. First Solar was the first company to develop CdTe solar cells with efficiency above 21% in 2014 and to manufacture CdTe modules with 19%

efficiency and large area (2.3 m^2) in 2019, but it is struggling to get a good market share in competition with c-Si modules [46].

4.2.3 Chalcopyrites and Kesterites

Copper indium diselenide ($CuInSe_2$, or CIS) has been developed since the 70s as a material with high potential for photovoltaic applications since it is a direct band gap semiconductor, with energy gap around 1eV and high optical absorption. Due to its properties, an active layer of a few μm could absorb as much as a silicon solar cell of hundreds of μm thickness, additionally, it can be doped both p-type and n-type. The material has a chalcopyrite crystalline structure, with grains of size around 1μm that tend to grow in columnar shape and therefore with most grain boundaries perpendicular to the p-n junction direction. Initially, p-n homojunction CIS cells were fabricated, but low power conversion efficiencies (less than 5%) were obtained; by using a n-type CdS layer, deposited on top of p-type CIS layer, p-n heterojunctions were fabricated. Due to the large energy gap (around 2.5 eV), it acts as a window layer which is highly doped so that surface recombination losses are reduced at the same time that transport of electrons from the junction is improved due to series resistance reduction. A further improvement was obtained by the partial substitution of indium by gallium in the active layer alloy ($CuIn_{1-x}Ga_xSe_2$, or CIGS), the addition of Ga increases the band gap to 1.3 eV and improves the electronic properties of the back contact [85, 94]. The typical CIGS cell structure is completed by adding the electrodes, which usually are a molybdenum layer on top of the glass substrate that acts as hole collecting electrode and a bilayer of intrinsic and Al-doped zinc oxide (ZnO) on top of a transparent conducting oxide acting as an electron collecting electrode (see energy level scheme in Fig. 4.7 from [20]).

The use of scarce and expensive indium in the CIGS solar cell motivated research on similar structures in which it can be replaced by abundant and cheaper materials.

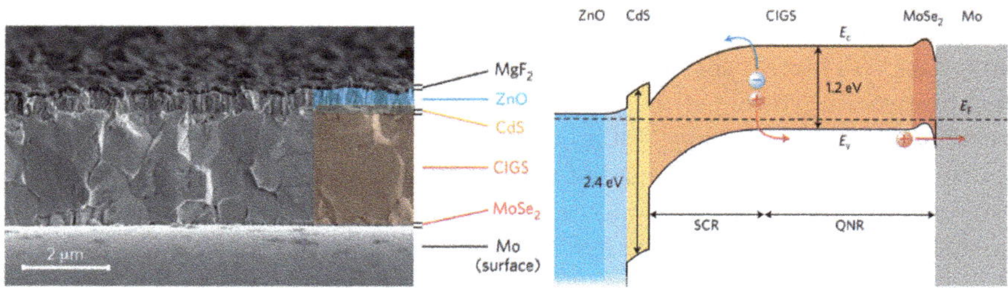

Fig. 4.7 Electronic and optical structure of CIGS solar cells. Left: Scanning electron micrograph of a cross section obtained from a complete CIGS device. Right: schematic band diagram of a CIGS solar cell under zero-bias voltage condition. Conduction band energy (EC), valence band energy (EV), Fermi level E_F, space charge region (SCR), and quasi-neutral region (QNR) are indicted as well (Reproduced with permission from [20])

This effort lead to the development of the kesterite solar cells, in which two group III atoms in the chalcopyrite are replaced with a group II and a group IV atom. The composition of the absorbing layer is therefore $Cu_2ZnSnSe_4$ (or CZTS). Both in CIGS chalcopyrite and CZTS kesterite solar cells, the selenium can be partially replaced by sulphur. The kesterite solar cells are expected to have similar behaviour since they are isovalent and have similar crystalline structure compared to chalcopyrite cells, but so far, best efficiencies with kesterite cells are still far from those obtained with the chalcopyrite cells [6, 101]. This is mainly due to the high density of defects both in the bulk and the interfaces, leading to very high recombination rates. A big research effort on the combination of materials for the kesterite alloys and modelling of the resulting properties is being carried out with the aim to improve the photovoltaic parameters of the kesterite solar cells with special focus on the understanding of the role of atomic disorder on the cation sub-lattice, as well as unwanted phase separation of $Cu_2ZnSn(S_x,Se_{1-x})_4$ alloys into ZnS, ZnSe, $CuSnS_3$ and $CuSnSe_3$ by the effect of thermal treatments after deposition of materials [54, 128].

Both chalcopyrite and kesterite solar cells are manufactured by vacuum deposition techniques following two main routes: (i) the co-evaporation of the elements either uniformly deposited or using the so-called three-stage process, or (ii) the deposition of the metallic precursor layers followed by selenization and/or sulphidization. The deposition stage is usually followed by thermal annealing, involving temperatures as high as 500 °C to enhance grain growth and recrystallization [85]. The CdS window layer and other anti-reflecting coating layers are deposited by chemical bath either in liquid or vapour phase. The structures with CdS/i-ZnO/n-ZnO window and buffer layers are good for finishing both chalcopyrite and kesterite devices, with ZnO doped with aluminium or boron and partial substitution of Zn by Mg in the ZnO layer. Other window layers acting as n-type material have been used, such as ZnS, ZnSe, In_2S_3, (Zn,In)Se, Zn(O,S) and $ZnMgO_6$ that can be deposited by wet chemistry or dry processes [115]. Recently, manufacture of chalcopyrite and kesterite solar cells by printing techniques are being explored with the aim to reduce process complexity and energy consumption by avoiding high temperature annealing processes (which will also permit the use of flexible plastic substrates). All solution-processed chalcopyrite and kesterite solar cells have been demonstrated with efficiencies slightly above 17 and 10% respectively, approaching the results of best vacuum processed cells (23.4 and 12.6% in 2021 [47]). The best results are obtained by routes that use hydrazine (N_2H_4) based solutions which can be either fully dissolved molecular species or a mixture of suspended nanoparticles and dissolved molecules. The properties of the inks and the printing methods have to be fine-tuned to control the growth of the different layers. Since hydrazine is a flammable and highly toxic material, effort has been devoted to explore non-hydrazine-based processing routes, although the fabricated cells show poorer performance than those based on hydrazine routes; thermal annealing steps in the printed processing routes are still broadly used, specially for a final selenization step, although in some limited cases temperatures have been reduced to as low as 200 °C opening the door to the use of cheap plastic substrates [6].

On top of the problem of defects as mentioned above (specially for kesterites), it is difficult to control the growth process since differences in the composition of the two materials in the p-n heterojunction may lead to the formation of unwanted species due to diffusion of atoms across the junction, such as $CuSe_2$ and CuS_2 in chalcopyrite cells and CuZn or SnZn in kesterite cells, creating narrow spikes in the conduction and valence bands near the junction that act as recombination centres and barriers to charge collection, since the carriers have to tunnel through the spike before it can be collected. Furthermore, the effect of alkali doping of the absorbing layer by Na or Li (or heavier K, Rb or Cs) ion migration from glass is still under discussion, but may have a positive effect on the improvement of stability and power conversion efficiency of both chalcopyrite and kesterite solar cells; also grading of composition by inclusion of Ga in the CIGS and Ge in the CZTS absorbing layers and postsurface sulfurization (or sulfurization after selenization, SAS), is being explored [54].

4.3 III-V Technologies

Also called third generation technology, it aims at producing solar electricity at a competitive cost with solar cells that are expensive and therefore they must have an *ultra high conversion efficiency* [42]. This can be achieved with tandem configurations in which several cells are monolithically connected in series, each having materials in the active layer that are good light absorbers with different spectral sensitivity. Ideally, an infinite stack of solar cells, each one absorbing photons of a given energy and using all incident photons to generate electrons (assuming that they can be extracted efficiently from the active layer of each cell) will have a power conversion efficiency close to the detailed balance thermodynamic limit (93% for a temperature difference between the Sun surface at 6000 K and solar cell at 300 K, while the limit calculated by Shockley and Queisser for a single junction solar cell is 30% [114]. For more practical designs with a limited number of cells within the tandem structure, the efficiencies that can be achieved depending on the combination of optimal band gap for each of the subcells in the tandem, ranging from 44.9% for two cells to 58.2% for incident spectral Sunlight AM1.5G, which can be further increased up to 68.5% under a $\times 1000$ light concentration factor [84].

The combination of elements of groups III and V of the periodic table produces a III-V semiconducting alloy with a band gap that can be engineered depending on the relative amount of the elements. In binary compounds, an equal number of elements of groups III and V are included in the unit cell of the material, which is a *zincblende* crystal structure (two interlocking face centred cubic lattices) [94]. Group III atoms contribute three valence electrons to bonding and group V contribute five, all the valence electrons are used in the bond and therefore an energy gap opens between both bands (at room temperature the valence band is mostly filled and the conduction band is mostly empty). When some of the atoms of group III are replaced by atoms of a different element of the same group, a ternary blend is obtained and the band gap can be modified in a controlled way depending on the amount of substitution.

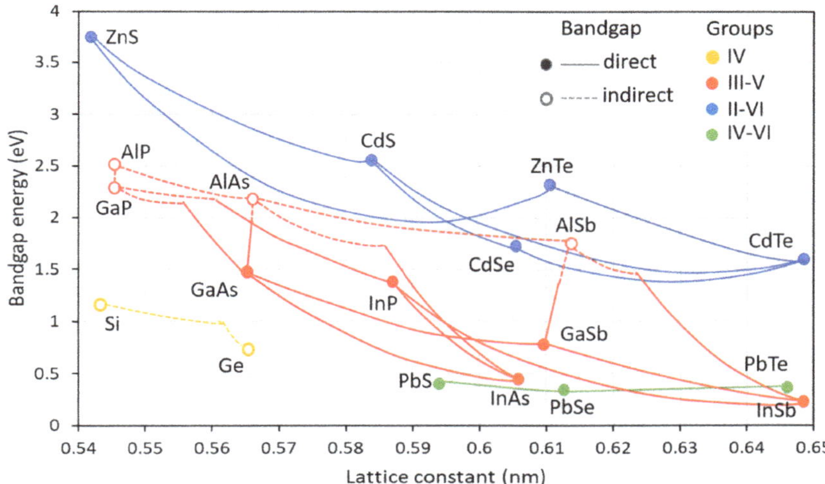

Fig. 4.8 Band gap energy and lattice constant of binary semiconducting alloys (symbols). The band gap can be engineered by partial substitution of elements leading to ternary semiconducting alloys, whose band gap and lattice constant is indicated by moving along the lines (dotted for indirect band gap and continuous for direct band gap)

In Fig. 4.8 some of the possible combinations are shown, the vertical axis indicates the band gap of a given alloy, the horizontal axis indicates the lattice constant of the unit cell; if two binary compounds have similar unit cells (mismatch lower than 5%) a ternary compound can be grown on top of a binary one and the bulk material still has good crystalline quality (with low strain and low defect density) and the band gap can be tuned accordingly if an epitaxial control of the doping and growth of the layers is achieved, delivering good quality p-n heterojunctions stacked in a tandem device. Doping in III-V semiconductors can be achieved by replacing one of the elements with another of different valence, acting as donors or acceptors in a similar way to doping in silicon.

For many of the possible combinations of III-V alloys, the gap is direct (gallium arsenide, GaAs, indium phosphide, InP, and gallium antimonide, GaSb). In particular, GaAs, with near-optimal direct band gap (1.42 eV) has an absorption coefficient ten times larger than silicon (with indirect 1.12 eV band gap) and therefore less material is required to manufacture a good absorbing layer. A single junction GaAs cell of a few μm thick could theoretically achieve 31% power conversion efficiency close to the thermodynamic limit for a single junction cell. Another advantage of GaAs compared to Si cells is a better temperature coefficient (temperature losses are lower) which is important for cells operating under concentration at high temperature. GaAs can be doped with silicon atoms, which act as donors when replacing some trivalent gallium atoms in the lattice, thus delivering a n-type material; tin is sometimes also used as tetravalent donor impurity to deliver n-type doping. For p-type doping carbon is the most widely used impurity, in this case, the carbon atom replaces an arsenic atom, which is pentavalent, creating a deficiency of valence electrons and thus acting as an acceptor impurity and delivering a p-type material. Alternatively, and arsenic atom can be replaced by a group II element, like beryllium, which creates an acceptor state and also delivers p-type material. Other relevant binary alloys are indium phosphide (InP) and gallium antimonide (GaSb). The ternary alloys can be fabricated with different fractions x of the third element that replaces some of the gallium atoms, for example, aluminium gallium arsenide ($Al_xGa_{1-x}As$), indium gallium arsenide ($In_xGa_{1-x}As$) and indium gallium phosphide ($In_xGa_{1-x}P$).

In the first approaches to III-V technology, n-type wafers from single crystals produced using either the liquid-encapsulated Czochralski (LEC) method or a Bridgman method were used [12, 24, 64]. But the best power conversion efficiencies were achieved with cells that are grown by a variety of epitaxial techniques: the most commonly used is molecular beam epitaxy (MBE), followed by others such as metal-organic chemical vapour deposition (MOCVD), metal-organic vapour phase epitaxy (MOVPE) and liquid phase epitaxy (LPE); the need to grow high-quality crystalline layers with very low impurity content almost make impossible the fabrication with cheaper processing routes already used in other technologies [85]. In all cases, the active layer of the solar cell is a single crystal. The slow growth of the layers (epitaxial atomic monolayers) enables a very high control of the atoms in each layer, which is used to carefully tune the composition of each layer. The doping impurities can be introduced during the growth of the layers or in a later stage by diffusion of chemical compounds. The configuration of single junction cells can be either a p-n design, or an n-p design, in both cases with a thin emitter (0.5 μm emitter for the p-n design and even thinner, 0.2μm for the n-p design) and a thicker base (2–5 μm), much thinner than in silicon. The tandem cells include a monolithical configuration with tunnel junctions between the stacked cells to create a series connection between each cell in the tandem. In Fig. 4.9, a schematic draft of a single junction cell (left) and a tandem with three cells (right) are shown. In all designs, the series resistance should be minimized, since it is a critical parameter when the cells are going to work under highly concentrated light.

A practical problem of III-V technology is the high cost of materials and the need to use cheaper substrates to grow the epitaxial layers that constitute the single junction or the tandem devices. This problem is still to be solved, recycling of GaAs wafers is one way to reduce cost; also germanium (Ge) substrates are commonly used, with a very good lattice constant match with GaAs, thus allowing epitaxial growth on Ge, although it is also a scarce and expensive material. Multijunction III-V cells have also been bonded to Si wafers, delivering efficiencies similar to those on Ge [18]. Also cells have been grown on GaAs substrates that are removed at a later stage of the process and recycled. When very thin substrates are used, the cells need to be glued to additional materials acting as supporting substrates to provide mechanical strength.

This third-generation technology was intended to provide solar electricity at a competitive cost; ten years ago it was expected that III-V tandem solar cells, once they achieved very high efficiency, will dominate the market [42]. In 2021, this technology holds the efficiency records for single junction cells: GaAs,

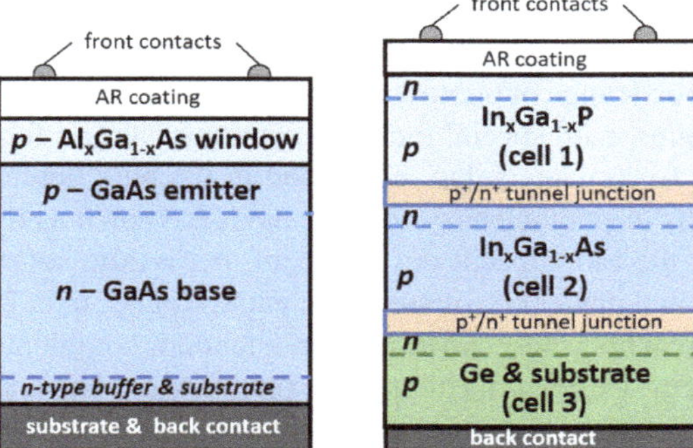

Fig. 4.9 Structure of III-V cells. Left: a single junction GaAs cell with an $Al_xGa_{1-x}As$ window layer. Right: a triple junction tandem cell, grown on a Ge substrate that also acts as third cell; in the monolithic tandem configuration with two contacts, tunnel junctions must be included between the cells (layer thickness not at scale)

29.1% under 1 Sun standard conditions (global AM1.5 spectrum at 1000 W/m^2 at 25 °C) and 30.5% under $\times$258 concentration [45, 65]; and for six junctions: AlGaInP/AlGaAs/GaAs/GaInAs($\times$3) tandem cells, 39.2% at 1 Sun and 47.1% under $\times$143 concentration [35]. Despite this progress, the deployment of III-V technologies for terrestrial applications has not occurred so far. In part it is explained because c-Si technology has reduced its cost below what could be expected a few years ago, and in part because the III-V tandem technology without concentration is still extremely expensive (cost per square meter of module) and working under concentration requires additional investment in the balance of system (BoS) components, modules with incorporated non-focusing Fresnel lenses and heat dissipation systems and biaxial solar trackers, which increases BoS cost and maintenance.

4.4 Organic and Hybrid Emerging Technologies

One of the most active areas in physics, chemistry and materials science in the past decade has been the development of organic and hybrid photovoltaic technologies. Thousands of articles have been published in scientific journals, hundreds of books in many publishers worldwide. Since the first demonstration devices, these emerging technologies were considered as one of the best alternatives to silicon and thin film technologies, capable to deliver very cheap solar electricity at a massive scale with low weight and flexible modules. This prediction has not been fulfilled yet, since crystalline silicon modules have continued reducing its price and have kept its market share; neither thin film inorganic technologies, nor emerging organic and hybrid have posed a real challenge to its market domination so far. Nevertheless, due to the large portfolio of materials and processing techniques that can be used to manufacture organic and hybrid photovoltaic modules, it is possible that a strong reduction in environmental impacts and economical costs will enable a disruptive penetration in the market for a broad range of photovoltaic applications. In this section, an overview of these technologies, with special focus on the materials and fabrication processes is presented, first for organic solar cells based in the bulk heterojunction concept, then a special class of photoelectrochemical devices known as dye sensitized solar cells, and finally, the most recent development, perovskite solar cells which have already demonstrated power conversion efficiencies higher than 25% in single junction devices and reaching 30% when used in a tandem combination of perovkskite and silicon junctions. Despite this impressive progress, organic and hybrid solar cells still do not have reached the market, although some companies are starting to sell modules for specific applications where flexibility and low weight provide a competitive advantage. The main drawback for a strong penetration in the market is the poor stability of the devices, due to degradation mechanisms still under investigation, the lifetime of these technologies is still too short to compete with crystalline silicon or thin film technology. A big effort in research is devoted to the extension of lifetime and impressive progress has been achieved in the past three years.

For a more in depth study of these technologies, there are excellent reviews and books with a detailed analysis of the physical and chemical phenomena involved in its operation and manufacturing procedures [9, 11, 49, 69, 95, 97, 98, 141, 146].

4.4.1 Organic Bulk Heterojunctions

Since the development of organic semiconducting polymers, its application to a broad range of electronic devices, previously manufactured with inorganic materials, has created a whole industry called "plastic electronics". Small organic molecules are also included in the class of materials used in plastic electronics. Organic light emitting diodes are already dominant in the screen industry, and growing in energy-saving lighting applications [32, 122], and organic thin film transistors are broadly applied in flexible and/or disposable electronic products [36]. The possibility of using solution processing techniques, including roll-to-roll printing of cheap and abundant materials prepared by organic chemistry recipes, opens the door to low-cost manufacturing of light weight and flexible organic photovoltaic modules [58, 95].

The first organic solar cell was prepared by creating a bilayer molecular heterojunction. A layer of copper phthalocyanine (CuPc) followed by a second layer of organic material, a perylene tetracarboxylic derivative (PV) and a final silver electrode were evaporated in vacuum on a glass substrate covered with indium tin oxide (ITO), a transparent conductive oxide. The cell had a low power conversion efficiency (around 1% at AM2 illumination), but already contained all fundamental ingredients for the development of organic photovoltaic devices: two organic materials with different electron affinity and two selective electrodes; interestingly, also a stability test was carried out during five days, in which J_{sc} and V_{oc} showed a degradation of $<2\%$

of its initial value, while FF degraded 30%, which was attributed to an increase in series resistance due to the degradation of the Ag electrode [120].

An important step was the use of conjugated polymers in the active layer of organic solar cells. The Nobel Prize in Chemistry in 2000 was awarded to Prof. A. J. Heeger, Prof. A. G. McDiarmid and Prof. H. Shirakawa for "*the discovery and development of electrically conductive polymers*". The prize recognizes a work of many years, since the 70s, to develop a special class of polymers with sp^2 hybridization which enabled to conduct electricity. The work started with polyacetylene, as the first polymer to show a conductivity which increased with temperature, like a semiconducting material [19, 113]; improvements in synthetic routes allowed the preparation of highly conductive thin films of polyacetylene and a broad class of sp^2 polymers that could be used in organic electronic devices and in particular in solar cells [112]. The conjugated polymers have a backbone of sp^2 hybridized carbon atoms and the overlapping of neighbouring p_z orbitals along the backbone creates delocalized π electronic systems which extend through several monomers of the backbone (lengths of a few nanometers). The energy difference between the highest occupied molecular orbital (HOMO) and the lowest unoccupied molecular orbital (LUMO) creates an energy gap of a few electronvolts (typically between 1 eV and 3 eV). Photons can be absorbed creating a neutral excited state called *exciton*, which can be considered as a bound pair of the photoexcited electron (in the LUMO) and a hole (in the HOMO) which is localized in a small volume of a few nm^3 of the molecular material or the polymer segment. Because of this localization, the main difference with inorganic semiconductors is that the exciton is bound by Coulomb interaction with an energy around 1 eV and it is difficult to dissociate into a separated electron and hole before it decays to the ground state. In inorganic materials, the exciton is extended to much higher volume, the Coulomb interaction is around 0.1 eV and the exciton is almost immediately separated into different charge carriers in the valence and conducting bands. In the conjugated polymer, the separation of the exciton can be assisted by adding a second organic molecule with higher electron affinity and thus acting as electron acceptor; the electron is transferred to the acceptor and the hole remains in the donor, both still bound by Coulomb interaction, they form a *geminate* pair that must be separated into different charge carriers that should travel to the respective electrode to be collected and delivered to the external circuit. Usually, the photogeneration process occurs in the donor material (a conjugated polymer, that can be considered as a p-type material by analogy to the inorganic case) and the acceptor (a molecule or another conjugated polymer with higher electron affinity, that can be considered as a n-type material). The difficulty is to separate the exciton into a geminate pair (still bound by Coulomb interaction) and then collect the carriers before they recombine, either before separation (in this case it is called geminate recombination) or after separation (called non-geminate recombination, occurring with an carrier generated in a different part of the material). For a good performance of the organic solar cells, the materials constituting the active layer must have energy levels carefully tuned to enable good light absorption and to facilitate charge separation after exciton generation, but also the layer morphology demands two important requirements: firstly, the size of the domains of donor an acceptor phases has to be of the order of the exciton diffusion length (a few nm), so the exciton created can find an interface between both phases and separated into a geminate pair before it decays; secondly, the donor and acceptor phases must have a percolation path to carry the separated charges to the different electrodes. On top of that, electron and hole transporting layers may be included to facilitate charge extraction from the active layer to the electrodes, they must be different: one with low work function (for electron collection) and another one with high work function (for hole collection).

The structure of the organic solar cell with bilayer structure in the active layer is not optimal since one planar interface between the donor and acceptor materials and with enough thickness to absorb light efficiently does not comply simultaneously with those requirements: a thickness of a few hundred nanometers is needed for light absorption, but then most of the photogenerated excitons will decay before they can be collected since they will be too far from the donor/acceptor interface. All inorganic solar cells, either single junction or tandem are designed with layered structures in which the p-n junction is planar; the organic equivalent was not delivering good efficiencies, although the possibilities of organic chemistry to synthesize a huge variety of compounds in principle granted multiple combinations for donor/acceptor interfaces. It took almost ten years to find a solution to this apparent contradiction: an active layer composed of two *interpenetrating polymer networks* which creates a distributed internal donor/acceptor heterojunctions; the so-called *bulk heterojunction* [50, 144]. The bulk heterojunction (BHJ) can be composed of two polymers with different electron affinity (donor and acceptor), or a polymer (usually a donor) and a molecular material (usually an acceptor). The other layers of the device have a planar architecture, either with traditional or inverted architecture as can be seen in Fig. 4.10. In the traditional or standard architecture, the electrons are collected by an electrode evaporated on top of the organic layers (usually Ca, or Al, electrode, with low work function and assisted by an electron transporting layer, ETL, acting as cathode interlayer) and the holes are collected by the transparent conductive oxide (usually indium doped tin oxide, or ITO, with a hole transporting layer, HTL), while in the inverted architecture, electrons are collected by the transparent conducting oxide (the ITO, but in this case with a zinc oxide ETL, while the holes are collected by a silver electrode (with higher work function and assisted by a HTL) [70]. The advantage of the inverted architecture is two-fold: on the one hand, it is more stable, since Ag is less reactive than Al or Ca (although Ag stability in oxygen is low) and on the other hand, it can be printed from a paste, avoiding vacuum evaporation processing. Furthermore, by including selective ETL and HTL layers, Ag can be used to fabricate both electrodes thus enabling an ITO-free all-solution fully printed roll-to-roll organic solar module manufacture [68].

Organic solar cell manufacture combines the possibility of designing and synthesizing a broad range of novel polymers and organic molecules for all the layers in the device, with the option of designing solar cells with different architectures, standard and inverted, or including tandem stacks, that can be manufactured by potentially low-cost processing routes based in all-solution printing methods that can be carried out in air, at low temperatures and on different substrates (glass, plastics, ceramics or metals). In 25 years, the portfolio of materials has increased enormously, from the

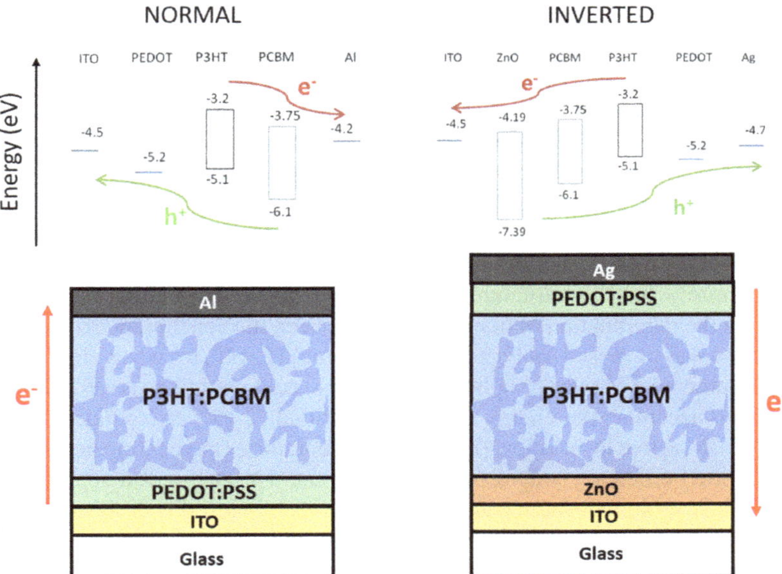

Fig. 4.10 Energy levels and architecture of normal and inverted organic bulk heterojunction (P3HT:PCBM) solar cells. In the upper plots, the HOMO, LUMO and work function of typical materials are shown, the organization of the layers changes the electron extracting electrode (opaque aluminium for the normal architecture or transparent ITO for the inverted architecture)

first demonstration devices using only one conjugated polymer, to the complex structures of tandem devices comprising several blended active layers and other organic and inorganic transporting layers and electrodes. A short summary of materials and device designs of plastic solar cells are presented below.

Photodiodes were fabricated with a thin film (100nm) of a conjugated polymer, poly-*p*-phenylene-vynilene (PPV), which is a good hole transporting material, on a glass substrate with an ITO transparent electrode and aluminium (or calcium or magnesium) as low work function opaque electrode evaporated on top of the structure [79]. This photodiode, initially considered as a photodetector device, achieved quantum efficiencies around 1% pointing to its possible use as a solar cell. Side chains were added to the PPV to increase solubility delivering poly[2-methoxy-5-(2'-ethyl-hexyloxy)-1,4-phenylene vinylene], MEH-PPV, and then a second component was blended in the active layer, the fullerene C_{60} used as "sensitizer" with higher electron affinity, thus creating for the first time a blend of donor and acceptor materials; another conjugated polymer, poly[3-octyl thiophene], P3OT, was also used in the photodetectors [145]. The films were spin cast from solution, delivering very thin active layers (100–300 nm). The road to create efficient plastic solar cells was open; all ingredients and recipes were already on the table.

The first bulk heterojunction solar cell had an active layer composed of a mixture of MEH-PPV and another modification of PPV, synthesized by adding a cyano group to increase its electron affinity in order to act as electron acceptor in the blend (CN-PPV). It was fabricated by spin casting on an ITO covered glass substrate and then annealed at 100 °C, a low temperature which could make it possible to fabricate solar cells on a plastic substrate [50]. The polymer mixture is segregated in different phases due to the low entropy of mixing, thus generating two interpenetrating networks which percolate throughout the volume of the active layer; the control of the morphology of these networks by thermal or solvent annealing is a technological tool to improve the nanostructure of the active layer at the required length scales. Very soon the combination of materials that has been the benchmark active layer for a long time was proposed: poly[3-hexyl-thiophene] (P3HT) acting as electron donor and a fullerene derivative [6,6]phenyl C_{61} butyric acid methyl ester (PCBM) acting as electron acceptor; the energy gap is not optimal (2 eV) and efficiency for this P3HT:PCBM active layer is low (hundreds of laboratories have fabricated cells, with most reported efficiency values between 3.8% and 4%, with a few reporting up to 5%), but it is very reproducible and many laboratories used this design as a starting point for further developments [25].

Several strategies were followed to increase light absorption by the polymers in the active layer, the phenyl units were replaced by thiophene units (as seen above for P3OT and P3HT) to planarize the backbone and extend the π conjugated domains; if π-stacked aggregates form, the energy gap is reduced. Bridging atoms have also been used to planarize the polymers [89]. The modification of the polymer backbone by combining electron donor and acceptor units enables a fine-tuning of the energy levels (energy gap, HOMO and LUMO) of the polymer in the blend, for example, including benzothiadiazole, carbazole, dithienobenzene or diketopyrrolopyrrole units in the copolymer backbone creating a so-called "push-pull" structure [13, 99]. The absorption of the electron acceptor component of the blend is low because the high symmetry of the C_{60} molecule forbids some optical transitions; if the molecule reduces its symmetry, the optical absorption is enhanced and therefore C_{70} with good blue and green absorption are currently used in combination with low band gap polymers; other modifications of PCBM, such adding more side chains or including endohedral atoms to slightly reduce its electron affinity and thus enhance the Voc of the cell have also been realized [55, 74, 104]. The use of non-fullerene acceptors (NFAs) in recent years has increased the absorption of the acceptor in the blend and therefore has boosted photocurrent generation, and at the same time increased V_{oc}, leading to cells with very good power conversion efficiency, reaching values comparable to inorganic cells (>18%) [77, 132, 133]. A large variety of NFAs have been used; they can be organized in three main groups; the first one based on Acceptor-Donor-Acceptor (A-D-A) calamitic small molecules, like fluorene, carbazole, indaceno-dithiophene and indaceno-dithieno-thiophene based acceptors; the second group based on perylene-3,4:9,10-tetracarboxylic acid diimides (PDIs), that may comprise one or more PDI monomer units in planar structures and a third group of polymer acceptors, such as polymeric naphthalene diimides, perylene diimides or terpolymers [127]. A large variety of soluble low band gap conjugated polymers acting as donor (and occasionally acceptors) and fullerene or non-fullerene molecules for acceptors have been synthesized and used in single junction or tandem organic solar cells [4].

Another task in organic solar cell manufacture is the control of the blend morphology, which should have phase segregated domains of a few nm width and should percolate in networks throughout all the volume of the active layer [33, 78] Solution processing can be tuned by using different solvents and by the addition of side chains to modify the polymer solubility in halogenated and non-halogenated solvents. Combination of solvents, and using solvent/anti-solvent combinations, processing additives, and thermal annealing procedures permits good control of the nanostructure for solution-processed devices, including tuning of optimal phase separation and vertical segregation of components in active layers that have a typical thickness between 100 nm and 300 nm [17, 126].

Electron transporting layers (ETL) and hole transporting layers (HTL) are usually included in the cell structure to improve charge carrier collection with good ohmic contact at the electrodes and to reduce surface recombination: low work function materials such as lithium fluoride (LiF) as ETL and high work function materials such as NiO, WO, MoO_3 and V_2O_5 as HTL [86]. The conjugated polymer poly[ethylene-dioxythiophene] doped with poly[styrene sulphonate] (PEDOT:PSS) can be used both as HTL (in standard cells) and as ETL with the help of an additional zinc oxide (ZnO) or titanium dioxide (TiO_2) layer which have low work function (in inverted cells). Other polymers have been included in transporting layers, like polyoxyethylene tridecyl ether, included between ITO and solution-processed layers TiO_x [118]; or the addition of conjugated poly-electrolyte interlayers [110].

The most used transparent electrodes are tin oxide doped with indium (ITO) or fluor (FTO); the opaque electrodes are a variety of metals, used depending on the architecture of the cell: low work function (Ca, Al, Mg) for standard cells and high work function (Ag, Au) for inverted cells; an inconvenience is the requirement of vacuum-based evaporation or sputtering process of this kind of electrodes (with the exception of Ag, that can be screen printed from inks or pastes of different composition). Other alternatives for cheaper and easily processable electrodes are silver grids embedded in PEDOT:PSS, silver nanowires, copper nanoparticle-based metal grids, carbon nanotubes, or graphene layers [26, 37, 66, 135]. The ITO, FTO or other transparent electrodes can be deposited on glass or on plastic substrates, being poly[ethylene-terephthalate] (PET) the most used plastic for this purpose. ETL, HTL and electrodes have typical thickness of 50nm to 100nm.

All these layers, with the exception of some electrode materials, are compatible with solution processing. The prepared inks can be printed, layer by layer, in a continuous roll-to-roll (R2R) or sheet-to-sheet (S2S) manufacture line, using a variety of methods that are already well known and optimized in the printing industry: slot die coating, spray coating, inkjet printing, flat bed and rotary screen printing, doctor blade, flexography, gravure, knife over edge, etc.... Some of these techniques have been successfully applied to manufacture large-size organic solar cells and modules. The number of patents filed for organic solar cell technology (materials and processes) is more than 5000 (2000–2018), with a peak of more than 400/year in 2013–2015 and a slight decline since then [69, 70, 105]. Several companies are already manufacturing OPV modules: Infinity PV (Denmark), Solarmer Energy Inc. (USA), Toshiba (Japan) and ZAE Bayern (Germany), which since 2019 holds the record of power conversion efficiency (12.6%) for an organic photovoltaic submodule of 26 cm^2 [47].

With this large portfolio of materials and manufacturing routes for each layer of the organic bulk heterojunction solar cell, it seems possible that an optimal combination with low environmental impact and low cost can be found. The materials are abundant with a few exceptions (like indium), but its chemical precursors and processing routes have impacts on human toxicity and may create environmental damages that should be reduced. Temperature process is usually low, and therefore devices with less embedded energy compared to inorganic cells may be manufactured. Life cycle assessment is the tool that will provide constraints to the broad range of possibilities and point to the alternatives that are worth to investigate in more depth.

4.4.2 Dye Sensitized

Inspired by early studies of artificial photosynthesis, and based on previous structures of photoelectrochemical cells, a breakthrough was achieved in 1991 when Brian O'Regan and Michael Grätzel demonstrated a power conversion efficiency of 7% with a new design of dye sensitized photovoltaic cell [7, 41, 92, 96]. The device consisted on a transparent 10 μm thin mesoporous film of titanium dioxide particles (TiO_2) a few nanometres in size, coated with a monolayer of a charge-transfer molecular dye to sensitize the film for light harvesting; a photon absorbed by the molecule excites an electron which is injected into the conduction band of the TiO_2. The regeneration of the dye is achieved by a redox reaction in a liquid electrolyte, an electron is delivered to the dye and the redox couple in solution is reduced in a counter electrode. The transfer of electrons from counter electrode to TiO_2 is mediated by the electrolyte and the photovoltage correspond to the energy difference between the Fermi level of the TiO_2 semiconductor (n-type) and the Nernst electrochemical potential of the redox couple in the electrolyte. This working mechanism, shown schematically in Fig. 4.11, is different from a solid-state inorganic (or organic) solar cell; the presence of a liquid electrolyte resembles the original electrochemical photovoltaic cell of Edmund Becquerel.

The success of Grätzel's cell in comparison with previous designs was the mesoporous quality of the titanium dioxide layer. It was deposited on a conducting glass sheet from a colloidal solution and then heated at 450 °C to create a percolative porous network by sintering the particles. The dye covers the sintered particles and the electrolyte penetrates the pores; the contact surface TiO_2/Dye/Electrolyte is hugely increased by the mesoporous quality of the film (×2000 increase in surface area for a 10μm layer of nanoparticles around 15 nm size assuming a cubic close packing, achieving around 50 m^2/g). Similarly to the bulk heterojunction cell, the morphology of the active layer is essential for an efficient operation of the solar cell.

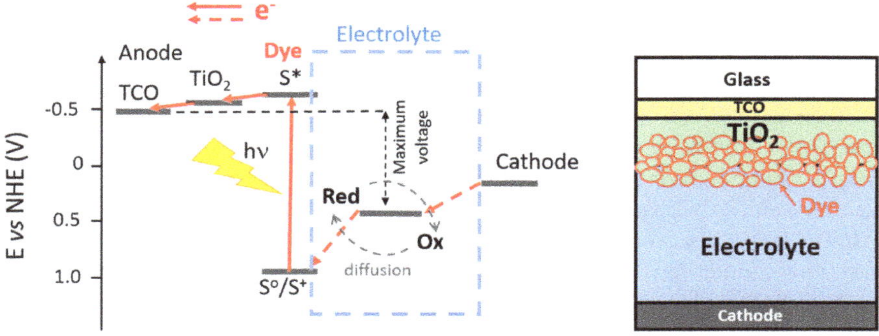

Fig. 4.11 Energy levels and architecture of a dye sensitized solar cell with liquid electrolyte showing how the intricate surface of a mesoporous TiO_2 layer is covered with dye molecules (the energy scale is referenced to Normal Hydrogen Electrode, NHE)

In the past 30 years, the power conversion efficiency of the dye sensitized solar cells (DSSCs) has almost doubled from the initial 7% to 13% in 2021. The original structure, composed of a stack of photoanode, dye sensitizer, liquid electrolyte with a redox couple and counter electrode has been kept constant although a very large variety of elements, compounds and molecules have been used for each of the layers (see Fig. 4.11).

The photoanode is the mesoporous nanostructured thin film deposited on a glass previously covered with a transparent conductive oxide, the most widely used material for the photoanode is anatase TiO_2 particles prepared by hydrolysis of a titanium precursor such as titanium(IV) alkoxide with excess water catalyzed by acid or base, followed by hydrothermal growth and crystallization [49]. The nanoparticles can be deposited from colloidal solutions by a variety of methods, such as sol-gel, spray pyrolysis, hydrothermal/solvothermal, always trying to reduce the temperature to find processing routes compatible with the use of plastic substrates. The application of a $TiCl_4$ treatment improves the quality of the films [93]. Other more complex routes like the growth of arrays of TiO_2 nanotubes or the use of alumina or polymeric templates to achieve the desired morphology were also tested, sometimes in combination with atomic layer deposition techniques [51, 88]. Also ZnO, SnO_2 and Nb_2O_5 have been used to fabricate photoanodes [141].

The dye, or sensitizer, is a critical component since it absorbs light to generate the photoexcited electron and inject it into the semiconductor anode; hundreds of molecules have been tested looking for the optimum HOMO/LUMO energy levels for effective charge injection, but also for good light absorption, good solubility and chemical stability, and good adsorption to the mesoporous photoanode (if possible avoiding surface recombination). All these tasks have been subject of intense research for the past 30 years and the results can be classified into the following groups: ruthenium polypyridyl dyes, metal-free organic dyes, porphyrin dyes and quantum dot sensitizers (alloys such as PbS, CdS, CdSe, Sb_2S_3), organic perovskites were also used and as will be seen in the next section, it was an important step towards the development of (solid state) perovskite solar cells. Detailed lists of sensitizers used in DSSCs can be found in reference [73].

The redox couple electrolyte must transfer electrons from the counter electrode to the oxidized dye without absorbing light; the iodide-triiodide (I^-/I_3^-) is easy to process, is low cost, has good reaction kinetics and it has been the most successful redox couple despite some inconveniences: it absorbs light at 430nm, it corrodes the counter electrode and its Nernst potential limits the maximum V_{oc} achievable by the solar cell to 0.9V [130]. Other couples have also been used, such as $Co^{(II/III)}$ polypyridyl complex, ferrocenium/ferrocene (Fc/Fc^+) couple, $Cu^{(I/II)}$ complex and thiolate/disulfide mediator [129, 131]. Quasi-solid electrolytes such as ionic liquids and polymer gels have also been used as electrolytes to avoid the problems of leakage or evaporation of the liquid, which nevertheless is still present in the quasi-solid electrolytes due to their thermodynamic instability. Solid electrolytes acting as hole transporting layers have been investigated; they range from inorganic p-type semiconductors (CuI/CuSCN and $CsSnI_3$) to polymers (PEDOT, P3HT) or organic molecules; the most successful has been 2,20,7,70-tetrakis (N,N-di- 4-methoxyphenylamino)-9,90-spirobifluorene (spiro-OMeTAD) [117]. The use of $CsSnI_3$, which has a perovskite crystalline structure, delivered the first solid-state DSSC with power conversion efficiency higher than 10% [21].

The counter electrode should create a good ohmic contact with the external circuit of the solar cell and inject electrons to reduce the redox couple (or transporting holes when considering a solid state electrolyte). Platinum (Pt) for liquid electrolytes and silver (Ag) or gold (Au) for quasi-solid or solid electrolytes, have been the best counter electrodes; but due to its high cost, other alternatives have been proposed: inorganic compounds (sulfides, phosphides, carbides, nitrides and metal oxides), conjugated polymers such as polyaniline (PANI), poly[3,4-ethylenedioxythiophene] (PEDOT), and polypyrrole (PPy), carbon materials (sp^2 carbon nanotubes or graphene and sp^3 carbon black or mesoporous carbon) or composites fabricated with combinations of any of them; a detailed analysis of all tested counter electrodes in dye sensitized solar cells can be found in [49, 71].

The multiple choice of processing routes is compatible with a low energy consumption, low cost, printable technology, but two main inconveniences hinder the practical application of DSSCs and have made difficult the penetration of this technology in the market: some of the required materials are expensive (for example, ruthenium in most used dyes and platinum as the best counter electrode) and the requirement of a liquid electrolyte posed some challenges for encapsulation and operational outdoor stability without leakage. Despite these difficulties, some companies like Solaronix SA (Switzerland), Greatcell Energy Pty Ltd (Australia) or G24 Power Ltd (United Kingdom) have successfully commercialized dye sensitized solar modules.

On the other hand, the intensive research to overcome these problems led to a new breakthrough: the development of the hybrid perovskite solar cell, that can be considered as a natural evolution of the dye sensitized solar cell, in which the search for a solid electrolyte, the inclusion of a solid good hole transport layer and the modification of the counter electrode led to a new solid-state hybrid technology.

4.4.3 Perovskites

The perovskite crystalline structure was well known for inorganic compounds with general formula ABX_3, where A and B are cations (A with larger size than B) and X is oxygen or an halogen atom. An inorganic perovskite containing layers of copper oxide delivered the first high-temperature superconductor material, which was discovered by Bernodz and Müller in 1986 and deserved the Nobel prize one year later [8]. Around a decade later, hybrid halide perovskites had shown an interesting transition from semiconducting to metallic behaviour and were used as layers to improve the performance of organic electronic devices [75, 87]. It took a long time until the hybrid metal(tin or lead)-halide perovskites were used as light absorbers in dye sensitized solar cells with liquid electrolytes, delivering a modest power conversion efficiency (3.8%) and poor stability [67]. Both efficiency (6.5%) and stability (a few hours) were improved very soon with perovskite "quantum dots" for liquid electrolyte cells, still resembling the dye sensitized approach [60]. The breakthrough came one year later, in 2012, when methyl ammonium lead iodide perovksite was used as sensitizer of a TiO_2 mesoporous structure with spiro-OMeTAD as hole transporting layer in solid-state cells which delivered power conversion efficiencies around 10% [72]. Importantly, all production steps were compatible with solution process and low temperature. This led to an impressive and very fast improvement of power conversion efficiency and lifetime for perovskite solar cells, reaching more than 25% since 2019 by several laboratories, Massachusetts Institute of Technology (MIT, USA), Korea Research Institute of Chemical Technology (KRICT, South Korea), Ulsan National Institute of Science and Technology (UNIST, South Korea), and the Swiss Federal Institute of Technology in Lausanne (EPFL, Switzerland), with operational stability of thousands of hours [47, 62, 63]. Remarkably, flexible perovskite solar cells on plastic substrates have also reached very high efficiency (19,51%) and retained 90% of its initial value after more than 1,000 hours in air (10% humidity) without encapsulation [59]. Perovskite solar cells in a two-terminal tandem structure on a crystalline silicon solar cell have been developed at University of Oxford by combining an infrared-tuned silicon heterojunction bottom cell with a caesium formamidinium lead halide perovskite [14]; this perovskite/silicon tandem approach is close to commercialization by Oxford PV (United Kingdom) and has achieved an impressive certified 29.5% power conversion efficiency in 2021, beating the record of the best GaAs single junction solar cell (without light concentration) [47].

The first perovskite solar cells had a structure very similar to dye sensitized solar cells with solid electrolyte (acting as hole transporting layer). Very soon, PSCs departed from its DSSC birth structure by adding electron transporting layers (ETLs), hole transporting layers (HTLs) and by modifying the mesoporous TiO_2 structure into a compact/mesoporous bilayer. The electrodes were changed to avoid the use of expensive metals (originally used as contraelectrodes in DSSC); also indium in the transparent conductive oxide (ITO) which cover glass or plastic substrates was replaced by fluor (in FTO) or by other cheaper materials. A schematic structure

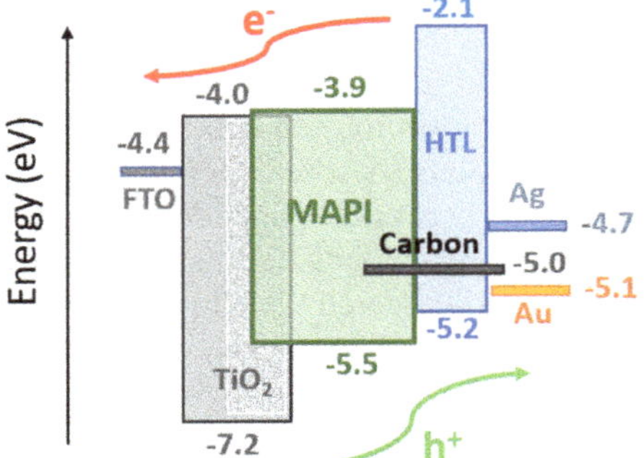

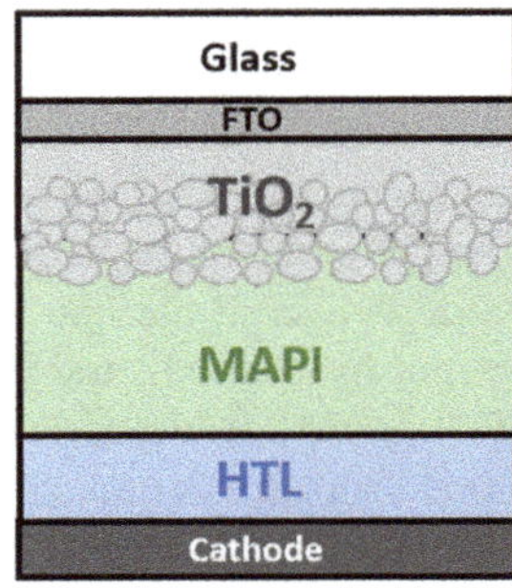

Fig. 4.12 Energy levels and architecture of a MAPI perovskite solar cell with a compact layer and a mesoporous layer of TiO_2. The energy diagram shows various options for cathode configuration, using a porous carbon layer, a hole transporting layer and two possible metals (Au and Ag)

of an advanced perovskite solar cell is shown in Fig. 4.12. The research effort has included a strong focus on increasing stability and thus providing longer lifetimes for outdoor operation of the PSCs; this focus has accompanied the aim of improving power conversion efficiency and at the same time reducing the environmental impact concerns that may arise by the use of lead and of some processing routes for organic compounds that include the use of toxic solvents or reactives (for example, spiro-OMeTAD synthesis has a strong environmental impact). With this combined purpose a large variation of materials (both organic and inorganic) have been used in all the layers of a PSC. A brief summary is presented below; a more detailed analysis of the advances in PSC technology is presented in recent books and reviews [97, 98, 125, 146].

The light absorbing material is the ABX_3 perovskite mentioned above; the most used so far is methyl ammonium lead iodide ($CH_3NH_3PbI_3$; MAPI), related ethylammonium lead iodide ($CH_3CH_2NH_3PbI_3$) and formamidinium lead iodide ($CH_5I_3N_2PbI_3$; FAPI) have also been used; other halides (Cl, Br) have totally or partially substituted iodine in the MAPI or FAPI compounds; the metal in the perovksite has been mostly lead, but also tin (or mixtures) have been used although generally producing lower efficiencies and worst stability. In order to improve the stability of the active layer, molecular additives have been used and mixed with the perovskite, the aminovaleric acid iodide ($HOOC(CH_2)4NH_3I$, AVAI) delivered an impressive extension of lifetimes up to more than 12,000 hrs by providing a 2D/3D interfacial structure [40]. For HTL, the most used compound is spiro-OMeTAD, but it has stability problems (on top of its complex synthetic route) and many HTLs have been developed with good results, small organic molecules, metal oxides (ZnO, NiO, NiO_x, Cu:NiO_x, $NiCo_2O_4$), polymers (PEDOT, PANI) and carbon materials (nanotubes, PCBM, carbon black). Several ETLs are being used, mainly compact and mesoporous TiO_2 bilayers, ZnO (layers or nanostructures), MgO, SnO_2, MoO_3/Ag structures, and again carbon materials (graphene, doped carbon nanotubes, PCBM,

$PC_{70}BM$, with recent approaches using a carbon black porous structure where the perovskite is infiltrated). The electrodes are similar to organic cells, where ITO or FTO covered glass or plastic is used as base for solution processing of the several layers and a final solution processed or evaporated metal electrode completes the structure (mainly Al and trying to avoid nobel metals, although silver and silver nanowires are still widely used).

It should be emphasized that a very good efficiency (22.7%) was obtained with P3HT as hole transporting layer, thus avoiding the use of spiro-OMeTAD, opening a new route for the development of cheap PSCs [63]; the recent use of P3HT, together with inclusion of PCBM layers, is a nice demonstration of crossover of technological research, where the organic bulk heterojunction classical P3HT:PCBM blend materials have come back to be merged with the perovskite technology [34, 38]. Further efforts towards better encapsulation, the development of tandem devices and better serial interconnection of cells in strings to fabricate large area modules are still going on. Several companies are working to manufacture modules with efficiencies above 20%, the best certified efficiency included in the NREL module efficiency chart is 17.9% for a module manufactured by Panasonic (area 800 cm^2) [47, 56]. For now, it seems that the most competitive commercial opportunity for perovskite solar cells will be the tandem structures on silicon developed by Oxford PV with 29.5% efficiency mentioned above.

4.5 From Cells to Modules

Once the solar cells have been manufactured they have to be connected to each other in order to increase the overall output power of the photovoltaic module. The manufacture of photovoltaic technologies has been explained in more detail in the previous section; the general concepts behind the jump from cell to module are briefly explained in the following paragraphs for the two main approaches: crystalline cells that are manufactured one by one and then connected into a single module or thin film cells manufactured simultaneously in a single block at module scale and then cut into single cells mostly by laser techniques; both methods are schematically shown in Fig. 4.13.

In the first case, the manufactured cells are connected serially in **strings** thus rising the output voltage of the module by adding the voltage delivered by each cell; a string with n cells will deliver a string voltage n times higher than the voltage of a single cell and will carry the same current as the one delivered by a single cell. The connection is carried out by a stripe of aluminium or copper that connects the busbars of the p-side of one cell to the busbars of the n-side of the next cell, and the same for all cells in the string. Ideally all the cells in the string should have the same photocurrent, if one of the cells is underperforming, it will act as a bottleneck for the current and limit the total current of the string to the value of this underperforming cell and the extra power generated by the other cells in the string will be dissipated as heat; roughly, this dissipated power is $n - 1$ times the difference between the

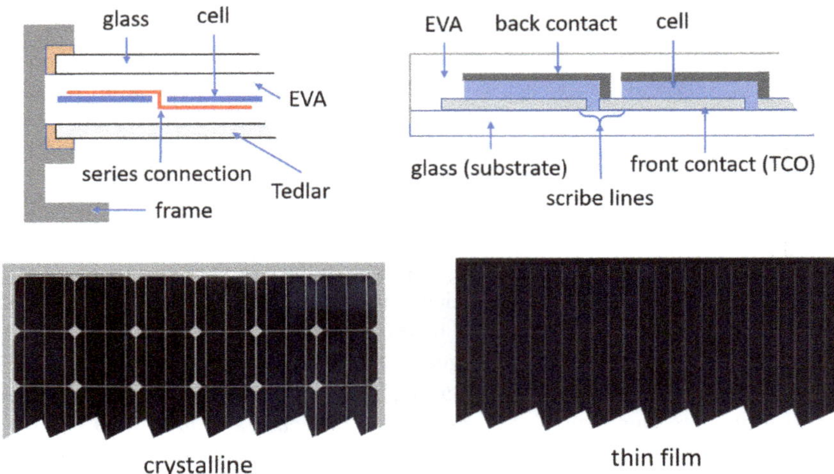

Fig. 4.13 Most common photovoltaic module structures: individual crystalline cells interconnected forming strings that are encapsulated in EVA and sandwiched between glass and Tedlar® (left) and thin film cells grown on a glass substrate and then scribed to form serially connected strings (right)

photocurrent delivered by each one of the good $n - 1$ cells and the photocurrent delivered by the bad cell, multiplied by the voltage of the cells under operation in the string. In some module architectures two or more strings are connected in parallel and therefore its current output is added. The trend in modern manufacture is to include a single string or even string of half-cells in order to increase the voltage and reduce the current delivered by each module (always the total power delivered by an ideal module with n cells will be n times the power delivered by a single cell).

The complete strings are then encapsulated in a plastic foil which in most cases is a thin film of ethylene-vinyl-acetate (EVA), around 0.5 mm thick and with a density of 0.92 g/cm^3; this process is called lamination and it is carried out at a temperature around 150 °C. The next step is adding the backsheet layer of the module, which is usually a Tedlar®/poly-ester/Tedlar® composite film. Tedlar® is the Dupont's registered trademark name for poly-vinyl fluoride film (PVF). Semitransparent Tedlar® films are being developed to be used as backsheet for bifacial modules. These backsheets are used in combination with poly-ethylene terephthalate (PET) films, delivering a total thickness which may vary from 0.25 mm to 1 mm. The module is completed with a toughened glass cover of around 3mm thick and an aluminium frame. Flat glass (SiO_2) for PV applications represent a growing industrial sector, and contributes to an important part of environmental impacts of PV modules and around 15–25% of its final cost in 2020 depending on the technology (its share depends on the specific technology as will be presented in Chap. 7). The soda-lime-silica (SLS) float glass industry for solar applications is highly developed and room for manufacture improvement is limited, although from the scientific point of view, still several improvements could be achieved and recent research points to functional glasses with photon management of UV radiation (most effective filtering to protect the polymeric components of the module, even with recycling UV photon capabilities that can be re-emitted in the visible part of the spectrum and delivered to the cell) and by chemical formulations to reduce ion migration (in order to avoid potential-induced degradation) and increase resistance to cracks [3]. The module manufacture is completed by a junction box, where the strings of the module are connected and acts as the output interface of the PV module by standard electrical connectors (usually models MC4 or MC5 rated for high voltage: 1500 V, and high current: 15 A). The junction box is glued to the back side of the module.

Thin film panels have a different approach to the internal connection of cells: they are not individual cells connected to each other by stripes as in the previous case; the cells in thin film modules are grown directly on a glass substrate of the same size of the module that is covered by the different layers (see Fig. 4.13). During the manufacture process, there are steps in which this film is cut creating scribe lines which divide the film into individual cells serially connected (usually a large stripe around 1cm width and with a length that goes from one side of the module to the other). The process of cutting requires using lasers or mechanical blades or in some emerging technologies a controlled displacement of the printing heads between the different printing steps in R2R or S2S processing (see Sect. 4.2 for details). The result is a series connection of the cells. The module is completed by a backsheet cover, a frame and a junction box similar to the previous case.

Recent models of photovoltaic modules may present variations from these standard descriptions. The most important is the absence of frame. These frameless modules reduce environmental impacts and economic cost (by strongly reducing aluminium consumption), but they require reinforcement of front glass and backsheet cover (in some cases they use double glass, specially in bifacial technologies) because they are more fragile and have an increased risk of breaking during transport and installation.

References

1. Aberle A, Neuhaus DH, Münzer A (2008) Industrial silicon wafer solar cells. Adv Opto-Electron 2007:024521. https://doi.org/10.1155/2007/24521. Publisher: Hindawi Publishing Corporation
2. Akbarnejad E, Ghorannevis Z, Mohammadi E, Fekriaval L (2019) Correlation between different CdTe nanostructures and the performances of solar cells based on CdTe/CdS heterojunction. J Electroanal Chem 849:113358
3. Allsopp BL, Orman R, Johnson SR, Baistow I, Sanderson G, Sundberg P, Stålhandske C, Grund L, Andersson A, Booth J, Bingham PA, Karlsson S (2020) Towards improved cover glasses for photovoltaic devices. Prog PhotovoltS: Res Appl 28(11):1187–1206. https://doi.org/10.1002/pip.3334. Publisher: John Wiley & Sons Ltd
4. Ameri T, Li N, Brabec CJ (2013) Highly efficient organic tandem solar cells: a follow up review. Energy Environ Sci 6(8):2390–2413. https://doi.org/10.1039/C3EE40388B
5. Aramoto T, Kumazawa S, Higuchi H, Arita T, Shibutani S, Nishio T, Nakajima J, Tsuji M, Hanafusa A, Hibino T, Omura K, Ohyama H, Murozono M (1997) 16.0% efficient thin-film

CdS/CdTe solar cells. Jpn J Appl Phys 36(Part 1, No. 10):6304–6305. https://doi.org/10.1143/jjap.36.6304, http://dx.doi.org/10.1143/JJAP.36.6304. Publisher: IOP Publishing

6. Azimi H, Hou Y, Brabec CJ (2014) Towards low-cost, environmentally friendly printed chalcopyrite and kesterite solar cells. Energy Environ Sci 7(6):1829–1849. https://doi.org/10.1039/C3EE43865A
7. Bard AJ (1980) Photo electro chemistry. Science 207(4427):139. https://doi.org/10.1126/science.207.4427.139
8. Bednorz JG, Müller KA (1986) Possible high T_c superconductivity in the Ba-La-Cu-O system. Zeitschrift für Physik B Condensed Matter 64(2):189–193. https://doi.org/10.1007/BF01303701
9. Bisquert J (2017) The physics of solar cells. perovskites, organics, and photovoltaic fundamentals. CRC Press (Francis & Taylor). https://www.routledge.com/The-Physics-of-Solar-Cells-Perovskites-Organics-and-Photovoltaic-Fundamentals/Bisquert/p/book/9781138099968
10. Bottosso C, Tao W, Wang X, Ma L, Galiazzo M (2013) Reliable metallization process for ultra fine line printing. In: Proceedings of the fourth workshop on metallization for crystalline silicon solar cells, vol 43, pp 80–85. https://doi.org/10.1016/j.egypro.2013.11.091
11. Brabec C, Scherf U, Dyakonov V (2014) Organic photovoltaics. Wiley-VCH Verlag GmbH & Co. https://doi.org/10.1002/9783527656912
12. Bridgman PW (1925) Certain physical properties of single crystals of tungsten, antimony, bismuth, tellurium, cadmium, zinc, and tin. Proc Am Acad Arts Sci 60(6):305–383. https://doi.org/10.2307/25130058
13. Bronstein H, Chen Z, Ashraf RS, Zhang W, Du J, Durrant JR, Shakya Tuladhar P, Song K, Watkins SE, Geerts Y, Wienk MM, Janssen RAJ, Anthopoulos T, Sirringhaus H, Heeney M, McCulloch I (2011) Thieno[3,2-b]thiophene-Diketopyrrolopyrrole-containing polymers for high-performance organic field-effect transistors and organic photovoltaic devices. J Am Chem Soc 133(10):3272–3275. https://doi.org/10.1021/ja110619k. Publisher: American Chemical Society
14. Bush KA, Palmstrom AF, Yu ZJ, Boccard M, Cheacharoen R, Mailoa JP, McMeekin DP, Hoye RLZ, Bailie CD, Leijtens T, Peters IM, Minichetti MC, Rolston N, Prasanna R, Sofia S, Harwood D, Ma W, Moghadam F, Snaith HJ, Buonassisi T, Holman ZC, Bent SF, McGehee MD (2017) 23.6%-efficient monolithic perovskite/silicon tandem solar cells with improved stability. Nat Energy 2(4):17009. https://doi.org/10.1038/nenergy.2017.9
15. Bye G, Ceccaroli B (2014) Solar grade silicon: technology status and industrial trends. Solar Energy Mater Solar Cells 130:634–646. https://doi.org/10.1016/j.solmat.2014.06.019
16. Campbell P, Green MA (2001) High performance light trapping textures for monocrystalline silicon solar cells. PVSEC 11 Part I 65(1):369–375. https://doi.org/10.1016/S0927-0248(00)00115-X, https://www.sciencedirect.com/science/article/pii/S092702480000115X
17. Campoy-Quiles M, Ferenczi T, Agostinelli T, Etchegoin PG, Kim Y, Anthopoulos TD, Stavrinou PN, Bradley DDC, Nelson J (2008) Morphology evolution via self-organization and lateral and vertical diffusion in polymer:fullerene solar cell blends. Nat Mater 7(2):158–164. https://doi.org/10.1038/nmat2102
18. Cariou R, Benick J, Feldmann F, Höhn O, Hauser H, Beutel P, Razek N, Wimplinger M, Bläsi B, Lackner D, Hermle M, Siefer G, Glunz SW, Bett AW, Dimroth F (2018) III-V-on-silicon solar cells reaching 33% photoconversion efficiency in two-terminal configuration. Nat Energy 3(4):326–333. https://doi.org/10.1038/s41560-018-0125-0
19. Chiang CK, Fincher CR, Park YW, Heeger AJ, Shirakawa H, Louis EJ, Gau SC, MacDiarmid AG (1977) Electrical conductivity in doped polyacetylene. Phys Rev Lett 39(17):1098–1101. https://doi.org/10.1103/PhysRevLett.39.1098
20. Chirila A, Buecheler S, Pianezzi F, Bloesch P, Gretener C, Uhl AR, Fella C, Kranz L, Perrenoud J, Seyrling S, Verma R, Nishiwaki S, Romanyuk YE, Bilger G, Tiwari AN (2011) Highly efficient Cu(In, Ga)Se2 solar cells grown on flexible polymer films. Nat Mater 10(11):857–861. https://doi.org/10.1038/nmat3122
21. Chung I, Lee B, He J, Chang RPH, Kanatzidis MG (2012) All-solid-state dye-sensitized solar cells with high efficiency. Nature 485(7399):486–489. https://doi.org/10.1038/nature11067

22. Conibeer G (2007) Third-generation photovoltaics. Mater Today 10(11):42–50. https://doi.org/10.1016/S1369-7021(07)70278-X
23. Cusano D (1963) CdTe solar cells and photovoltaic heterojunctions in II-VI compounds. Solid-State Electro 6(3):217–232. https://doi.org/10.1016/0038-1101(63)90078-9
24. Czochralski J (1918) Ein neues Verfahren zur Messung der Kristallisationsgeschwindigkeit der Metalle. Zeitschrift für Physikalische Chemie 92U(1):219–221. https://doi.org/10.1515/zpch-1918-9212
25. Dang MT, Hirsch L, Wantz G (2011) P3HT:PCBM, Best seller in polymer photovoltaic research. Adv Mater 23(31):3597–3602. https://doi.org/10.1002/adma.201100792. Publisher: John Wiley & Sons Ltd
26. Emmott CJM, Urbina A, Nelson J (2012) Environmental and economic assessment of ITO-free electrodes for organic solar cells. Solar Energy Mater Solar Cells 97:14–21. https://doi.org/10.1016/j.solmat.2011.09.024. Go to ISI://WOS:000300653800003, type: Journal Article
27. Eyer A, Schillinger N, Reis I, Räuber A (1990) Silicon sheets for solar cells grown from silicon powder by the SSP technique. J Cryst Growth 104(1):119–125. https://doi.org/10.1016/0022-0248(90)90319-G
28. Fan Z, Razavi H, Jw Do, Moriwaki A, Ergen O, Chueh YL, Leu PW, Ho JC, Takahashi T, Reichertz LA, Neale S, Yu K, Wu M, Ager JW, Javey A (2009) Three-dimensional nanopillar-array photovoltaics on low-cost and flexible substrates. Nat Mater 8(8):648–653. https://doi.org/10.1038/nmat2493
29. Fan Z, Ruebusch DJ, Rathore AA, Kapadia R, Ergen O, Leu PW, Javey A (2009) Challenges and prospects of nanopillar-based solar cells. Nano Res 2(11):829. https://doi.org/10.1007/s12274-009-9091-y
30. Filtvedt W, Javidi M, Holt A, Melaaen M, Marstein E, Tathgar H, Ramachandran P (2010) Development of fluidized bed reactors for silicon production. Solar Energy Mater Solar Cells 94(12):1980–1995. https://doi.org/10.1016/j.solmat.2010.07.027
31. Filtvedt W, Holt A, Ramachandran P, Melaaen M (2012) Chemical vapor deposition of silicon from silane: review of growth mechanisms and modeling/scaleup of fluidized bed reactors. Solar Energy Mater Solar Cells 107:188–200. https://doi.org/10.1016/j.solmat.2012.08.014
32. Friend RH, Gymer RW, Holmes AB, Burroughes JH, Marks RN, Taliani C, Bradley DDC, Santos DAD, Brédas JL, Lögdlund M, Salaneck WR (1999) Electroluminescence in conjugated polymers. Nature 397(6715):121–128. https://doi.org/10.1038/16393
33. Frost JM, Cheynis F, Tuladhar SM, Nelson J (2006) Influence of polymer-blend morphology on charge transport and photocurrent generation in donor-acceptor polymer blends. Nano Lett 6(8):1674–1681. https://doi.org/10.1021/nl0608386. Publisher: American Chemical Society
34. Galatopoulos F, Papadas IT, Armatas GS, Choulis SA (2018) Long thermal stability of inverted perovskite photovoltaics incorporating fullerene-based diffusion blocking layer. Adv Mater Interfaces 5(20):1800280. https://doi.org/10.1002/admi.201800280. Publisher: John Wiley & Sons Ltd
35. Geisz JF, France RM, Schulte KL, Steiner MA, Norman AG, Guthrey HL, Young MR, Song T, Moriarty T (2020) Six-junction III-V solar cells with 47.1% conversion efficiency under 143$\times$Suns concentration. Nat Energy 5(4):326–335. https://doi.org/10.1038/s41560-020-0598-5
36. Gelinck G, Heremans P, Nomoto K, Anthopoulos TD (2010) Organic transistors in optical displays and microelectronic applications. Adv Mater 22(34):3778–3798. https://doi.org/10.1002/adma.200903559. Publisher: John Wiley & Sons Ltd
37. Georgiou E, Choulis SA, Hermerschmidt F, Pozov SM, Burgués-Ceballos I, Christodoulou C, Schider G, Kreissl S, Ward R, List-Kratochvil EJW, Boeffel C (2018) Printed copper nanoparticle metal grids for cost-effective ito-free solution processed solar cells. Solar RRL 2(3):1700192. https://doi.org/10.1002/solr.201700192. Publisher: John Wiley & Sons Ltd
38. Giacomo FD, Razza S, Matteocci F, D'Epifanio A, Licoccia S, Brown TM, Carlo AD (2014) High efficiency $(ch_3nh_3pbi)_{(3--x)}cl_x$ perovskite solar cells with poly(3-hexylthiophene) hole transport layer. J Power Sour 251:152–156. https://doi.org/10.1016/j.jpowsour.2013.11.053

39. Goodrich A, Hacke P, Wang Q, Sopori B, Margolis R, James TL, Woodhouse M (2013) A wafer-based monocrystalline silicon photovoltaics road map: utilizing known technology improvement opportunities for further reductions in manufacturing costs. Solar Energy Mater Solar Cells 114:110–135. https://doi.org/10.1016/j.solmat.2013.01.030
40. Grancini G, Roldán-Carmona C, Zimmermann I, Mosconi E, Lee X, Martineau D, Narbey S, Oswald F, De Angelis F, Graetzel M, Nazeeruddin MK (2017) One-year stable perovskite solar cells by 2D/3D interface engineering. Nat CommunD 8:15684. https://doi.org/10.1038/ncomms15684
41. Grätzel M (1981) Artificial photosynthesis: water cleavage into hydrogen and oxygen by visible light. Acc Chem Res 14(12):376–384. https://doi.org/10.1021/ar00072a003. Publisher: American Chemical Society
42. Green MA (2001) Third generation photovoltaics: ultra-high conversion efficiency at low cost. Prog PhotovoltS: Res Appl 9(2):123–135. https://doi.org/10.1002/pip.360. Publisher: John Wiley & Sons Ltd
43. Green MA (2009) The path to 25% silicon solar cell efficiency: history of silicon cell evolution. Prog PhotovoltS: Res Appl 17(3):183–189. https://doi.org/10.1002/pip.892. Publisher: John Wiley & Sons Ltd
44. Green MA (2015) The Passivated Emitter and Rear Cell (PERC): from conception to mass production. Solar Energy Mater Solar Cells 143:190–197. https://doi.org/10.1016/j.solmat.2015.06.055
45. Green MA, Hishikawa Y, Dunlop ED, Levi DH, Hohl-Ebinger J, Yoshita M, Ho-Baillie AW (2019) Solar cell efficiency tables (Version 53). Prog PhotovoltS: Res Appl 27(1):3–12. https://doi.org/10.1002/pip.3102. Publisher: John Wiley & Sons Ltd
46. Green MA, Dunlop ED, Hohl-Ebinger J, Yoshita M, Kopidakis N, Hao X (2020) Solar cell efficiency tables (version 56). Prog PhotovoltS: Res Appl 28(7):629–638. https://doi.org/10.1002/pip.3303. Publisher: John Wiley & Sons Ltd
47. Green MA, Dunlop ED, Hohl-Ebinger J, Yoshita M, Kopidakis N, Hao X (2021) Solar cell efficiency tables (Version 58). Prog PhotovoltS: Res Appl 29(7):657–667. https://doi.org/10.1002/pip.3444. pubLisher: John Wiley & Sons Ltd
48. Haase F, Hollemann C, Schäfer S, Merkle A, Rienäcker M, Krügener J, Brendel R, Peibst R (2018) Laser contact openings for local poly-Si-metal contacts enabling 26.1%-efficient POLO-IBC solar cells. Solar Energy Mater Solar Cells 186:184–193. https://doi.org/10.1016/j.solmat.2018.06.020
49. Hagfeldt A, Boschloo G, Sun L, Kloo L, Pettersson H (2010) Dye-sensitized solar cells. Chem Rev 110(11):6595–6663. https://doi.org/10.1021/cr900356p. Publisher: American Chemical Society
50. Halls JJM, Walsh CA, Greenham NC, Marseglia EA, Friend RH, Moratti SC, Holmes AB (1995) Efficient photodiodes from interpenetrating polymer networks. Nature 376(6540):498–500. https://doi.org/10.1038/376498a0
51. Hamann TW, Martinson ABF, Elam JW, Pellin MJ, Hupp JT (2008) Atomic layer deposition of TiO2 on aerogel templates: new photoanodes for dye-sensitized solar cells. J Phys Chem C 112(27):10303–10307. https://doi.org/10.1021/jp802216p. Publisher: American Chemical Society
52. Hands ADP, Ryden KA, Meredith NP, Glauert SA, Horne RB (2018) Radiation effects on satellites during extreme space weather events. Space Weather 16(9):1216–1226. https://doi.org/10.1029/2018SW001913. Publisher: John Wiley & Sons Ltd
53. Hauser A, Hahn G, Spiegel M, Fath P, Bucher E, Feist H, Breitenstein O, Rakotoniaina JP (2001) Comparison of different techniques for edge isolation. In: Proceedings of the seventeenth european photovoltaic solar energy conference, WIP-Renewable Energies, Munich, pp 1739–1742. https://www.hahn.uni-konstanz.de/typo3temp/secure_downloads/80509/0/e5ae247f78559f02e330c754c4b20fc80216de0a/VC3_11_AH.pdf
54. He M, Yan C, Li J, Suryawanshi MP, Kim J, Green MA, Hao X (2021) Kesterite solar cells: insights into current strategies and challenges. Adv Sci n/a(n/a):2004313. https://doi.org/10.1002/advs.202004313. Publisher: John Wiley & Sons, Ltd

55. He Y, Chen HY, Hou J, Li Y (2010) Indene-c$_{60}$ bisadduct: a new acceptor for high-performance polymer solar cells. J Am Chem Soc 132(15):5532. https://doi.org/10.1021/ja101780s. Publisher: American Chemical Society
56. Higuchi H, Negami T (2018) Largest highly efficient 203×203 mm^2 $CH_3NH_3PbI_3$ perovskite solar modules. Jpn J Appl Phys 57(8S3):08RE11. https://doi.org/10.7567/jjap.57.08re11, http://dx.doi.org/10.7567/JJAP.57.08RE11. Publisher: IOP Publishing
57. Hollemann C, Haase F, Rienäcker M, Barnscheidt V, Krügener J, Folchert N, Brendel R, Richter S, Großer S, Sauter E, Hübner J, Oestreich M, Peibst R (2020) Separating the two polarities of the POLO contacts of an 26.1%-efficient IBC solar cell. Sci Rep 10(1):658. https://doi.org/10.1038/s41598-019-57310-0
58. Hoppe H, Sariciftci NS (2004) Organic solar cells: an overview. JMater Res 19(7):1924–1945. https://doi.org/10.1557/JMR.2004.0252
59. Huang K, Peng Y, Gao Y, Shi J, Li H, Mo X, Huang H, Gao Y, Ding L, Yang J (2019) High-performance flexible perovskite solar cells via precise control of electron transport layer. Adv Energy Mater 9(44):1901419. https://doi.org/10.1002/aenm.201901419. Publisher: John Wiley & Sons Ltd
60. Im JH, Lee CR, Lee JW, Park SW, Park NG (2011) 6.5% efficient perovskite quantum-dot-sensitized solar cell. Nanoscale 3(10):4088–4093. https://doi.org/10.1039/C1NR10867K, http://dx.doi.org/10.1039/C1NR10867K
61. Isshiki M, Wang J (2017) II-IV semiconductors for optoelectronics: CdS, CdSe, CdTe. In: Kasap S, Capper P (eds) Springer handbook of electronic and photonic materials. Springer International Publishing, Cham, p 1. https://doi.org/10.1007/978-3-319-48933-9_33
62. Jeong M, Choi IW, Go EM, Cho Y, Kim M, Lee B, Jeong S, Jo Y, Choi HW, Lee J, Bae JH, Kwak SK, Kim DS, Yang C (2020) Stable perovskite solar cells with efficiency exceeding 24.8% and 0.3-V voltage loss. Science 369(6511):1615. https://doi.org/10.1126/science.abb7167, http://science.sciencemag.org/content/369/6511/1615.abstract
63. Jung EH, Jeon NJ, Park EY, Moon CS, Shin TJ, Yang TY, Noh JH, Seo J (2019) Efficient, stable and scalable perovskite solar cells using poly(3-hexylthiophene). Nature 567(7749):511–515. https://doi.org/10.1038/s41586-019-1036-3
64. Jurisch M, Eichler S, Bruder M (2015) 9 - vertical bridgman growth of binary compound semiconductors. In: Rudolph P (ed) Handbook of crystal growth (2nd edn). Elsevier, Boston, pp 331–372. https://doi.org/10.1016/B978-0-444-63303-3.00009-2, https://www.sciencedirect.com/science/article/pii/B9780444633033000092
65. Kayes BM, Nie H, Twist R, Spruytte SG, Reinhardt F, Kizilyalli IC, Higashi GS (2011) 27.6% Conversion efficiency, a new record for single-junction solar cells under 1 sun illumination. In: 2011 37th IEEE photovoltaic specialists conference, pp 000004–000008. https://doi.org/10.1109/PVSC.2011.6185831
66. Kim S, Yim J, Wang X, Bradley DD, Lee S, deMello JC (2010) Spin- and spray-deposited single-walled carbon-nanotube electrodes for organic solar cells. Adv Funct Mater 20(14):2310–2316. https://doi.org/10.1002/adfm.200902369. Publisher: John Wiley & Sons Ltd
67. Kojima A, Teshima K, Shirai Y, Miyasaka T (2009) Organometal halide perovskites as visible-light sensitizers for photovoltaic cells. J Am Chem Soc 131(17):6050–6051. https://doi.org/10.1021/ja809598r
68. Krebs FC (2009) All solution roll-to-roll processed polymer solar cells free from indium-tin-oxide and vacuum coating steps. Organ Electron 10(5):761–768. https://doi.org/10.1016/j.orgel.2009.03.009
69. Krebs FC (2010) Polymeric solar cells. DEStech Publications Inc, Lancaster, PA, U.S.A, Materials, Design, Manufacture
70. Krebs FC, Gevorgyan SA, Alstrup J (2009) A roll-to-roll process to flexible polymer solar cells: model studies, manufacture and operational stability studies. J Mater Chem 19(30):5442–5451. https://doi.org/10.1039/B823001C
71. Lee CP, Li CT, Ho KC (2017) Use of organic materials in dye-sensitized solar cells. Mater Today 20(5):267–283. https://doi.org/10.1016/j.mattod.2017.01.012

72. Lee MM, Teuscher J, Miyasaka T, Murakami TN, Snaith HJ (2012) Efficient hybrid solar cells based on meso-superstructured organometal halide perovskites. Science 338(6107):643. https://doi.org/10.1126/science.1228604
73. Lee TD, Ebong AU (2017) A review of thin film solar cell technologies and challenges. Renew Sustain Energy Rev 70:1286–1297. https://doi.org/10.1016/j.rser.2016.12.028
74. Lenes M, Morana M, Brabec CJ, Blom PWM (2009) Recombination-limited photocurrents in low bandgap polymer/fullerene solar cells. Adv Funct Mater 19(7):1106–1111. https://doi.org/10.1002/adfm.200801514. Publisher: John Wiley & Sons Ltd
75. Liang K, Mitzi DB, Prikas MT (1998) Synthesis and characterization of organic-inorganic perovskite thin films prepared using a versatile two-step dipping technique. Chem Mater 10(1):403–411. https://doi.org/10.1021/cm970568f
76. Little RG, Nowlan MJ (1997) Crystalline silicon photovoltaics: the hurdle for thin films. Progress Photovoltaics: Res Appl 5(5):309–315. https://doi.org/10.1002/(SICI)1099-159X(199709/10)5:5 309::AID-PIP180 3.0.CO;2-X. Publisher: John Wiley & Sons Ltd
77. Liu Q, Jiang Y, Jin K, Qin J, Xu J, Li W, Xiong J, Liu J, Xiao Z, Sun K, Yang S, Zhang X, Ding L (2020) 18% Efficiency organic solar cells. Sci Bull 65(4):272–275. https://doi.org/10.1016/j.scib.2020.01.001
78. Loos J (2010) Volume morphology of printable solar cells. Mater Today 13(10):14–20. https://doi.org/10.1016/S1369-7021(10)70182-6
79. Marks RN, Halls JJM, Bradley DDC, Friend RH, Holmes AB (1994) The photovoltaic response in poly(p-phenylene vinylene) thin-film devices. J Phys Condens Matter 6(7):1379–1394. https://doi.org/10.1088/0953-8984/6/7/009
80. Meier J, Dubail S, Golay S, Kroll U, FaÿS, Vallat-Sauvain E, Feitknecht L, Dubail J, Shah A (2002) Microcrystalline silicon and the impact on micromorph tandem solar cells. PVSEC 12 Part I 74(1):457–467. https://doi.org/10.1016/S0927-0248(02)00111-3, https://www.sciencedirect.com/science/article/pii/S0927024802001113
81. Meier J, Spitznagel J, Kroll U, Bucher C, FaÿS, Moriarty T, Shah A (2004) Potential of amorphous and microcrystalline silicon solar cells. In: Proceedings of symposium D on thin film and nano-structured materials for photovoltaics, of the E-MRS 2003 spring conference 451-452:518–524. https://doi.org/10.1016/j.tsf.2003.11.014, https://www.sciencedirect.com/science/article/pii/S0040609003015475
82. Meija J, Coplen TB, Berglund M, Brand WA, De Biévre P, Gröning M, Holden NE, Irrgeher J, Loss RD, Walczyk T, Prohaska T (2016) Atomic weights of the elements 2013 (IUPAC Technical Report). Pure Appl Chem 88(3):265–291. https://doi.org/10.1515/pac-2015-0305
83. Messenger SR, Summers GP, Burke EA, Walters RJ, Xapsos MA (2001) Modeling solar cell degradation in space: a comparison of the NRL displacement damage dose and the JPL equivalent fluence approaches. Progress Photovoltaics: Res Appl 9(2):103–121. https://doi.org/10.1002/pip.357. Publisher: John Wiley & Sons Ltd
84. Micha DN, Silvares Junior RT (2019) The influence of solar spectrum and concentration factor on the material choice and the efficiency of multijunction solar cells. Sci Rep 9(1):20055. https://doi.org/10.1038/s41598-019-56457-0
85. Miles RW, Zoppi G, Forbes I (2007) Inorganic photovoltaic cells. Mater Today 10(11):20–27. https://doi.org/10.1016/S1369-7021(07)70275-4
86. Mingorance A, Xie H, Kim HS, Wang Z, Balsells M, Morales-Melgares A, Domingo N, Kazuteru N, Tress W, Fraxedas J, Vlachopoulos N, Hagfeldt A, Lira-Cantu M (2018) Interfacial engineering of metal oxides for highly stable halide perovskite solar cells. Adv Mater Interfaces 5(22):1800367. https://doi.org/10.1002/admi.201800367. Publisher: John Wiley & Sons Ltd
87. Mitzi DB, Feild CA, Harrison WTA, Guloy AM (1994) Conducting tin halides with a layered organic-based perovskite structure. Nature 369(6480):467–469. https://doi.org/10.1038/369467a0
88. Mor GK, Varghese OK, Paulose M, Shankar K, Grimes CA (2006) A review on highly ordered, vertically oriented TiO2 nanotube arrays: Fabrication, material properties, and solar energy applications. Solar Energy Mater Solar Cells 90(14):2011–2075. https://doi.org/10.1016/j.solmat.2006.04.007

89. Mühlbacher D, Scharber M, Morana M, Zhu Z, Waller D, Gaudiana R, Brabec C (2006) High photovoltaic performance of a low-bandgap polymer. Adv Mater 18(21):2884–2889. https://doi.org/10.1002/adma.200600160. Publisher: John Wiley & Sons Ltd
90. Nakayama N, Matsumoto H, Yamaguchi K, Ikegami S, Hioki Y (1976) Ceramic thin film CdTe solar cell. Jpn J Appl Phys 15(11):2281–2282. https://doi.org/10.1143/jjap.15.2281
91. Nampalli N, Hallam B, Chan C, Abbott M, Wenham S (2015) Evidence for the role of hydrogen in the stabilization of minority carrier lifetime in boron-doped Czochralski silicon. Appl Phys Lett 106(17):173501. https://doi.org/10.1063/1.4919385. Publisher: American Institute of Physics
92. Nazeeruddin MK, Liska P, Moser J, Vlachopoulos N, Grätzel M (1990) Conversion of light into electricity with trinuclear ruthenium complexes adsorbed on textured TiO2 films. Helvetica Chimica Acta 73(6):1788–1803. https://doi.org/10.1002/hlca.19900730624. Publisher: John Wiley & Sons Ltd
93. Nazeeruddin MK, Kay A, Rodicio I, Humphry-Baker R, Mueller E, Liska P, Vlachopoulos N, Graetzel M (1993) Conversion of light to electricity by cis-X2bis(2,2'-bipyridyl-4,4'-dicarboxylate)ruthenium(II) charge-transfer sensitizers (X = Cl^-, Br^-, I^-, CN^-, and SCN^-) on nanocrystalline titanium dioxide electrodes. J Am Chem Soc 115(14):6382–6390. https://doi.org/10.1021/ja00067a063. Publisher: American Chemical Society
94. Nelson J (2003) The physics of solar cells. Imperial College Press and distributed by World Scientific Publishing Co. https://doi.org/10.1142/p276, https://www.worldscientific.com/doi/abs/10.1142/p276, _eprint: https://www.worldscientific.com/doi/pdf/10.1142/p276
95. Nelson J (2011) Polymer:fullerene bulk heterojunction solar cells. Mater Today 14(10):462–470. https://doi.org/10.1016/S1369-7021(11)70210-3
96. O'Regan B, Grätzel M (1991) A low-cost, high-efficiency solar cell based on dye-sensitized colloidal TiO2 films. Nature 353(6346):737–740. https://doi.org/10.1038/353737a0
97. Park NG (2015) Perovskite solar cells: an emerging photovoltaic technology. Mater Today 18(2):65–72. https://doi.org/10.1016/j.mattod.2014.07.007
98. Park NG, Graetzel M, Miyasaka T (2016) Organic-inorganic halide perovskite photovoltaics: from fundamentals to device architectures. Springer, Berlin
99. Peet J, Kim JY, Coates NE, Ma WL, Moses D, Heeger AJ, Bazan GC (2007) Efficiency enhancement in low-bandgap polymer solar cells by processing with alkane dithiols. Nat Mater 6(7):497–500. https://doi.org/10.1038/nmat1928
100. Podestá A, Armani N, Salviati G, Romeo N, Bosio A, Prato M (2006) Influence of the fluorine doping on the optical properties of CdS thin films for photovoltaic applications. In: EMSR 2005—proceedings of symposium F on thin film and nanostructured materials for photovoltaics 511–512:448–452. https://doi.org/10.1016/j.tsf.2005.11.069, https://www.sciencedirect.com/science/article/pii/S0040609005022868
101. Polizzotti A, Repins IL, Noufi R, Wei SH, Mitzi DB (2013) The state and future prospects of kesterite photovoltaics. Energy Environ Sci 6(11):3171–3182. https://doi.org/10.1039/C3EE41781F
102. Raut HK, Ganesh VA, Nair AS, Ramakrishna S (2011) Anti-reflective coatings: A critical, in-depth review. Energy Environ Sci 4(10):3779–3804. https://doi.org/10.1039/C1EE01297E
103. Romeo N, Bosio A, Romeo A (2010) An innovative process suitable to produce high-efficiency CdTe/CdS thin-film modules. 17th Int Mater Res Congress 94(1):2–7. https://doi.org/10.1016/j.solmat.2009.06.001, https://www.sciencedirect.com/science/article/pii/S0927024809002141
104. Ross RB, Cardona CM, Guldi DM, Sankaranarayanan SG, Reese MO, Kopidakis N, Peet J, Walker B, Bazan GC, Van Keuren E, Holloway BC, Drees M (2009) Endohedral fullerenes for organic photovoltaic devices. Nat Mater 8(3):208–212. https://doi.org/10.1038/nmat2379
105. Sadula A, Azzopardi B, Chircop J (2018) Innovation updates for organic and perovskites solar cells. In: 2018 IEEE 7th World Conference on Photovoltaic Energy Conversion (WCPEC) (A Joint Conference of 45th IEEE PVSC. In: 28th PVSEC & 34th EU PVSEC), pp 1090–1094. https://doi.org/10.1109/PVSC.2018.8547864; Journal Abbreviation: 2018 IEEE 7th World Conference on Photovoltaic Energy Conversion (WCPEC) (A Joint Conference of 45th IEEE PVSC, 28th PVSEC & 34th EU PVSEC)

106. Schiff EA, Hegedus S, Deng X (2010) Amorphous silicon-based solar cells. In: Handbook of photovoltaic science and engineering. Wiley, New York, pp 487–545. https://doi.org/10.1002/9780470974704.ch12, https://onlinelibrary.wiley.com/doi/abs/10.1002/9780470974704.ch12. Section: 12 _eprint: https://onlinelibrary.wiley.com/doi/pdf/10.1002/9780470974704.ch12
107. Schneiderlöchner E, Preu R, Lüdemann R, Glunz SW (2002) Laser-fired rear contacts for crystalline silicon solar cells. Progress Photovolt Res Appl 10(1):29–34. https://doi.org/10.1002/pip.422. Publisher: John Wiley & Sons Ltd
108. Schreurs D, Nagels S, Cardinaletti I, Vangerven T, Cornelissen R, Vodnik J, Hruby J, Deferme W, Manca JV (2018) Methodology of the first combined in-flight and ex situ stability assessment of organic-based solar cells for space applications. J Mater Res 33(13):1841–1852. https://doi.org/10.1557/jmr.2018.156
109. Schüttauf JW, Niesen B, Löfgren L, Bonnet-Eymard M, Stuckelberger M, Hänni S, Boccard M, Bugnon G, Despeisse M, Haug FJ, Meillaud F, Ballif C (2015) Amorphous silicon-germanium for triple and quadruple junction thin-film silicon based solar cells. Solar Energy Mater Solar Cells 133:163–169. https://doi.org/10.1016/j.solmat.2014.11.006
110. Seo JH, Gutacker A, Sun Y, Wu H, Huang F, Cao Y, Scherf U, Heeger AJ, Bazan GC (2011) Improved high-efficiency organic solar cells via incorporation of a conjugated polyelectrolyte interlayer. J Am Chem Soc 133(22):8416–8419. https://doi.org/10.1021/ja2037673. Publisher: American Chemical Society
111. Shimura F (2007) Single-crystal silicon: growth and properties. In: Kasap S, Capper P (eds) Springer handbook of electronic and photonic materials. Springer US, Boston, MA, pp 255–269. https://doi.org/10.1007/978-0-387-29185-7_13
112. Shirakawa H (1995) Synthesis and characterization of highly conducting polyacetylene. Proc Int Conf Sci Technol Synthetic Metals 69(1):3–8. https://doi.org/10.1016/0379-6779(94)02340-5
113. Shirakawa H, Louis EJ, MacDiarmid AG, Chiang CK, Heeger AJ (1977) Synthesis of electrically conducting organic polymers: halogen derivatives of polyacetylene, (CH). J Chem Soc Chem Commun 16:578–580. https://doi.org/10.1039/C39770000578
114. Shockley W, Queisser HJ (1961) Detailed Balance Limit of Efficiency of p-n Junction Solar Cells. J Appl Phys 32(3):510–519. https://doi.org/10.1063/1.1736034. Publisher: American Institute of Physics
115. Siebentritt S (2004) Alternative buffers for chalcopyrite solar cells. Thin Film PV 77(6):767–775. https://doi.org/10.1016/j.solener.2004.06.018
116. Sinton RA, Cuevas A (1996) Contactless determination of current-voltage characteristics and minority carrier lifetimes in semiconductors from quasi-steady state photoconductance data. Appl Phys Letts 69(17):2510–2512. https://doi.org/10.1063/1.117723. Publisher: American Institute of Physics
117. Snaith H, Grätzel M (2007) Electron and hole transport through mesoporous TiO2 infiltrated with Spiro-MeOTAD. Adv Mater 19(21):3643–3647. https://doi.org/10.1002/adma.200602085. Publisher: John Wiley & Sons Ltd
118. Steim R, Choulis SA, Schilinsky P, Brabec CJ (2008) Interface modification for highly efficient organic photovoltaics. Appl Phys Lett 92(9):093303. https://doi.org/10.1063/1.2885724. Publisher: American Institute of Physics
119. Swirhun JS, Sinton RA, Forsyth MK, Mankad T (2011) Contactless measurement of minority carrier lifetime in silicon ingots and bricks. Progress in Photovoltaics: Res Appl 19(3):313–319. https://doi.org/10.1002/pip.1029. Publisher: John Wiley & Sons Ltd
120. Tang CW (1986) Two-layer organic photovoltaic cell. Appl Phys Lett 48(2):183–185. https://doi.org/10.1063/1.96937. Publisher: American Institute of Physics
121. Tang J, Huo Z, Brittman S, Gao H, Yang P (2011) Solution-processed core-shell nanowires for efficient photovoltaic cells. Nat. Nanotechnol. 6(9):568–572. https://doi.org/10.1038/nnano.2011.139
122. Thejo-Kalyani N, Dhoble S (2012) Organic light emitting diodes: Energy saving lighting technology-A review. Renew Sustain Energy Rev 16(5):2696–2723. https://doi.org/10.1016/j.rser.2012.02.021

123. Tsunomura Y, Yoshimine Y, Taguchi M, Baba T, Kinoshita T, Kanno H, Sakata H, Maruyama E, Tanaka M (2009) Twenty-two percent efficiency HIT solar cell. 17th Int Photovolt Sci Eng Conf 93(6):670–673. https://doi.org/10.1016/j.solmat.2008.02.037, https://www.sciencedirect.com/science/article/pii/S0927024808000834
124. United States Geological Survey (2021) Mineral commodity summaries 2021. Report, Reston, VA. https://doi.org/10.3133/mcs2021
125. Urbina A (2020) The balance between efficiency, stability and environmental impacts in perovskite solar cells: a review. J Phys: Energy 2(2):022001. https://doi.org/10.1088/2515-7655/ab5eee
126. Urbina A, Abad J, Fernandez Romero AJ, Lacasa JS, Colchero J, Gonzalez-Martinez JF, Rubio-Zuazo J, Castro GR, Gutfreund P (2019) Neutron reflectometry and hard X-ray photoelectron spectroscopy study of the vertical segregation of PCBM in organic solar cells. Solar Energy Mater Solar Cells 191:62–70. https://doi.org/10.1016/j.solmat.2018.10.004, Go to ISI://WOS:000456640000009, type: Journal Article
127. Wadsworth A, Moser M, Marks A, Little MS, Gasparini N, Brabec CJ, Baran D, McCulloch I (2019) Critical review of the molecular design progress in non-fullerene electron acceptors towards commercially viable organic solar cells. Chem Soc Rev 48(6):1596–1625. https://doi.org/10.1039/C7CS00892A
128. Walsh A, Chen S, Wei SH, Gong XG (2012) Kesterite thin-film solar cells: advances in materials modelling of Cu2ZnSnS4. Adv Energy Mater 2(4):400–409. https://doi.org/10.1002/aenm.201100630. Publisher: John Wiley & Sons Ltd
129. Wang HQ, Li N, Guldal NS, Brabec CJ (2012) Nanocrystal V_2O_5 thin film as hole-extraction layer in normal architecture organic solar cells. Org Electron 13(12):3014–3021. https://doi.org/10.1016/j.orgel.2012.08.007
130. Wang M, Chamberland N, Breau L, Moser JE, Humphry-Baker R, Marsan B, Zakeeruddin SM, Grätzel M (2010) An organic redox electrolyte to rival triiodide/iodide in dye-sensitized solar cells. Nat Chem 2(5):385–389. https://doi.org/10.1038/nchem.610
131. Wang M, Grätzel C, Zakeeruddin SM, Grätzel M (2012) Recent developments in redox electrolytes for dye-sensitized solar cells. Energy Environ Sci 5(11):9394–9405. https://doi.org/10.1039/C2EE23081J
132. Wang R, Yao Y, Zhang C, Zhang Y, Bin H, Xue L, Zhang ZG, Xie X, Ma H, Wang X, Li Y, Xiao M (2019) Ultrafast hole transfer mediated by polaron pairs in all-polymer photovoltaic blends. Nat Commun 10(1):398. https://doi.org/10.1038/s41467-019-08361-4
133. Wang R, Xu J, Fu L, Zhang C, Li Q, Yao J, Li X, Sun C, Zhang ZG, Wang X, Li Y, Ma J, Xiao M (2021) Nonradiative Triplet Loss Suppressed in Organic Photovoltaic Blends with Fluoridated Nonfullerene Acceptors. J Am Chem Soc 143(11):4359–4366. https://doi.org/10.1021/jacs.0c13352. Publisher: American Chemical Society
134. Wronski CR, Carlson DE (2001) Amorphous silicon solar cells. In: Archer M, Hill R (eds) Clean electricity from photovoltaics, series on photoconversion of solar energy, vol 1. Imperial College Press and World Scientific, pp 199–243. https://doi.org/10.1142/9781848161504_0005,
135. Wu J, Becerril HA, Bao Z, Liu Z, Chen Y, Peumans P (2008) Organic solar cells with solution-processed graphene transparent electrodes. Appl Phys Lett 92(26):263302. https://doi.org/10.1063/1.2924771. Publisher: American Institute of Physics
136. Yadav S, Chattopadhyay K, Singh CV (2017) Solar grade silicon production: a review of kinetic, thermodynamic and fluid dynamics based continuum scale modeling. Renew Sustain Energy Rev 78:1288–1314. https://doi.org/10.1016/j.rser.2017.05.019
137. Yamaguchi M (2001) Radiation-resistant solar cells for space use. Solar Cells Space 68(1):31–53. https://doi.org/10.1016/S0927-0248(00)00344-5
138. Yan D, Cuevas A, Bullock J, Wan Y, Samundsett C (2015) Phosphorus-diffused polysilicon contacts for solar cells. In: Proceedings of the 5th international conference on crystalline silicon photovoltaics (SiliconPV 2015) 142:75–82. https://doi.org/10.1016/j.solmat.2015.06.001, https://www.sciencedirect.com/science/article/pii/S0927024815002706

139. Yasuda K, Morita K, Okabe TH (2014) Processes for production of solar-grade silicon using hydrogen reduction and/or thermal decomposition. Energy Technol 2(2):141–154. https://doi.org/10.1002/ente.201300131. Publisher: John Wiley & Sons Ltd
140. Yaws CL, Jelen FC, Li KY, Patel P, Fang C (1979) New technologies for solar energy silicon: cost analysis of UCC Silane Process. Solar Energy 22(6):547–553. https://doi.org/10.1016/0038-092X(79)90027-6
141. Ye M, Wen X, Wang M, Iocozzia J, Zhang N, Lin C, Lin Z (2015) Recent advances in dye-sensitized solar cells: from photoanodes, sensitizers and electrolytes to counter electrodes. Mater Today 18(3):155–162. https://doi.org/10.1016/j.mattod.2014.09.001
142. Yoshikawa K, Kawasaki H, Yoshida W, Irie T, Konishi K, Nakano K, Uto T, Adachi D, Kanematsu M, Uzu H, Yamamoto K (2017) Silicon heterojunction solar cell with interdigitated back contacts for a photoconversion efficiency over 26%. Nat Energy 2(5):17032. https://doi.org/10.1038/nenergy.2017.32
143. Yoshikawa K, Yoshida W, Irie T, Kawasaki H, Konishi K, Ishibashi H, Asatani T, Adachi D, Kanematsu M, Uzu H, Yamamoto K (2017b) Exceeding conversion efficiency of 26% by heterojunction interdigitated back contact solar cell with thin film Si technology. In: Proceedings of the 7th international conference on crystalline silicon photovoltaics 173:37–42. https://doi.org/10.1016/j.solmat.2017.06.024, https://www.sciencedirect.com/science/article/pii/S092702481730332X
144. Yu G, Heeger AJ (1995) Charge separation and photovoltaic conversion in polymer composites with internal donor/acceptor heterojunctions. J Appl Phys 78(7):4510–4515. https://doi.org/10.1063/1.359792 Publisher: American Institute of Physics
145. Yu G, Pakbaz K, Heeger AJ (1994) Semiconducting polymer diodes: large size, low cost photodetectors with excellent visible-ultraviolet sensitivity. Appl Phys Lett 64(25):3422–3424. https://doi.org/10.1063/1.111260. Publisher: American Institute of Physics
146. Zhang J, Zhang W, Cheng HM, Silva SRP (2020) Critical review of recent progress of flexible perovskite solar cells. Mater Today 39:66–88. https://doi.org/10.1016/j.mattod.2020.05.002, https://www.sciencedirect.com/science/article/pii/S1369702120301747

Chapter 5
The Limits of Raw Materials Embedded in PV Modules

In this chapter, the limits of raw materials consumption in the fabrication of solar modules will be analysed. Two kinds of limits can be considered. First, the limits that arise from the ultimately recoverable reserves of a given element required to manufacture the module; secondly, the limits that arise from the impacts on human health and the environment from mining, processing and eventually recycling of the materials embedded in the module and used in the manufacturing process.

In the first case, the philosophy behind the limitation is similar to fossil fuels where the production of the fuel obeys a logistic curve which will peak at some time and then decrease [23]. This simple hypothesis, when applied to fossil fuels, predicts a future time were production from a given exploitation site will peak, which together with the rate of production, the ultimately recoverable reserves can be extrapolated. Depending on the symmetry of the logistic curve (the rising and decaying parts of the peak can be different) some uncertainty is introduced in the calculation, but it has been successful to predict the decay of production in important oil fields [22]. Similarly, the limitations arising from element scarcity may pose a risk on photovoltaic module production if a scarce material is used in some specific technology and the required production rate per year becomes important compared to the known reserves.

In the second case, the limitation is created by the health and environmental impacts due to the use of the material within the module and during the manufacturing process. Again, a similarity with fossil fuels can be observed: greenhouse gas emissions emitted by fuel burning have important environmental impacts, in particular in climate change, and it makes clear that consumption of fossil fuels must be reduced and eventually stopped before the ultimately recoverable fossil fuel reserves are burned; it has been calculated that in order to limit global warming to 2 °C, a third of oil reserves, half of gas reserves and over eighty per cent of current coal reserves should remain unused from 2010 to 2050 [28]. In renewable energy technologies there is no fossil fuel consumption, but environmental impacts arising from use of raw materials may generate important risks that could limit a massive deployment

A. Urbina, *Sustainable Solar Electricity*, Green Energy and Technology,
https://doi.org/10.1007/978-3-030-91771-5_5

of installed capacity. The quantification of impacts obtained by LCA methodology, and in particular, some impact categories, are specially focused on raw materials depletion and other associated categories associated to consumption of materials (see Chap. 3). On the other hand, material processing (from mining, purification, use in production lines of different photovoltaic technologies, recycling and landfilling) also generate impacts on many other categories, such as cumulative energy demand and associated emissions (analysed in Chap. 6) and many other categories such as climate change (warming potential), ecotoxicity or human health impacts, etc…that will be presented in Chap. 7.

The detailed analysis of mineral resource use impacts and in particular materials depletion in life cycle impact assessment (LCIA) methodologies has been discussed by the scientific community since long time ago. No LCIA method has a globally accepted sets of impact categories or characterization factors, leading to variations in depletion potential results across models [11, 21]. An expert group formed by 62 members and called "task force mineral resources" was created by the Life Cycle Initiative hosted by the United Nations. They accomplished several tasks: (i) defining the safeguard subject for mineral resources within the area of protection natural resources; (ii) formulating seven key questions regarding the consequences of mineral resource (for example, current resource use leading to changes in opportunities for future users to use resources, or potential restrictions of resource availability for current resource users); (iii) reviewing 27 different life cycle impact assessment methods for mineral resource use in the "natural resources" area of protection; and (iv) recommending seven existing LCIA methods to evaluate mineral resource depletion (ADP_{UR}, SOP_{URR}, $LIME2_{endpoint}$, CEENE, ADP_{ER}, ESSENZ, and GeoPolRisk), adding several suggestions to improve the methods [3, 35].

The methods can be grouped in four categories: depletion methods, future efforts methods, thermodynamic accounting methods, and supply risk methods. The first two cases provide the basic assessment about resources while the other two provide complementary information. The depletion methods quantify the decrease in mineral resource stocks due to extraction; between abiotic resource depletion potential methods, ADP_{UR} is the one that provides most robust relative potential of long-term depletion of natural stocks of mineral resources, and its assessment is constant over time since Earth crustal mineral content estimates have been quite stable over time [7, 35]. Some uncertainty is contained in future effort methods since they quantify the additional societal efforts required in the future as a result of current extraction, in these methods ore mining is normally assumed to occur from the highest to the lowest grade of mineral content, which is not generally the case since different grade ores are exploited in parallel in different parts of the world according to technological, economical and even political considerations; despite these uncertainties, the SOP method provides reliable results, while ORI or SCP relies excessively on volatile mineral demand price [11]. Thermodynamic accounting methods, such as CEENE, calculate the exergy difference between the mineral resource as found in nature and a reference compound in the natural environment, they provide complementary information to LCIA methods, but are still to be integrated in a global LCA framework. Finally, supply risk methods (like GeoPolRisk) are more focused on highlighting the supply risk between countries based on trading relationships, country stability and international politics, its conclusions may vary over time since the scenarios and assumptions also vary over time and are sometimes subject to unexpected shocks (as it was the case of oil crisis in the 70s), also provisioning local stocks to guarantee supply chains for products and services affect the risk assessment [10]; this focus lays beyond the conventional LCA framework and will be discussed in more detail in Part IV of this book.

Recently, Arvidsson et al. proposed a midpoint-level mineral resource impact assessment method called the crustal scarcity indicator (CSI), with characterization factors called crustal scarcity potentials (CSPs) measured as kg silicon equivalents per kg element [2]. For assessment of mineral resource depletion associated with photovoltaic technologies, this method is interesting because it relates silicon (with CSP = 1 by definition since it is used as the reference value) to all other materials, which are assessed relative to silicon. The CSPs proposed for each element can be multiplied by the amount of mass extracted from Earth to obtain the CSI indicator for a certain product. Furthermore, since crustal concentrations of elements have been suggested to correlate with several important resource metrics (reserves, reserve base, reserves plus cumulative production, and ore deposits), they constitute good proxies for long-term global elemental scarcity. The CSI method has been compared to other LCIA methods (such as abiotic depletion, the surplus ore, the cumulative exergy demand and the environmental priority strategies in product development methods); the results of the comparison indicate that CSI is reliable and robust for long-term global elemental scarcity calculation while requiring few assumptions and input parameters.

The characterization factor for element i in the CSI method is called crustal scarcity potential and it is defined as:

$$\mathrm{CSP}_i = \frac{1/C_i}{1/C_{\mathrm{Si}}}, \tag{5.1}$$

where C_i is the crustal concentration (in ppm) of element i, and silicon is considered the reference element. The CSPs can be applied in a LCIA method for resource depletion by multiplying the mass of element i extracted from the crust (m_i, in kg required for the manufacture of the functional unit used in the LCA, data collected from life cycle inventory) by the corresponding CSP_i as indicated in the following equation:

$$\mathrm{CSI} = \sum_i m_i \times \mathrm{CSP}_i. \tag{5.2}$$

Data on crustal concentrations of elements have been evaluated since more than one century ago and most recent standard data are available for upper, middle and lower Earth's crust; also an average crust is calculated with a mix in proportions of 31.7%, 29.6% and 38.8%, respectively [33]. In some cases, crustal concentrations are given for the corresponding oxides (silicon, titanium, aluminium, iron, manganese, magnesium, calcium, sodium, potassium and phosphorous) and the concentrations of the pure elements are calculated based on their molar shares of the respective oxides in order to obtain their CSPs [2].

An important clarification about the dynamic nature of the concept of production, reserves and resources should be made before a detailed analysis of potential limitations for future PV production due to access to raw materials is carried out. In most of the studies, the risk of supply is analysed in terms of annual production of a given element and the trend in production growth that is extrapolated in different scenarios; therefore, its value depend on past trends and on choices for the scenarios' definition. At a second level, the reserves are used for risk evaluation; but reserves data are also dynamic, they may be reduced by constant mining of a raw material, or they can be increased if new deposits are discovered and may be exploited with known technology in an economically viable way (or expected to be viable with a reasonable technological development in short or medium term). The reserves can be considered as an aggregated working inventory of all mining companies' supplies of an economically extractable mineral commodity; the CSI definition provided by Eqs. 5.1 and 5.2 correlates well with reserves (including base, marginal and inferred reserves). The resource is defined by the U. S. Geological Survey as [38].

> Resource is a concentration of naturally occurring solid, liquid or gaseous material in or on the Earth's crust in such form and amount that economic extraction of a commodity from the concentration is currently or potentially feasible.

Again, the definition is linked to technical and economic considerations, although in this case it is projected into future "potential" developments. The quantification of resources involves a certain degree of uncertainty and therefore several categories have been established: identified (location, grade, quality and quantity are known or estimated from geologic evidence), demonstrated (quantity and quality computed from sampling in specific sites) and inferred (estimates from geologic evidence that may or may not be supported by samples or measurements). The extraction of a specific material may be restricted by laws or regulations and in this case they are considered "restricted resources" (it may also apply to reserves).

In this chapter, a list of required elements for each photovoltaic technology is provided by the numerous LCA studies which include a detailed inventory. The annual production and estimated reserves are obtained from the most recent mineral commodity summaries of the United States Geological Survey [38], the world mineral production report of the British Geological Survey [6] and the reports on critical metals for strategic technologies and sectors of the Joint Research Centre of the European Commission [9, 14, 29]. The resource depletion impacts are calculated following the CSI methodology described above.

5.1 Silicon Feedstock and Other Raw Materials Embedded in the PV Cells

The materials embedded in solar cells for different technologies are considered in this section. It refers to the materials that are included as constituent part in the final product. The list of considered minerals that are the raw material at the starting point of the whole manufacturing process are listed in Table 5.1, where data from different sources have been included. A very large list is obtained and not all elements will have the same impact since they contribute with different weights to the composition of the cell and the world production of each technology is dominated by silicon technology. Hydrogen and oxygen are not included when embedded elements are considered since they are effectively inexhaustible and their use will not have impact on long-term resource depletion, furthermore, data of elementary or molecular hydrogen and oxygen abundance on Earth are usually not provided in reports [2].

A list of materials is provided in Table 5.1 in alphabetical order. All included elements are related to one or more PV technologies. The main contribution of each element to the different parts of the solar cell is detailed in the fourth column. This list will be used as reference for the subsequent analysis of raw material production, reserves and demand projections carried out in this section.

The list of elements required for the PV technologies is long, but its relative importance is mediated by two factors: the amount of each element required per functional unit of photovoltaic electricity and the market share of each technology. Today, the market share for crystalline silicon is higher than 90% (see Chap. 3) and it is expected to remain dominant in the following decades.

Therefore, silicon is the element with the higher potential impact on the future of PV technology. Fortunately, silicon is the most abundant element on Earth's crust and it is taken as reference to calculate the crustal scarcity potentials of the other elements shown in Table 5.1 [2]. But silicon technology could be affected by scarcity of production capacity of crystalline ingots for wafers, as it happened for two years (2007–2008) provoking a transitory rise in silicon wafer price. Today industrial capacity is able to provide demand for the forthcoming years (even without considering silicon recycling, which is low).

Other technologies can be more affected by scarcity of elements required for its active layer. It is the case for thin film materials, that although they require less material per functional unit, the elements are scarce. For CdTe technology, specially tellurium may pose a supply risk; for CIGS technology, indium, gallium and selenium are also scarce minerals, with indium competing as a reference material for transparent conducting oxides used in organic and hybrid technologies. In this case, tensions between demand and supply could appear for production volumes in the range of a few hundreds GW_p [8, 43].

Material usage for silicon cells has been reduced significantly during the last 15 years from around 16 g/W_p to about 3.6 g/W_p due to increased efficiencies and thinner wafers, which have evolved from 300 μm in 2004 to 175 μm in 2020, although it has remained stable for the past ten years [15]. For thin film materials,

Table 5.1 List of elements (listed by element alphabetical order) used for manufacturing solar cells of all technologies

Element	Symbol	Technology	Cell component
Aluminium	Al	III-V and others (as electrode)	Active layer, electrode
Arsenic	As	III-V	Active layer
Boron	B	c-Si, m-Si, a-Si:H	p-type dopant
Cadmium	Cd	CdTe,	Active layer, window layer
Calcium	Ca	Organic	Electrode
Carbon black	C	Organic and hybrid	ETL, HTL, electrode
Carbon (Graphite)	C	Organic and hybrid	Electrode
Copper	Cu	CIGS and others (as electrode)	Active layer, electrode
Fluorine	F	Organic and hybrid	Dopant in TCO
Gallium	Ga	III-V	Active layer
Germanium	Ge	Si/Ge and Ge/CdTe, III-V	Active layer, substrate (in III-V)
Indium	In	CIGS, III-V, and others (as TCO)	Active layer, dopant in TCOs
Iron	Fe	Organic and hybrid	ETL, HTL, ferroelectric layers
Lead	Pb	Hybrid	Active layer
Lithium	Li	Organic	ETL
Magnesium	Mg	Organic and hybrid	ETL, HTL
Molybdenum	Mo	CIGS, organic and hybrid	ETL, back contact
Nickel	Ni	Organic and hybrid	ETL, HTL
Phosphorus	P	c-Si, a-Si:H	n-type dopant
Ruthenium	Ru	Organic and hybrid	Sensitizer (DSSC)
Selenium	Se	CIGS	Active layer
Silicon	Si	c-Si, a-Si:H, Si/Ge, Si/PSC, III-V	Active layer, substrate (in III-V)
Silver	Ag	All	Electrode
Sulphur	S	CdTe, organic	Window layer (CdS) active layer (organic)
Tellurium	Te	CdTe	Active layer, window layer
Tin	Sn	All	Active layer, electrode, TCO
Titanium	Ti	Organic and hybrid	ETL, electrode
Vanadium	V	Organic and hybrid	ETL
Zinc	Zn	CdTe, organic and hybrid	ETL, HTL, electrode

Table 5.2 c-Si technology: material requirements for a solar cell in kg/MW$_p$ from several references indicated in first row and calculated average. It includes metals required for frames, soldering and cables of typical module

c-Si	Reference 1	Reference 2	Reference 3	Reference 4	Average
Si	3653			5377.53	4515.26
Al	10593			12511	11552
Cu	2741	2194.1	7597.5	3554	4021.65
Sn	577	463.1		442	494.03
Ag	24	19.2	355.9	113.08	128.04
Mg	53.5			45.84	49.67
Ni			1.1	0.94	1.02

1: [29]
2: [30]
3: [12]
4: [39, 40]

Table 5.3 CIGS technology: material requirements for a solar cell in kg/MW$_p$ from several references indicated in first row and calculated average

CIGS	Ref. 1	Ref. 2	Ref. 3	Ref. 4	Ref. 5	Ref. 6	Ref. 7	Average
Cu	21.2	21					16.9	19.7
In	18.99	18.9	27.4	15.5	22.5	27.4	27.4	22.58
Ga	2.34	2.3	5		7.5	5	5	4.52
Se	9.56	9.6	45.3		45	45.3		30.95
In (in TCO)	44.29						94.3	69.25
Sn (in TCO)	5.95						85.8	45.8

1: [29]
2: [30]
3: [12]
4: [16]
5: [5]
6: [1]
7: [8]

where thickness of active layer has been kept almost constant during a long time of development, the material use is already highly optimized around a few μm, but efficiency has been increased and therefore the material use per W$_p$ has been also reduced; in particular average values for use of main elements in the active layers are: 0.068 g/W$_p$ silicon in a-Si:H technology, 0.064 g/W$_p$ cadmium and 0.067 g/W$_p$ tellurium in CdTe technology and 0.019 g/W$_p$ copper, 0.022 g/W$_p$ indium, 0.004 g/W$_p$ gallium and 0.031 g/W$_p$ selenium in CIGS technology (calculated as average of reported values provided in Tables 5.2, 5.3, 5.4 and 5.5; with the limiting material (or "restricted metal") being indium and tellurium for CIGS and CdTe technologies respectively [16]. Indium is also used in transparent conducting oxide electrodes

Table 5.4 CdTe technology: material requirements for a solar cell in kg/MW_p from several references indicated in first row and calculated average

CdTe	Ref. 1	Ref. 2	Ref. 3	Ref. 4	Ref. 5	Ref. 6	Ref. 7	Ref. 8	Average
Cd		61.1	63.3			85	63.3	49.2	64.38
Te	93.3	47.2	61.9	55		97.5		47.2	67.01
Cu					42.8				42.8
In (in TCO)	15.9	15.9							15.9
Sn (in TCO)	21.4							6.6	14

1: [29]
2: [30]
3: [12]
4: [16]
5: [4]
6: [5]
7: [1]
8: [8]

Table 5.5 a-Si technology: material requirements for a solar cell in kg/MW_p from several references indicated in first row and calculated average

a-Si	Reference 1	Reference 2	Reference 3	Reference 4	Average
Si					68.55[a]
In	5.32				5.32
Sn	0.714				0.714
Ge (in a-Si/Ge)		6.9	4.4	6.9	6.06

[a]Calculated from thickness and electrical parameters for typical a-Si:H cell mentioned in text.
1: [29]
2: [12]
3: [16]
4: [1]

(indium doped tin oxide, ITO, with around 0.022 g/W_p indium and 0.045 g/W_p tin) and could be a restriction to massive deployment unless it is replaced by fluorine doped tin oxide (FTO)

The difference between the material finally embedded in the cell and the initial input required for cell production, that is, the utilization rate for c-Si is currently 50% and may be strongly improved up to 90% in optimistic scenarios by 2040; for a-Si is already at 90%, while for other thin films there is still room for improvement from current 60 to 90% by 2040 [43].

The values provided in the tables refer to typical solar cells with the following parameters: c-Si cell thickness 170 μm, PCE = 15.8% [42]; and for thin film technologies: CIGS cell thickness 3 μm, PCE = 12% , CdTe thickness 3 μm, PCE = 10%, a-Si thickness 5 μm, PCE = 10% [30]; or small differences in performance (given in the corresponding references). For TCOs, it has been assumed a 100 nm layer thickness with composition In_2O_3 (90%) and Sn (10%). Some data are originally provided in g/m^2 and it has been converted to kg/MW$_p$, assuming the cell power conversion efficiency and layer thickness mentioned above.

5.2 Glass, Plastics and Frames for the PV Modules

The content of glass, plastic and metallic frames of PV modules is linked to the technology under consideration. The most common crystalline silicon modules have a weight per nominal capacity around 80 g/W$_p$ to 100 g/ W$_p$, with glass accounting for more than 70% of total weight and frame around 10%. These contributions have evolved in time with a tendency to increase the share of glass in total weight and reduce the contribution of frame leading to the manufacture of frameless modules. This tendency is also observed in thin film technologies, where frameless modules are more commonly manufactured. In Table 5.6 the composition of a typical crystalline silicon solar cell with 15% power conversion efficiency is presented, these values have slowly evolved by reducing frame weight (and share), glass weight (although its share in total weight increased), and also reducing silicon layer thickness (with limited impact in its share of total weight); the nominal peak power per module between 2000 and 2010 was around 200 W$_p$ but since 2010, the peak power per module have increased to peak powers per module around 350 W$_p$ to (more recently) almost 500 W$_p$, with reduced glass thickness and lighter frames, leading to slightly different weight percentage of each component as shown in Table 5.7. For the purpose of comparison, both modules have been considered to have the same power conversion efficiency (15%).

Table 5.6 Example for the composition of a c-Si standard module (215 W$_p$) up to 2010 [41]

Component	Share (%, 2003)	Share (%, 2007)	Weight (g/W_p, 2007)
Solar cells	4	3.48	3.6
Glass	62.7	74.16	77.3
Frames (e.g. AlMgSi0,5)	22	10.3	10.7
EVA	7.5	6.55	6.8
Backing film (Tedlar)	2.5	3.6	3.8
Junction box	1.2	1.1	1.2
Adhesive, potting compound	1.15	1.16	1.1
Total Weight (g/W_p)			104.5

Table 5.7 Example for the composition of a c-Si module (450 W_p) after 2010 [42]

Component	Material	g/W_p	%
Cover	Glass	59.9	76.22
Encapsulant	EVA	4.5	5.75
Backsheet	PET	3	3.77
Frame	Al	6.1	7.82
Cells and Ribbon	Si	3.7	4.7
	Ag	0.032	0.04
	Cu	0.58	0.74
	Sn	0.056	0.07
	Pb	0.033	0.04
Sealant, Potting, Compound	PIB, TPT, silicone, other	0.67	0.85

Table 5.8 Composition of CdTe and CIGS modules [4]

Material	Layer thickness (μm)	CdTe mass (kg)	Mass fraction (%)	Layer thickness (μm)	CIGS mass (kg)	Mass fraction (%)
Glass	6400	11.40	96.7	6000	11.67	96.9
EVA	450	0.356	3.1	450	0.356	2.96
CdTe	3–4	0.012–0.016	0.11–0.13			
Cu	0.4	0.003	0.02			
SnO_2	0.5	0.002	0.02			
CdS	0.4	0.001	0.01	0.05	0.000037	0.0003
ZnO				1.0	0.0042	0.035
$CuInS_2$				1.5	0.005	0.04
Mo				0.4	0.0031	0.025

For thin film cells, the glass amount is more than 95% of total weight in frameless modules, where EVA backsheet is the next major contributor with 3% and solar cells not reaching 1% as expected for thin film technologies where the thickness of active layers are lower than 5 μm. In Table 5.8 a summary of weight distribution for CdTe and CIGS modules is presented, as reported in a study by Berger et al about end-of-life of modules and the possibility of recovery of materials by using a novel approach for recycling [4].

In all technologies, front glass contribution to weight and to LCA category impacts is important. Reduction of glass thickness is limited due to structural considerations and in 2020 module production had reached a glass thickness of 3 mm in 85% of manufactured modules worldwide, and between 2 mm and 3 mm for the remaining 15%; this ratio is expected to change to 55/42% by 2030 with a few modules (around

3%) using glass thinner than 2 mm [25]. Furthermore, a recent LCA study has shown that single-crystalline silicon modules with glass-glass encapsulation have around 10% lower impacts in several categories compared to glass-backsheet encapsulation [31]. This is independent of total absolute emissions which are different according to the country of production; for example, regarding greenhouse gas emissions (carbon footprint), the impact is reduced 8% in China, 10% in Germany and 12.5% in average European Union, a result confirming that bifacial modules (with glass in the two sides) could potentially have lower impacts than conventional modules.

If the materials embedded in the solar cells and balance of system components (including supporting structures) are added, a fair comparison with other energy alternatives (renewable or non-renewable) can be shown in terms of demand of materials per unit of delivered energy along the lifetime of energy systems. In Fig. 5.1, the comparison of four common materials whose demand is presented (cement, iron, aluminium and copper); the bars in the figures spans from higher to lower estimates using collected data (2010) and the extrapolation to possible reduction in material demand due to technological improvements that may be achieved by 2050 (data from [37]). Cement and iron demand is similar for photovoltaic technology and non-renewable energy technologies, while the most demanding are large hydro, wind and concentrated solar power. Regarding aluminium and copper, this trend is reversed, with photovoltaic technology on top of demand but with high uncertainty in potential material savings due to future technological development (the only exception is CdTe with has lower aluminium demand).

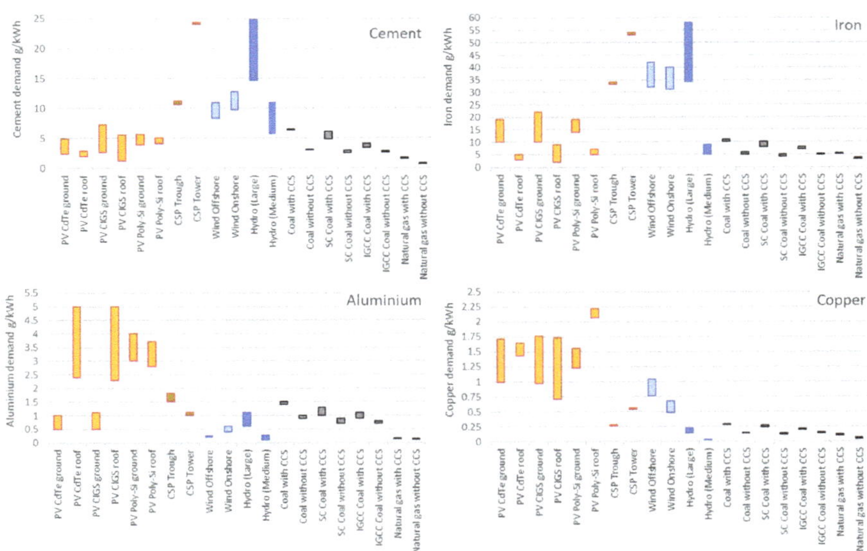

Fig. 5.1 Bulk material (g) requirements per unit energy (kWh) produced by different energy technologies [37]

5.3 Strategic and Scarce Materials Embedded in PV Modules

When considering the concepts of abundance (or scarcity) of an element and the classification of the material as strategic, not only its abundance on Earth's crust and the estimated reserves are considered, also the availability of the resource according to the geographical location of mines and the geopolitical considerations of the countries which supply significant shares of annual mineral production. The focus of this section is to present the data of production trends of all materials involved in any of the current or emerging PV technologies. Data of production and reserves are obtained from [6, 38] and they are compared with estimated demand for PV industry in the coming years as well as the crustal scarcity potential calculated by Rickard Arvidsson [2]. A more detailed analysis of the geopolitical risks is provided in Chap. 12, Sect. 12.5.

In Table 5.9 all elements are listed in alphabetical order, for the details of which technology (and part of cell) uses a given material, see Table 5.1. Silicon is the most abundant element in Earth's crust and has been kept in the list since its CSP is used as reference (hence its value is 1 for Si), other four abundant elements have not been included although they are also used for PV module fabrication, because its production is very high compared with PV industry demand, they are carbon (black or coal), with CSP = 14, iron (CSP = 5.4), calcium (CSP = 6.2) and sulphur (CSP = 700); a few comments on these four elements are included below. On the other hand, metals such as aluminium, titanium or magnesium, which are abundant, have been included in the table because its world production although being high is one order of magnitude lower than the other four.

The values provided in Table 5.9 are a present-time global view of world annual production of all elements related to any PV technology. Values for 2019 production from BGS and USGS are coincident in most materials, with slight deviations due to data collection from some countries or mineral definition. Values from USGS for 2020 may be affected by COVID-19 pandemia, since some mining sites were closed for a period of time in major producer countries. Silicon production both for ferro-silicon and silicon metal (>99% Si content) have been included in the table, although there are other silicon-containing minerals that can be also used as a starting point for a purification process for solar grade or electronic grade silicon. Silicon does not pose any risk for future availability of the element, and only purification and crystallization capacity could potentially suppose a bottleneck for solar cell manufacture for wafer-based Si technology that represented 95% of market share in 2020 (130 GW_p of installed capacity in 2019 as reported by [15]); but nowadays, and up to 2030 production capacity is able to match demand at the current trend of production and the only risk is the excessive concentration of industrial capacity in China, which accounts for high shares of mineral production (53%), Si processing (50%), Si cell manufacture (86%) and module assembly (70%) in 2019 [9, 14]; and therefore the main recommendation for the rest of the world regarding Si supply and

Table 5.9 Crustal scarcity potential (CSP), production, production growth and estimated reserves of elements required by PV industry

Material	Symbol	CSP kg Si_{eq}/kg	Annual Production (tonnes)				World reserves
			BGS	USGS	USGS	Growth[a]	USGS Million tonnes
			2019	2019	2020	%	
Aluminium	Al	3.4	62,850,158	63,200,000	65,200,000	51.61	55,000–75,000
Arsenic	As	110,000	57,585	32,300	32,000	38.96	0.64
Boron (borates)	B	26,000	6,889,063	3,632,000	3,630,000	17.48	1,200
Cadmium	Cd	3,500,000	27,526	24,400	23,000	18.12	0.57
Carbon (Graphite)	C	140	1,132,328	1,100,000	1,100,000	–46.81	320
Copper (mine)	Cu	10,000	20,676,784	20,400,000	20,000,000	28.41	870[b]
Fluorine[c]	F	510	6,478,841	7,460,000	7,600,000	–9.50	320
Gallium	Ga	18,000	380	351	300	156.76	1 (estimated[d])
Germanium	Ge	22,000	95	131	130	–22.13	Not available[e]
Indium	In	5,400,000	851	968	900	32.35	Not available[f]
Lead (mine)	Pb	26,000	4,684,075	4,720,000	4,400,000	7.43	2,000
Lithium	Li	18,000	1,906,494[g]	86,000	82,000	240.93	86
Magnesium	Mg	10	1,059,736	1,120,000	1,100,000	39.83	Unlimited
Molybdenum	Mo	350,000	275,187	294,000	300,000	12.29	25.4
Nickel (mine)	Ni	4,800	2,702,428	2,610,000	2,500,000	68.36	>300
Phosphorus	P	650	226,162,769	227,000,000	223,000,000	25.56	>300,000
Ruthenium	Ru	470,000,000	60	30	30	–1.79	0.005
Selenium	Se	2,200,000	4,264	2,880	2,900	69.88	0.1[h]
Silicon (ferro-Si)	Si	1	1,807,840	8,410,000[9]	8,000,000[i]	–34.66	Unlimited
Silicon (metal)	Si	1	3,010,606			52.35	Unlimited
Silver	Ag	5,100,000	26,261	26,500	25,000	12.27	0.5
Tellurium	Te	57,000,000	625	520	490	359.56	0.031
Tin	Sn	170,000	304,572	296,000	270,000	–7.15	4.3

(continued)

Table 5.9 (continued)

Material	Symbol	CSP	Annual Production (tonnes)				World reserves
			BGS	USGS	USGS	Growth[a]	USGS
		kg Si_{eq}/kg	2019	2019	2020	%	Million tonnes
Titanium	Ti	67	13,262,262	16,300,000	16,010,000	18.53	2,000
Vanadium	V	2,000	81,031	86,800	86,000	19.24	63
Zinc (mine)	Zn	3,900	12,333,887	12,700,000	12,000,000	–1.24	1,900

[a]Production growth in ten years 2010–2019 calculated with BGS data.
[b]Copper reserves: identified 2.1 billon tonnes; estimated undiscovered 3.5 billion tonnes.
[c]Fluorine production and reserves refer to fluorspar (CaF_2).
[d]Gallium: estimated reserves as by-product of bauxite (50 ppm) and Zn (50 ppm) but only 10% potentially recoverable.
[e]Germanium: data on recoverable Ge from Zn ores not available; globally only 3% of Ge content is recovered.
[f]Indium: most commonly recovered from zinc-sulphide ore mineral sphalerite (1–100 ppm).
[g]Lithium: data from BGS for Li includes all minerals and in particular 1.6 million tonnes produced by Australia (Spodumene).
[h]Selenium: estimated from 40 to 900 ppm in copper reserves, and uneconomical recovery from 0.5 to 12 ppm from coal reserves.
[i]Silicon data from USGS includes ferro-silicon and silicon metal.

silicon-based technologies is diversification of the supply chain. This geopolitical recommendation will be analysed in more detail in Chap. 12.

Other required elements have not been included in the table because they are abundant and its annual production is much higher than its global demand; furthermore, its use in the PV industry is low. They are carbon (coal) and iron (used for steel manufacture of BoS components for electronic components, framing and support or tracking devices for modules either in plants or roof-top or BIPV systems), with annual production >8 and >3 billion tonnes in 2019 respectively, and estimated world reserves in excess of 800 billion tonnes for both minerals [6]. Similarly, calcium and sulphur are produced from a broad range of minerals of which the most abundant are gypsum and anhydrite, providing an almost unlimited amount of reserves, and they are scarcely used in PV technologies [38].

Other metals included in the table require a more detailed analysis. Copper is used mainly for BoS components (power electronics and cabling), its annual world production is high (>20 million tonnes in 2020) and reserves amount to 2.1 billion tonnes identified and an extra estimated undiscovered 3.5 billion tonnes. But price of copper is unstable with a sustained increasing trend, which has led to high shares of recycled copper in the supply chain; for example, copper recovered from scrap contributed about 38% of the U.S. copper supply [38]. Aluminium (second most abundant element in Earth's crust, with annual production >60 million tonnes) is a metal widely used in the PV industry and it is not considered a bottleneck for future supply [14]. On the other hand silver could pose a potential risk of supply, its annual production is 26,000 tonnes and although silver was a principal product at several mines, silver is primarily obtained as a by-product from lead-zinc mines, copper mines, and gold mines (reserves estimated around 500,000 tonnes as reported by [38]); silver content is being reduced in alloys for contacts in all PV technologies, with increasing use of tin and aluminium as substitutes. Silver is the only element in silicon technology that could require a replacement in the medium term, specially if an impact on final price of system is to be avoided, in competition with other energy-related technologies (since silver is itself a replacement for more scarce and expensive metals, for example as catalyst in batteries). Currently around 30% of silver is recovered and recycled worldwide, with an annual growth of 0.6% [39, 40].

A very different picture arises for thin film technologies. Ideally they were considered a good alternative to silicon because they require reduced amounts of material in the active layer (see Tables 5.3 and 5.4 for CIGS and CdTe technologies). But despite this fact, several elements have been identified as potential bottlenecks for a massive production of PV modules, which may limit future cumulative manufacture of PV modules of these technologies, although it has been estimated that production could peak around 100 GW_p/year by 2030, then be reduced again but with a potential recovery that in any scenario will be lower than 200 GW_p/year by 2100 [8, 17, 19, 27]. Market trends have reduced the share of thin film technologies to around 5% of global market which represents around 7.5 GW_p of annual installed capacity (2019), well below the possible limits posed by mineral supply and it is expected to remain in these values [15, 24, 25]. Nevertheless, six metals have been identified which could generate a potential supply risk for thin film technologies, they are scarce in Earth's crust and only produced as by-products of other primary commodities:

Cadmium is used in CdTe technology. It is produced as a by-product of processing sphalerite, a zinc sulphide ore containing varying amounts of cadmium (average about 0.03%). With production in the range of 20,000 tonnes per year, a few thousand tonnes are not recovered every year because it is actually uneconomical, but it could be recovered if high demand for cadmium pushes prices up. Moreover, around 25% of world production is obtained from recycled NiCd batteries [5]; therefore the risk for CdTe technology arising from Cd is much lower than the risk from Te.

Tellurium pose a risk for CdTe technology. It is produced as a by-product of copper mining, and only 490 tonnes were produced in 2020, in ten years (2010–2019), the production has grown 360% despite a small reduction in the past two years [6]; reserves are only 31,000 tonnes (estimated as recoverable from slimes of copper mining with 100ppm of Te content and without including data from Russia, which for this mineral are not available in the USGS [38]). A very small amount of tellurium is recovered from scrapped photoreceptors employed in old photocopy-machines in Europe and some pilot plants are starting to recover tellurium from recycled CdTe solar cells (see Sect. 8.2).

Indium is obtained as a by-product of zinc mining (also mainly from spharelite mineral, in ranges from 1ppm to 100ppm). Its use in many electronic devices as transparent conducting oxide (50% of current indium production is used in ITO coatings) creates a very competitive market which has generated high prices for indium with a production around 900 tonnes in 2020, an increment of 32%

in the past ten years. The recovery of indium from reprocessing mining wastes (tailings and slags) is now being practised on a limited scale but could become a significant source of future indium supply if high demand pulls prices up and justifies investment. As secondary source, indium is recovered from electronic scrap waste and processed domestically in many countries; in 2017 around 37.5% of indium was recycled worldwide, half of it reprocessed in China; this recycling industry could be a significant source of indium in the near future [5, 39, 40]. Scarcity and very high prices could pose a risk on CIGS and III-V technologies, which rely on indium for its active layer [26], but also on other organic and hybrid emerging technologies which require indium for the TCO contacts, in this later case, replacement with fluorine doped tin oxide (FTO) and the use of alternative electrodes is a good solution (see Sect. 4.4).

Gallium is a component of the active layer of CIGS and III-V technologies and it has been considered an element that could hinder the massive deployment of both technologies. Most of gallium production (380 tonnes in 2019 [6]) comes as by-product of bauxite ore processing for aluminium, and minor amounts are obtained from residues produced during the process to recover primary zinc metal; in both cases the gallium content is around 50 ppm and only 10% is potentially recoverable with total reserves estimated around 1 million tonnes. Considering that gallium production has grown in two steps, from less than 20 tonnes before 2010 to 237 tonnes in 2011, and more than doubled in seven years to be reduced since then, it is difficult to make estimations of future production. Very high power conversion efficiency III-V solar cells for PV space applications and high mobility transistors for electronic devices rely on this material and it is difficult that CIGS technology could compete in a horizon of high prices for this commodity although supply is enough to meet the actual demand; still very little gallium is recovered from recycled obsolete electronic materials and there are increasing commercial tensions: China produces 80% of gallium and a small reduction in Chinese primary production lead to a 32% increase in 2020 price [38].

Selenium is refined from primary copper ores. Around 4,000 tonnes are currently produced every year by reprocessing anode slimes resulting from electrolytic purification of copper mining ores, with a ten years increment of 69% [6]. Reserves of 100,000 tonnes are estimated from 40 to 900 ppm in copper reserves, and uneconomical recovery from 0.5 to 12 ppm from coal reserves. Although it is a scarce material, it is not considered that it could pose constraints to CIGS technology since search for primary deposits is progressing fast and recent technological improvements for its extraction from copper, coal, zinc, lead and gold mining residues and in oil refineries could result in an increase in supply if prices justify recovery. Only 5% of selenium is currently recycled, with an increasing trend that could double by 2050 [39].

Germanium production of around 100 tonnes per year is obtained from two main sources: refining of zinc mining residues (60%) and from leaching of fly ash, a waste product of coal combustion (30%). Until 2010 germanium supply was strongly dependant on zinc production and when zinc demand was low but germanium demand remained high, a reduction in Ge production lead to supply shortages and sudden price hikes (around 20% in 2007 and 2008); since then, an despite a constant or slow reduction in annual production, prices have been stable an even they have been reduced in recent years. World reserves are undisclosed, but could be considered high if technology for recovery of germanium from coal combustion is improved. Furthermore, worldwide around 30% of the total consumed germanium is produced from recycled materials, a figure which is increasing, specially in the infra-red optics industry, where 60% of the used germanium metal is routinely recycled as new scrap. This trend indicates that although germanium production is low, it will not pose a significant risk to the two PV technologies where it is mostly used, a-Si/Ge tandem cells and as substrate for III-V technology.

The European Union regularly assess the criticality of raw materials applying a methodology which is a combination of indicators about mineral global supply risk (internal production, import reliance and recycling), economical considerations (capital cost, value added) and potential substitutes for each application (evaluating trade-off of cost and performance). The first assessment carried out in 2011 identified 14 critical raw materials (CRM); it has been updated in 2014, 2017 and 2020 which has increased the list up to 30 CRMs. Interestingly, silicon metal has been considered a CRM by the EU since 2014, and in the final list of 2020 the following materials related to different photovoltaic technologies (from Table 5.1) were included: bauxite (for aluminium production), borates (for boron production), coking coal, fluorspar (for fluor production), gallium, germanium, indium, lithium, magnesium, natural graphite, phosphorus, silicon metal, titanium and vanadium [14]. A long list that emphasizes European vulnerability of its own photovoltaic industry (in 2021 almost completely delocalized outside Europe with the exception of some emerging technologies).

As a summary of global worldwide potential risks for solar electricity sustainability arising from the use of raw materials, two main conclusions can be pointed out: firstly, the c-Si (wafer based) photovoltaic technology, currently representing 95% of the market (and expected to remain high in the near future) is not threatened by any supply risk; the only possibility of a supply shortage could arise from silver, which is an expensive metal with many industrial uses, but silver could be replaced by aluminium, copper and tin without compromising power conversion efficiency or stability of c-Si PV modules. Secondly, two thin film technologies already in the market are constrained by high supply risk of tellurium (CdTe technology) and indium (CIGS technology); despite its small market share (around 5%) and the small amount of material used in the cell layers ($<5\ \mu m$), its manufacture at a rate of a few hundreds of GW_p cannot be achieved in the future; furthermore, the competition for indium against other PV technologies (III-V) and the electronics industry (TCOs) could push prices up and reduce more its market share. Indium scarcity could also represent a problem for other emerging organic and hybrid technologies, since they require transparent electrodes fabricated so far with indium tin oxide (ITO), but this risk is also low, since there are effective replacements for ITO to be used as electrodes.

Other materials, such as gallium, selenium and germanium, could also pose a certain risk, but it is considered low in most reports, with the other two metals (tellurium and indium) representing the real bottleneck for massive production in the affected technologies [5, 9, 14, 29]. The copper zinc tin sulphide (CZTS) kesterite approach to thin film technology aims to replace all scarce materials used in CIGS technology with Earth's crust abundant materials, which together with potentially low-cost printing technologies used in its production, could open a good market share for this thin film technology in competition with c-Si and other emerging technologies [27].

Finally, although used in very small quantities in dye sensitized solar cells (DSSC), ruthenium deserves a special mention: it is included in the "platinum-group metals" (PGM) production lists, with a stable production reported between 30 and 60 tonnes and estimated reserves of only 5,000 tonnes, it has seen a strong price increment of 136% in 2020 due to increasing demand [6, 38]. The most efficient molecular sensitizer in DSSC technology includes a ruthenium atom, but research is intensive to develop sensitizers that does not include critical metals. Nevertheless, forecasts for DSSC does not seem to include a significant share in future PV market, since its solid-state evolution led to the perovskite approach which seems to have much brighter future.

5.4 Polluting and Toxic Materials Embedded in PV Modules and Used in Its Manufacturing Process

A list of materials and substances embedded in the PV modules and used during its manufacturing process, including a brief comment on environmental hazards and health risk during manufacture are listed in Table 5.10 and commented below. The detailed analysis of impacts on several categories (including human health or ecotoxicity, between others) of embedded materials in solar cells and processing or end-of-life waste are included in the LCA studies presented in Chap. 7. In this later case, the relative importance of impacts is linked to the amount of materials embedded in the final product (cell or module) and the amount of substances required during processing; futhermore, they can be partially recovered, reused or recycled. Since characterization factors and a LCIA method should also be considered to calculate quantitatively the impacts, only a detailed life cycle assessment provides a realistic vision of impacts and risks, which in general are low despite the quite impressive list provided in Table 5.10.

The list is a long one and it emphasizes the need to carry out a detailed analysis of health and safety precautions in all manufacturing lines for all PV technologies. Most of the potential hazards may occur during fabrication (and during recycling, see Chap. 8). Nevertheless, after a PV module is fabricated, including its encapsulation, glass, backsheet and frame, the toxic elements content of the module is very low and therefore risks for human health and environment during installation, operation and

Table 5.10 List of polluting and toxic materials and substances used during manufacture of PV modules

Hazardous material	Used in	Health issues
Arsine	III-V CVD	Accidental risk, blood, kidney
Arsenic compounds	III-V CVD MBE	Cancer, lung
Ammonia	CdS deposition	Lung desease, irritant, environmental damage
Cadmium	CdTe and CdS deposition	Extremely toxic, Cancer, kidney
Cadmium chloride	CdTe deposition	Cd release, environmental damage
Cadmium sulphate	CdTe deposition	Cd release, environmental damage
Carbon tetrachloride	Etchant	Liver cancer, greenhouse gas
Chlorosilanes	a-Si and x-Si deposition	Irritant
Diborane	a-Si dopant	Central nervous system, pulmonary
Germane	a-Si dopant, Si/Ge tandem	Blood, central nervous system, kidney
Hydrogen	a-Si deposition	Fire hazard
Hydrogen fluoride	Etchant	Irritant, burns, bone, teeth
Hydrogen selenide	CIGS sputtering	Irritant, gastrointestinal, flammable
Hydrogen sulphide	CIGS sputtering	Irritant, central nervous system, flammable
Hydrochloric acid	Reduction chamber	Irritant, corrosive
Indium compounds	CIGS deposition	Pulmonary, bone, gastrointestinal
Lead	Soldering	Central nervous system, gastrointestinal, blood, kidney, reproductive
Nitric acid	Wafer cleaning	Irritant, corrosive
Phosphine	a-Si dopant	Irritant, central nervous system, gastrointestinal, flammable
Phosphorous oxychloride	x-Si dopant	Irritant, kidney
Potassium hydroxide	c-Si, Wafer treatment	Irritant, eye, lung, and skin damage
Selenium dioxide	CIGS deposition	Irritant, tissue poison
Silane	a-Si and x-Si deposition	Irritant, spontaneous explosion in air
Silica	Si Mining	Silicosis, a severe lung disease
Silicon tetrafluoride	a-Si deposition	Corrosive, irritant to eye, lung, and skin
Silicon tetrachloride	c-Si	React with water and cause environmental hazard
Sulphur hexafluoride	Clean the reactors	Potent greenhouse gas (25000$\times$ CO_{2eq})
Sulphur dioxide	Within chamber	Acid rain
Sodium hydroxide	c-Si, Wafer treatment	Irritant, eye, lung, and skin damage
Tellurium compounds	CIGS deposition	Central nervous system, cyanosis, liver
Thiourea	CdS deposition	Carcinogen, damage to bone marrow, skin allergy

(continued)

Table 5.10 (continued)

Hazardous Material	Used in	Health Issues
Trichloroethylene	III-V as solvent	Carcinogen
Triethyl gallium	III-V	Severe eye and skin burns, fatal if swallowed or inhaled, highly flammable, espontaneous explosion in air
Trimethyl arsenic	III-V	Releases arsenic and has effects on lung, liver, immune, and blood
Trimethyl gallium	III-V	Severe eye and skin burns, respiratory tract burns, spontaneous explosion in air

dismantling is low, even in the case of accidental damage (for example fire events or floods). Several agencies worldwide have carried out a detailed analysis of those risks and published recommendations for the handling of the PV components, in particular the United States Environmental Protection Agency (EPA) and Occupational Safety and Health Administration (OSHA) and the European Chemicals Agency (ECHA). In the following sections, special mention to four of the potentially most hazardous materials related to photovoltaic technologies are described.

5.4.1 Silicon Mining and Processing Risks

Silicon is the most abundant element in Earth's crust, but the mining of silica creates an important human health problem: silicosis, a severe lung disease that cannot be cured. It occurs after long exposure to silica dust; once the lung is damaged, the sickness cannot be reversed and therefore prevention is the only mitigation measure. Treatment of silicosis is focused on slowing down the progression of the disease and relieving the symptoms. In the past, incidence of the sickness was high, but strict rules on working conditions in mines, glass factories and building industry were able to reduce the incidence in twenty-first century, although contrary to expectations, it is increasing again specially in emerging economies of developing countries. Many experts consider that silicosis could become a pressing global health issue and that it should be considered an "epidemic in the making" [32]. PV industry is a high consumer of silica, mostly for the manufacturing of glass for the modules and should implement regulations that prevent using silica that has not been mined with the highest standards of workers safety.

Several chemical substances that are used in the silicon processing, specially in the chloride-hydride purification stage, present high toxicity according to EPA, OSHA and ECHA. Between them $SiHCl_3$, SiH_2Cl_2, SiH_4, $SiCl_4$, and $SiHBr_3$ are the most dangerous for human health and environment. Nevertheless, they are used in small quantities and in very controlled working environments within high-tech factories where overall health risk for workers is kept under control. Special mention requires silane (SiH_4), which present a very high explosion risk because it explodes spontaneously in air in a broad range of concentrations; it has also been coded by OSHA as HE4, meaning acute toxicity and short-term high-risk effects. High economical investment is required for chemical security in all factories using silane, chloro-silane(s) and its derivatives for silicon based technologies, specially a-Si:H.

5.4.2 Cadmium Toxicity

Cadmium is considered "extremely toxic" by EPA, OSHA and ECHA. It is a carcinogenic substance that may affect kidney, liver, bone, and can lead to blood damage from ingestion and lung cancer from inhalation. It is suspected of causing genetic defects, of damaging fertility and of badly affecting the unborn child. It can cause severe environmental damage, and it is considered very toxic for the aquatic life. Additionally, in pure form, it is pyrophoric and catches fire spontaneously if exposed to air. In a manufacturing centre for CdTe modules, workers should not be exposed to an airborne concentration of cadmium in excess of five micrograms per cubic meter of air (5 $\mu m/m^3$), calculated as an eight-hour time-weighted average exposure (TWA) according to OSHA.

The risk of cadmium release can be produced during manufacturing process with risk to exposed workers, and to the environment by accidental destruction of photovoltaic CdTe modules during its lifetime or due to a bad waste treatment at the end of life. Nevertheless, risk during fire events have been evaluated and considered low since its evaporation temperature is higher than the one that can be reached in a fire event [16, 18]. On the other hand, cadmium is produced as a by-product of zinc mining and can either be collected for several uses or discharged to the environment posing another risk. The use of Cd in the photovoltaic industry can make Cd treatment much safer than other current Cd uses, for example in Ni-Cd batteries, where it is used less efficiently and the handling of the substance is less controlled throughout its lifetime; CdTe PV technology is considered a safer way to sequestrate cadmium and therefore avoiding its release to environment as a by-product of zinc mining.

5.4.3 Lead Toxicity

If exposed to lead through accidental inhalation or ingestion, workers may develop a variety of ailments, such as neurological effects, gastrointestinal effects, anemia and kidney disease, and it may also damage fertility. The maximum accepted levels of lead in drinking water and air were set to 15 and 0.15 $\mu g/L$, respectively, according to EPA. The OSHA establish a permissible exposure limit of lead at 50 $\mu g/m^3$an eight-hour TWA (ten times higher than cadmium limits); nevertheless, if levels go beyond 30 $\mu g/m^3$ in manufacturing centres, specific compliance activities must be carried out, including regular medical tests for workers.

Lead can also be released to the environment, where it remains as dust indefinitely and it can result in decreased growth and reproduction in plants and animals, and neurological effects in vertebrates. Perovskite photovoltaic technology contains lead in the most efficient active layers. Rain with different pH on perovskite photovoltaic modules may induce lead release into the environment; this effect was experimentally evaluated in detail and the results find that even in the case of catastrophic module destruction, the release to the environment is extremely low [20]. The total content of lead in perovskite cells is very small and health detailed life cycle assessment have found that the impact of lead on several categories is low and in any case lower than other processes including in perovskite manufacture, like spiro-OMeTAD production [13, 34].

5.4.4 Sulphur Hexafluoride Environmental Damage

On top of those mentioned above, between all the substances listed in Table 5.10, the risks of releasing sulphur hexafluoride (SF_6) should be emphasized. According to the Intergovernmental Panel of Climate Change (IPCC), it is rated as one of the most potent greenhouse gases (GHGs) per molecule; as one tonne of SF_6 is equivalent to 25,000 tonnes of CO_2. Furthermore, if released to atmosphere, it can create acid rain when reacting with silicon to generate silicon tetrafluoride (SiF_4) and sulphur difluoride (SF_2), or be reduced to tetrafluorosilane (SiF_4) and sulphur dioxide (SO_2) [36].

In this chapter, the risks arising from the materials used in PV technologies have shown that future deployment of photovoltaic capacity may reach the Tera-Watt scale without facing insurmountable risks. Nevertheless, the impacts on human health and the environment of the production routes of the embedded materials in the modules and the substances used during its manufacture should be carefully evaluated. The tool to do so is Life Cycle Assessment and its results will be presented and analysed in Chap. 7.

References

1. Andersson BA, Jacobsson S (2000) Monitoring and assessing technology choice: the case of solar cells. Viability Photovolt 28(14):1037–1049
2. Arvidsson R, Söderman ML, Sandén BA, Nordelöf A, André H, Tillman AM (2020) A crustal scarcity indicator for long-term global elemental resource assessment in LCA. Int J Life Cycle Assess 25(9):1805–1817
3. Berger M, Sonderegger T, Alvarenga R, Bach V, Cimprich A, Dewulf J, Frischknecht R, Guinée J, Helbig C, Huppertz T, Jolliet O, Motoshita M, Northey S, Peña CA, Rugani B, Sahnoune A, Schrijvers D, Schulze R, Sonnemann G, Valero A, Weidema BP, Young SB (2020) Mineral resources in life cycle impact assessment: Part II- recommendations on application-dependent use of existing methods and on futuremethod development needs. Int J Life Cycle Assess 25(4):798–813

4. Berger W, Simon FG, Weimann K, Alsema EA (2010) A novel approach for the recycling of thin film photovoltaic modules. Resour Conserv Recycl 54(10):711–718
5. Bleiwas DI (2010) Byproduct mineral commodities used for the production of photovoltaic cells. Tech. Rep. USGS Circular 1365, USGS United States Geological Survey. https://pubs.usgs.gov/circ/1365/
6. British Geological Survey (2021) World Mineral Production 2015-2019. Tech. rep., British Geological Survey. https://www2.bgs.ac.uk/mineralsuk/statistics/wms.cfc?method=searchWMS, iSBN 978-0-85272-790-4
7. Calvo G, Mudd G, Valero A, Valero A (2016) Decreasing ore grades in global metallic mining: a theoretical issue or a global reality? Resources 5(4). https://doi.org/10.3390/resources5040036
8. Candelise C, Speirs JF, Gross RJ (2011) Materials availability for thin film (TF) PV technologies development: a real concern? Renew Sustain Energy Rev 15(9):4972–4981
9. Carrara S, Alves Días P, Plazzota B, Pavel C (2020) Raw materials demand for wind and solar PV technologies in the transition towards a decarbonised energy system. Tech. Rep. JRC119941, Joint Research Centre (European Commision). https://doi.org/10.2760/160859, https://publications.jrc.ec.europa.eu/repository/handle/JRC119941, iSBN: 978-92-76-16225-4 ISSN: 1831-9424
10. Dewulf J, Benini L, Mancini L, Sala S, Blengini GA, Ardente F, Recchioni M, Maes J, Pant R, Pennington D (2015) Rethinking the area of protection "Natural Resources" in life cycle assessment. Environ Sci Technol 49(9):5310–5317
11. Drielsma JA, Russell-Vaccari AJ, Drnek T, Brady T, Weihed P, Mistry M, Simbor LP (2016) Mineral resources in life cycle impact assessment-defining the path forward. Int J Life Cycle Assess 21(1):85–105
12. Elshkaki A, Graedel T (2013) Dynamic analysis of the global metals flows and stocks in electricity generation technologies. J Clean Prod 59:260–273
13. Espinosa N, Serrano-Luján L, Urbina A, Krebs FC (2015) Solution and vapour deposited lead perovskite solar cells: ecotoxicity from a life cycle assessment perspective. Solar Energy Mater Solar Cells 137:303–310
14. European Commission (2020) Study on the EUâs list of critical raw materials—final report (2020). Print ISBN 978-92-76-21050-4. https://doi.org/10.2873/904613ET-01-20-491-EN-C PDF ISBN 978-92-76-21049-8. https://doi.org/10.2873/11619ET-01-20-491-EN-N
15. Fraunhofer-ISE (2021) Photovoltaics Report 2021. Tech. rep., Fraunhofer Institute for Solar Energy Systems, ISE, Germany. https://www.ise.fraunhofer.de/content/dam/ise/de/documents/publications/studies/Photovoltaics-Report.pdf
16. Fthenakis V (2009) Sustainability of photovoltaics: the case for thin-film solar cells. Renew Sustain Energy Rev 13(9):2746–2750
17. Fthenakis V, Wang W, Kim HC (2009) Life cycle inventory analysis of the production of metals used in photovoltaics. Renew Sustain Energy Rev 13(3):493–517
18. Fthenakis VM (2004) Life cycle impact analysis of cadmium in CdTe PV production. Renew Sustain Energy Rev 8(4):303–334
19. Grandell L, Höök M (2015) Assessing rare metal availability challenges for solar energy technologies. Sustainability 7(9). https://doi.org/10.3390/su70911818
20. Hailegnaw B, Kirmayer S, Edri E, Hodes G, Cahen D (2015) Rain on methylammonium lead iodide based perovskites: possible environmental effects of perovskite solar cells. J Phys Chem Lett 6(9):1543–1547
21. Hauschild MZ, Goedkoop M, Guinée J, Heijungs R, Huijbregts M, Jolliet O, Margni M, De Schryver A, Humbert S, Laurent A, Sala S, Pant R (2013) Identifying best existing practice for characterization modeling in life cycle impact assessment. Int J Life Cycle Assess 18(3):683–697
22. Höök M, Hirsch R, Aleklett K (2009) Giant oil field decline rates and their influence on world oil production. China Energy Effic 37(6):2262–2272
23. Hubbert MK (1956) Nuclear energy and the fossil fuels. In: Publication 95. Shell Development Company, San Antonio, Texas, USA

24. IRENA (2020) Global Renewables Outlook: Energy transformation 2050. Tech. rep., International Renewable Energy Agency (IRENA), iSBN 978-92-9260-238-3
25. ITRPV (2021) International Technology Roadmap for Photovoltaic. Results 2020. Tech. rep. https://itrpv.vdma.org/documents/27094228/29066965/20210ITRPV/08ccda3a-585e-6a58-6afa-6c20e436cf41
26. Jean J, Brown PR, Jaffe RL, Buonassisi T, Bulović V (2015) Pathways for solar photovoltaics. Energy Environ Sci 8(4):1200–1219
27. Lee TD, Ebong AU (2017) A review of thin film solar cell technologies and challenges. Renew Sustain Energy Rev 70:1286–1297
28. McGlade C, Ekins P (2015) The geographical distribution of fossil fuels unused when limiting global warming to 2°C. Nature 517(7533):187–190
29. Moss R, Tzimas E, Kara H, Willis P, Kooroshy J (2011) Critical Metals in Strategic Energy Technologies Assessing Rare Metals as Supply-Chain Bottlenecks in Low-Carbon Energy Technologies. Tech. Rep. 978-92-79-20698-6, Joint Research Centre (European Commision), Netherlands. https://doi.org/10.2790/35716, https://publications.jrc.ec.europa.eu/repository/handle/JRC65592, eUR–24884-EN-2011 JRC65592 ISBN: 978-92-79-20699-3. https://publications.jrc.ec.europa.eu/repository/handle/JRC65592
30. Moss R, Tzimas E, Kara H, Willis P, Kooroshy J (2013) The potential risks from metals bottlenecks to the deployment of strategic energy technologies. Spec Sect: Run TransitS Sustain Econ Struct Eur Union 55:556–564
31. Müller A, Friedrich L, Reichel C, Herceg S, Mittag M, Neuhaus DH (2021) A comparative life cycle assessment of silicon PV modules: impact of module design, manufacturing location and inventory. Solar Energy Mater Solar Cells 230:111277
32. Rosental PA (ed) (2017) Silicosis. A World History. Johns Hopkins University Press. https://jhupbooks.press.jhu.edu/title/silicosis
33. Rudnick R, Gao S (2014) Composition of the continental crust 4.1. In: Holland HD, Turekian KK (eds) Treatise on geochemistry (2nd edn). Elsevier, Oxford, pp 1–51. https://doi.org/10.1016/B978-0-08-095975-7.00301-6, https://www.sciencedirect.com/science/article/pii/B9780080959757003016
34. Serrano-Luján L, Espinosa N, Larsen-Olsen TT, Abad J, Urbina A, Krebs FC (2015) Tin- and lead-based perovskite solar cells under scrutiny: an environmental perspective. Adv Energy Mater 5(20):1501119
35. Sonderegger T, Berger M, Alvarenga R, Bach V, Cimprich A, Dewulf J, Frischknecht R, Guinée J, Helbig C, Huppertz T, Jolliet O, Motoshita M, Northey S, Rugani B, Schrijvers D, Schulze R, Sonnemann G, Valero A, Weidema BP, Young SB (2020) Mineral resources in life cycle impact assessment-Part I: a critical review of existing methods. Int J Life Cycle Assess 25(4):784–797
36. Sundaram S, Benson D, Mallick TK (2016) Solar photovoltaic technology production. Elsevier, Academic, Potential Environmental Impacts and Implications for Governance
37. UNEP (2016) Green Energy Choices: The benefits, risks and trade-offs of low-carbon technologies for electricity production. Tech. rep., United Nations Environment Programme. https://wedocs.unep.org/handle/20.500.11822/7694, iSBN number: 978-92-807-3490-4
38. United States Geological Survey (2021) Mineral commodity summaries 2021. Report, Reston, VA,
39. Valero A, Valero A, Calvo G, Ortego A (2018a) Material bottlenecks in the future development of green technologies. Renew Sustain Energy Rev 93:178–200. https://doi.org/10.1016/j.rser.2018.05.041, https://www.sciencedirect.com/science/article/pii/S1364032118303861
40. Valero A, Valero A, Calvo G, Ortego A, Ascaso S, Palacios JL (2018b) Global material requirements for the energy transition. An exergy flow analysis of decarbonisation pathways. Energy 159:1175–1184. https://doi.org/10.1016/j.energy.2018.06.149, https://www.sciencedirect.com/science/article/pii/S0360544218312143
41. Wambach K, Sander K (2015) Perspectives on management of end-of-life photovoltaic modules. In: Proceedings of the 31st European photovoltaic solar energy conference and exhibition. Hamburg, Germany, https://doi.org/10.4229/EUPVSEC20152015-7EO.2.5, https://www.eupvsec-proceedings.com/proceedings?paper=33471

42. Wambach K, Heath G, Libby C (2017) Life Cycle Inventory of Current Photovoltaic Module Recycling Processes in Europe. Tech. Rep. Report IEA-PVPS T12-12:2017, IEA PVPS Task12, Subtask 2, LCA, iSBN 978-3-906042-67-1
43. Zuser A, Rechberger H (2011) Considerations of resource availability in technology development strategies: the case study of photovoltaics. Resour Conserv Recycl 56(1):56–65

Chapter 6
The Energy Balance of Solar Electricity

The discussion of the energy balance of a photovoltaic system during its lifetime started at the beginning of PV systems deployment in the early 80s. The critics often argued that a photovoltaic system never produced more energy than the required to manufacture it. The discussion was finished many years ago, when the energy embedded in the PV systems were compared with the electricity produced during its lifetime, resulting in an overwhelmingly positive balance. This balance, often measured as the energy payback time (EPBT) or as energy return on (energy) investment (EROI), requires the careful compilation of the energy embedded in the manufacture of cells, modules and balance of system components. Therefore, this chapter is organized in three main sections: one devoted to the calculation of embedded energy in different photovoltaic technologies, another one devoted to the calculation of the energy produced by a photovoltaic system during its lifetime, which is strongly dependant on the performance ratio of the system, a parameter that informs about all possible losses that an operational PV system may have in an given location; and finally the analysis of the balance provided by the EPBT and EROI parameters, which enables a fair comparison between PV technologies and with other energy technologies (renewable or non-renewable).

6.1 Embedded Energy in Photovoltaic Systems

The compilation of the embedded energy, that is, the cumulative energy demand (CED), of a photovoltaic system is part of the life cycle inventory. It is considered as an input for the production process of the product (the cell, or module, or system) or the service (the provided electricity). The energy input may arise from different contributions that are analysed in detail in this section: the energy embedded in the

A. Urbina, *Sustainable Solar Electricity*, Green Energy and Technology,
https://doi.org/10.1007/978-3-030-91771-5_6

processing of materials, in the manufacturing of modules and in the transport and building of the system. In many studies, the energy required for end-of-life operations are also included in the embedded energy calculation. This provides a comparison of embedded energy in different PV technologies. The balance of system life cycle inventory and impact assessment will be presented in Chap. 9.

6.1.1 Embedded Energy in the Processing of Materials

The cumulative energy demand embedded in PV module production has been calculated in detail using LCA inventories. An aggregation of the energy demand for each group of processes is shown in Tables 6.1 and 6.2 for two examples of crystalline silicon technologies, together comprising more than 95% of actual module production. The calculations were carried out by Erik Alsema, Vasilis Fthenakis, Marco Raugei and many others since late 90s; the values shown in the tables are a cap to actual CED values; on the one hand, the material requirements per module have been reduced (see Chap. 5) and industrial processes optimized; on the other hand the values originally provided are primary energy per area of module (MJ/m^2) which refers to a functional unit for the LCA study of 1 m^2; one step forward is to compute the CED for a functional unit of the nominal peak power (1 W_p) of module. This information has the disadvantage that the quality of the module (influenced by the processing method) is considered indirectly in the evaluation of the CED: a manufacturing route that demands the same CED per square meter of module may deliver better modules, thus reducing the CED per Watt peak without improving the energy efficiency of the process. In Tables 6.1, 6.2 and 6.3 both values are provided, in the second column, with MJ/W_p assuming the efficiencies indicated in the Table caption. Again, the efficiencies of commercial modules are today better than those

Table 6.1 Cumulative energy demand (CED) breakdown for single-crystal silicon PV modules, calculated with data from inventories by Alsema et al. [1, 19] (MJ/m^2 module area, and converted to MJ/W_p assuming module PCE = 14%)

Process sc-Si	MJ/m^2	MJ/W_p	Share (%)
Silicon production	450	3.21	7.3
Silicon purification	1800	12.86	29.0
Crystallization, contouring	2300	16.43	37.1
Wafering	250	1.79	4.0
Cell processing	550	3.93	8.9
Frame (Al)	500	3.57	8.1
Module assembly	350	2.50	5.6
Total module	6200	44.29	100.0

Table 6.2 Cumulative energy demand (CED) breakdown for multi-crystalline silicon PV modules, calculated with data from inventories by Alsema et al. [1, 19] (MJ/m^2 module area, and converted to MJ/W_p assuming module PCE = 13%)

Process mc-Si	MJ/m^2	MJ/W_p	Share (%)
Silicon production	450	3.46	9.6
Silicon purification	1800	13.85	38.3
Crystallization, contouring	750	5.77	16.0
Wafering	250	1.92	5.3
Cell processing	600	4.62	12.8
Frame (Al)	500	3.85	10.6
Module assembly	350	2.69	7.4
Total module	4700	36.15	100.0

Table 6.3 Cumulative energy demand (CED) breakdown for amorphous silicon PV modules, calculated with data from inventories by Alsema et al. [1, 19] (MJ/m^2 module area, and converted to MJ/W_p assuming module PCE = 7%)

Process a-Si	MJ/m^2	MJ/W_p	Share (%)
Cell material	50	0.71	4.2
Substrate + encapsulation material	350	5.00	29.2
Cell/module processing	400	5.71	33.3
Overhead operations	250	3.57	20.8
Equipment manufacturing	150	2.14	12.5
Total module	1200	17.14	100.0

considered for the calculation, and therefore, the results can be considered a cap for CED values.

For both crystalline types of module (single crystal and multi-crystalline), the most energy consuming step is the purification process nearly followed by the crystallization of the silicon material. This process is nowadays highly optimized and it is difficult to expect big improvements in energy saving during the coming years. The energy consumption rate in the Siemens process has reached 140 kWh/kg of Si, 50 kWh/kg for the production of high purity $SiHCl_3$ and 90 kWh/kg for the H_2 reduction and/or thermal decomposition [40].

For thin film amorphous silicon, the most energy consuming process is the cell manufacture, while the most important energy embedded in materials is the encapsulant and substrate, and not the materials of the photovoltaic cell active and transporting layers, a characteristic which is common to all thin film technologies and can be seen in Table 6.3 for a-Si modules.

The cumulative energy demand of emerging technologies requires a different approach. Again, the energy embedded in the processing of materials is important, but focussed on chemical processing instead of mining and purification techniques (typical of inorganic PV technologies). Due to the very diverse chemical routes used to synthesise organic molecules, the range of reported values is wide.

When tandem of mixed technologies are considered, the lifetime of the cells monolithically connected is an important factor. The most successful perovskite/silicon tandem, currently with record power conversion efficiency (29.5%), combines two cells, one with long lifetime (Si, with $T_{80} > 25$ years) and one with much shorter lifetime (perovskite, with T_{80} still below a few years). The final energy balance between cumulative energy demand and the solar electricity produced throughout its lifetime is strongly affected by the time at which the perovskite cell reduces its functionality and the transparency of the material once it has reached its lifetime since the perovskite is on top of the silicon cell which will be operational several years more. Although it may be a major drawback for a reduction in the final balance, the cumulative energy demand is still low [33].

The sputtering of transparent conducting oxides is a process which demands high energy, either in inorganic commercial technologies or in organic/hybrid emerging technologies, with values ranging from 6.3kWh per m^2 of module area, in contrast with printing technologies which require much lower energy, for example, for screen printing it ranges from 0.02 to 0.41 kWh/m^2 [29].

6.1.2 Embedded Energy in the Manufacturing of Modules

The energy required for cell processing is around 550, 600 and 400 MJ/m^2 for single crystal, multi-crystalline and amorphous silicon respectively (from Tables 6.1, 6.2 and 6.3); which can be converted to embedded energy per nominal capacity, delivering values of 3.93, 4.62 and 5.71 MJ/W_p. Note the impact of power conversion efficiency used for the calculation, which are the same as reported by the researchers at the time of their work; since PCE has improved steadily, and assuming that energy process is kept constant, with present day best PCE at module level, the values reported above are reduced to 2.25 MJ/W_p, 2.94 MJ/W_p and 4.08 MJ/W_p respectively.

These values for energy embedded in the cell are similar absolute quantities per square meter of module, but its share with respect energy embedded in the whole module is very different, being only 8.9 and 12.8% for sc-Si and mc-Si, but increasing to 33.3% for a-Si, emphasizing the different strategies that are required to reduce CED in wafer-based versus thin film technologies: in the first case reducing the CED embedded in material processing (already difficult and only achievable because the thickness of the wafers are reduced) while for thin film, the roadmap is to reduce the CED of the cell and module processing, where there is still plenty of reduction possibilities thanks to the large diversity of processing methods for deposition and structuring of materials embedded in the thin film cells. Cumulative energy demand embedded in PV module manufacture has been constantly reduced in the past few years, and currently the most advanced silicon heterojunction solar cells (SHJ) have a best value of CED = 0.8 MJ/W_p [29].

The three commercial technologies (CdTe, CIGS and III-V tandems) are based on material deposition on substrates by different methods that require vacuum chambers, the methods are being improved and energy consumption is slowly reduced. But a breakthrough in energy embedded in the thin film processing has been achieved with emerging technologies where the film deposition method is based on printing technologies from solution, the CED for thin film technologies varies from 894 MJ/m^2 to less than 200 MJ/m^2, which together with the variation in reported power conversion efficiencies translates into values ranging from 2–5 MJ/W_p, overlapping with the values obtained for crystalline silicon solar cells reported above; several authors recommend to use CED calculations per square meter of module (FU = 1 m^2) when comparing different technologies to avoid excessive dependence on value-choice; the use of harmonization procedures narrows the wide dispersion of reported data [5]. For small molecules and polymer photovoltaic technologies, the values have a similar order of magnitude and were found to range between 2.9 and 5.7 MJ/W_p [3]. Nevertheless, the reduction of cell processing energy in thin film and emerging technologies can be potentially higher, which together with the small amount of material required in thin films (of a few hundreds of nanometers thick), have constantly reduced the CED of the devices in the past few years and as will be presented in Sect. 6.3 have led to energy payback times lower than one year.

For all PV technologies, the energy embedded in the frame of modules is important because aluminium is a material which requires high energy processing. The typical energy embedded in aluminium framing is about 500 MJ/m^2 of module, independently of the technology under consideration [1, 2, 19]. In crystalline silicon cells, an Al frame is still included by most manufacturers, although there is a trend to manufacture frameless modules that could reduce aluminium consumption (lower impacts on mineral resources) and the energy required for aluminium and frame processing, thus bringing down the module CED around −10%. Thin film modules are now commonly manufactured without frame.

Table 6.4 presents the top contributor of primary energy demand at module level for different PV technologies, in the table a distinction between either process energy or energy embedded in materials has been carried out, data for the table have been collected from references indicated in each row, the compilation in reference [9], and from the recent report "Solar photovoltaic modules, inverters and systems: options and feasibility of EU Ecolabel and Green Public Procurement criteria" by the Joint Research Centre (European Commission) [11].

When renewable and non-renewable energy technologies are compared, fossil technologies have high cumulative non-renewable energy demand (CED), which is typical of energy technologies with very different life cycles, since the consumption of fuel in fossil and nuclear electricity technologies provides the biggest contributions of energy demand from the manufacturing stage (including raw materials extraction, equipment manufacture and building of the production plants) to the operational phase when the fuel is consumed. This very different approach has risen some criticism on the direct comparison of cumulative energy demand of renewable

Table 6.4 Top contributor to cumulative energy demand (CED, primary) at module level classifying between either process energy or energy embedded in materials

Technology	Electricity intensive process	Energy embedded in material	Module component	Reference
sc-Si		Si crystallization	Active layer	Alsema [1]
sc-Si		Si wafer growth (FZ)	Active layer	Celik et al. [8]
mc-Si		Si purification	Active layer	Alsema [1]
a-Si	Cell process		Active layer	Alsema [1]
a-Si/nc-Si	PECVD[a]		Active layer	Kim and Fthenakis [27]
a-Si/nc-Si	PECVD[a]		Active layer	Mohr et al. [32]
CdTe		Al	Frame	Kato et al. [26]
CdTe, Zn_3P_2	Substrate cleaning in heated ultrasonic bath cleaning drying with N2			Collier et al. [10]
CIGS	Co-evaporation of Cu,In,Ga and selenisation		Active layer	Collier et al. [10]
CZTS	Co-sputtering of Cu, Zn, Sn and sulphurisation		Active layer	Collier et al. [10]
GaAs	MOVPE[b]		Cell stack	Meijer et al. [30]
GaAs	MOVPE[b]		Cell stack	Mohr et al. [31]
OPV	N_2 glovebox		Active layer, back electrode, encapsulation	García-Valverde et al. [21]
OPV	ITO sputtering		Transparent electrode	Espinosa et al. [13]
OPV	Al/Cr sputtering		Back electrode	Espinosa et al. [14]
OPV	PEDOT:PSS slot-die coating and drying		Hole-transport layer	Emmott et al. [12]
OPV	ITO sputtering	Transparent electrode		Anctil et al. [3]
OPV [ITO-free]		PET film	Substrate and encapsulation barriers	Espinosa et al. [15]
PSC [TiO_2]		Au	Back electrode	Gong et al. [22]
PSC [TiO_2]		TiO_2 annealing	Transporting layer	Celik et al. [8]
PSC [ZnO]	ITO sputtering		Transparent electrode	Gong et al. [22]
QDPV		Al, ETFE[c], EVA[d]	Encapsulation	Sengül and Theis [37]

[a]Plasma-enhanced chemical vapour deposition.
[b]Metal-organic vapour phase epitaxy.
[c]Ethylene tetra-fluoro-ethylene.
[d]Ethylene vinyl acetate

and non-renewable technologies; a parameter that considers the balance of energy throughout the whole lifetime will provide a fairer comparison although it is not included in standard LCA approaches: this is the energy payback time, which is presented in Sect. 6.3.

6.2 Solar Electricity Production of a Photovoltaic System

A photovoltaic system will generate a certain amount of electricity during its lifetime "use phase" that depends on two main groups of parameters: technical parameters and environmental parameters. The first group defines the characteristics of the module at the gate of the factory and depends on the PV technology under consideration, they have been presented in Chap. 2. The second group is comprised of operational parameters that will influence the PV system output during a long period of time and which mainly depend on environmental parameters (irradiance, temperature, wind and humidity) of the geographical location where the PV system is built; but also good design of the system and good practice during the operational phase have a strong impact on the electricity output during the system lifetime; monitorization methods and maintenance practice should be taken into account for an evaluation of a system performance.

In this chapter, the methods to calculate the energy output of a photovoltaic system are briefly presented and then analyzed from the Life Cycle Assessment perspective, focussing on what could be considered a "gate to grave" scope. The environmental impacts of operational and the end-of-life phases are evaluated and, together with the "cradle to gate" results of Chap. 5 comprise a full LCA scope of PV technologies. First, the tools to calculate the electricity production of a PV system with a given nominal power and installed in a specific geographical location are provided. Then an overview of the large variety of PV system applications is presented, with examples of case studies of yield and performance, including some recommendations of exploitation and maintenance best practices. The end of life of the PV systems are considered. Finally, the issue of size dependant impacts of PV systems is presented, with a comparison of roof-top, BIPV or large plants and the possibility of multifunctional use of the required space is discussed, with special focus on a new important trend for double use of land in more landscape-integrated PV systems: agrivoltaics.

6.2.1 Electricity Production and Yield

The electricity output of a photovoltaic system with a given nominal power is called yield. Depending on PV technology, system design, environmental conditions and good maintenance practice, the yield relates energy and power, it informs about how much electricity can be obtained from a PV system in a certain location during a certain period of time. There are three definitions of photovoltaic yield, regulated by the standard IEC 61724 [25]: PV module (array) energy yield ($Y_{A,t}$), final yield ($Y_{F,t}$) and reference yield ($Y_{R,t}$). All yields are measured in hours, but it is recommended to indicate the ratio of units from which it is calculated in order to avoid confusions.

The PV module **array energy yield** is the ratio of energy (DC electricity) produced by the array of modules (the generator) to the nominal power (measured in STC conditions). In other words, it is the generated DC electricity (kWh)/kW$_p$ of installed PV. It is calculated for a certain period of time (hourly, daily, monthly or annual)

$$Y_{A,t} = \frac{E_{G_{\mathrm{DC}},t}}{P_{G,\mathrm{STC}}} \left(\frac{\mathrm{kWh}}{\mathrm{kW}_p} \rightarrow h \right). \tag{6.1}$$

The PV system **final energy yield** is the ratio of energy (AC electricity) produced by the PV system to the nominal power (measured in STC conditions); it is the generated AC electricity (kWh)/kW$_p$ of installed PV. It is calculated for a certain period of time (hourly, daily, monthly or annual)

$$Y_{F,t} = \frac{E_{S_{\mathrm{AC}},t}}{P_{G,\mathrm{STC}}} \left(\frac{\mathrm{kWh}}{\mathrm{kW}_p} \rightarrow h \right). \tag{6.2}$$

The **reference yield** is the ratio of the total in-plane irradiation (or insolation) per square meter to the standard conditions irradiance (STC, 1 kW/m^2). It provides information about the available solar resource in a geographical location, and can be expressed as "Sun equivalent hours" because it is equivalent to receive STC irradiance during the calculated reference yield *hours* for a certain period of time (hourly, daily, monthly or anual, but it is most often used on a monthly or annual basis). It is the amount of theoretically available solar resource in a geographical location

$$Y_{R,t} = \frac{E_{H_{\mathrm{theo}},t}}{G_{\mathrm{STC}}} \left(\frac{\mathrm{kWh/m^2}}{\mathrm{kW/m^2}} \rightarrow h \right). \tag{6.3}$$

When calculated or measured, the three yield definitions deliver different values and the differences between them have seasonal variations. The **performance ratio (PR)** is considered as the best parameter to aggregate the diversity of effects that may affect the yield of a PV system, its definition is simple and at the same time powerful: the PR is the ratio of the final yield to the reference yield

$$\mathrm{PR} = \frac{Y_{F,t}}{Y_{R,t}}. \tag{6.4}$$

The performance ratio is expressed as % (or with a value between 0 and 1) and indicates the amount of useful energy that is delivered by the system as a % of what could be ideally achieved for a given PV technology in a specific geographical location. It is an aggregated indicator of losses of the PV system, and provides information about the quality of the system design, construction and operation during a period of time. The PR has many contributions: temperature losses, low irradiance, spectral effects, angular losses, soiling, shadowing, parameter dispersion (from manufacturers), maximum power point tracker and power electronic efficiencies, losses in wires, degradation of modules, etc…

The PR is a combined measurement of how good a PV system has been designed (minimization of intrinsic losses by design: for example, inverters efficiency or cable

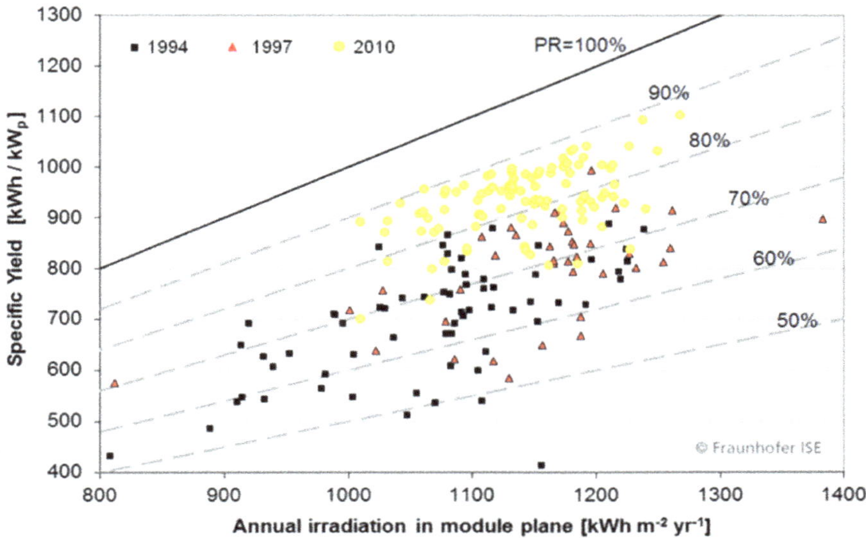

Fig. 6.1 Performance Ratio (PR) development for PV systems in Germany, with data from Fraunhofer ISE "1000 Dächer Jahresbericht" 1994, 1997 and 2011 (system evaluation) (Reproduced with permission from [17])

thickness optimized to reduce voltage losses) and how good it is performing in real operation conditions (cleaning, maintenance, quick failure detection and fixing). The PR has evolved in time from typical values around 70% in the 90s to values around 90% after 2010, thirty years of knowledge which has allowed to reach high levels of performance for any PV system. An example of this evolution are the PR values measured in Germany from 1994 to 2010 shown in Fig. 6.1 for similar geographical conditions and maintenance policies. This values may vary around the world, but are indicative of the ultimate possibilities of improvement in PR values: 90% with low variance is a limit that is very difficult to surpass.

In an inverse approach, if the reference yield of a location and the performance ratio of a system is known, the calculation of the produced useful energy (delivered AC electricity) is straightforward, starting with the definition of the energy delivered in a period of time t (Eq. 6.2)

$$E_t = P_{G_{\mathrm{STC}}} \times Y_{F,t}, \tag{6.5}$$

and substituting the definition for final yield using Eq. 6.4:

$$E_t = P_{G_{\mathrm{STC}}} \times Y_{R,t} \times \mathrm{PR}. \tag{6.6}$$

For the calculation of energy produced during the lifetime of the PV system, a careful modelling (or experimental measurement) of the PR must be carried out, it is

Table 6.5 Empirically determined coefficients used to predict module power at maximum power point in environmental conditions different from standard test conditions (STC) for three different PV technologies ([24])

Coefficient	c-Si	CIGS	CdTe
k_1	−0.017237	−0.005554	−0.046689
k_2	−0.040465	−0.038724	−0.072844
k_3	−0.004702	−0.003723	−0.002262
k_4	0.000149	−0.000905	0.000276
k_5	0.000170	−0.001256	0.000159
k_6	0.000005	0.000001	−0.000006

difficult to make predictive calculations since degradation of the PV modules must be taken into account, and the environmental data required to calculate the reference yield also varies in the long term due to climate change. The JRC recommendations for LCA applied to photovoltaic systems is to use a site-specific PR value or the following default values: PR = 0.75 for roof-top installations and PR = 0.8 for ground-mounted utility installations [18, 19]; degradation is included in these values that must be considered as time averages. For annually calculated energy production, PR should be updated every year to check for changes in parameters contributing to PR calculation, especially the degradation of the modules. In 2021, most manufacturers of PV modules guarantee degradation coefficients lower than 0.5% per year (that is a 0.5% reduction in nominal peak power per year); this leads to T_{80} times longer than 25 years, although the modules will be operational much longer times.

An alternative approach is used by the Joint Research Centre (European Commission) PVGIS on-line calculation tool. It is based on an empirical power rating method proposed by Thomas Huld in which the power of the system is calculated for different irradiance and temperature conditions by a correction to the power conversion efficiency provided by manufacturers for standard test conditions [23, 24]

$$P(G', T') = G'(P_{\mathrm{STC}} + k_1 \ln(G') + k_2 \ln(G')^2 + k_3 T' + k_4 T' \ln(G') + k_5 T' \ln(G')^2 + k_6 (T')^2), \tag{6.7}$$

where the normalized in-plane irradiance G' and temperature T' are defined as

$$G' \equiv G/G_{\mathrm{STC}}, \tag{6.8}$$

$$T' \equiv T_{\mathrm{mod}} - T_{\mathrm{STC}}. \tag{6.9}$$

The coefficients k_1 to k_6 are empirically obtained for each PV technology by fitting to measured data. The coefficients used in PVGIS are based on measurements performed at the European Solar Test Installation (JRC-ESTI) and are given in Table 6.5.

Besides the detailed calculation of the electricity production of a PV system in order to obtain the energy balance, the **production phase** of a PV system has an important contribution to impacts in any LCA study. The production of PV modules of different technologies have been analyzed in detail in Chap. 4 and for Balance of System (BoS) components will be presented in Chap. 9. The following aspects are taken into consideration in order to calculate the contribution of the **construction phase** and **use phase** of the PV system to the different impact categories [18].

- Construction phase
  - Transports to the PV power plant site (where the PV plant will be operated);
  - Construction and installation, including foundation, supporting structures and fencing.
- Use phase
  - Auxiliary electricity demand;
  - Cleaning of panels;
  - Maintenance;
  - Repair and replacements, if any.

A special mention should be made for building integrated PV systems (BIPV), where the double functionality of the PV modules must be included in the LCA study: the modules produce electricity and are also a structural part of the building providing additional services like weather protection, thermal insulation or shading. The best FU for this case is the amount of AC electricity produced by the system (quantified in kWh for a certain period of time) and the boundary must include all BoS which may have special characteristics for the integration in the building structure (mounting systems, microinverters). The calculation of electricity production of BIPV systems require a site-specific analysis since module orientation and electrical connections will depend more on the building structure and architecture design requirements and in most cases will not be the optimal for PV production; similarly lifetime should be adapted (usually larger than 25–30 years unless the modules are replaced before the end of building lifetime wich is usually much longer) and operation and maintenance activities strongly differ with respect to small home-systems (roof-top attached) and large PV plants. Furthermore, the quantification of the double functionality is often very difficult to define, but LCA studies will be more useful if they are able to quantify and compare the global performance of the combined building+PV system with the separated cases of building and PV system whose impacts are added. On the other hand, the focus of the LCA may be the building itself, and BIPV is considered as a modification of the default characteristics of the building.

6.2.2 *Lifetime of Photovoltaic Systems*

The IEA-PVPS Task 12 report on Methodology Guidelines on Life Cycle Assessment of Photovoltaic recommends the following life expectancy considerations for the LCA study of operational PV systems throughout its full lifetime [18].

Modules: 30 years for mature module technologies (glass-glass or glass-Tedlar backsheet) used in ground mounted, building attached and building integrated PV modules; life expectancy may be lower for foil-only encapsulation; this life expectancy is based on typical PV module warranties (T_{80}, for 20% or less efficiency degradation after 25 years) and the expectation that modules last beyond their warranties. In BIPV systems, since modules are integrated in the building structure (façades, roofs), the lifetime is longer and degradation losses after T_{80} should be evaluated according to manufacturer's recommendations and included in PR calculation.

Inverters: 15 years for small plants (residential PV); 30 years with 10% part replacement every 10 years (the parts that are assumed to be replaced need to be specified) for large size plants utility PV;

Transformers: 30 years;

Mounting and supporting structures: 30 years for building attached roof-top and façade installations, and between 30 and 60 years for building integrated installations and for ground-mount installations on metal supports. Sensitivity analyses should be carried out by varying the service life of the ground-mount supporting structures within the same time span;

Cabling: 30 years;

Manufacturing plants (capital equipment): The lifetime may be shorter than 30 years due to the rapid development of technology.

All assumptions need to be listed clearly in the final report of any LCA study.

6.3 Energy Payback Time and Energy Return on (Energy) Investment

A Net Energy Analysis (NEA) methodology was developed since the early 70s to complement economical prospective analysis. The oil crisis was the shock that prompted a careful analysis that was focussed on quantifying the energy returns of any economic investment, and soon the analysis was closing the loop on energy itself: How much energy must be invested to harvest an energy return? Is it possible to get an net energy gain (surplus) delivered to the end user?

The aim of the methodological discussions was to measure the relation of energy diverted from society to make energy available to society and the metric used for the calculation should be applicable to any energy technology. The "energy payback time" and the "energy return on (energy) investment" are the two main tools developed to answer these questions.

6.3.1 *Energy Payback Time Definition*

The Energy Payback Time (EPBT) is the period of time required by a renewable energy system to generate the same amount of energy that was used to produce the system itself. It is usually quantified in equivalent primary energy using a conversion efficiency factor. It can also be defined as the ratio of cumulative energy demand (CED) to mean net energy generated annually

$$\mathrm{EPBT} = \frac{\mathrm{CED}}{E_{G,a}/\eta_{\mathrm{grid}} - E_{O\&M}}, \tag{6.10}$$

where:

- CED is the cumulative primary energy demand (in MJ oil-equivalent);
- $E_{G,a}$ is the mean energy generated annually (kWh electricity);
- η_{grid} is the grid efficiency, that is, the primary energy to electricity conversion efficiency at the demand side (kWh electricity per MJ oil-equivalent) for the grid of a specific country or region where the PV system is deployed, it is calculated as the ratio of the yearly electricity output of the entire grid to the total primary energy harvested from the environment for the operation of the grid in the same year;
- $E_{O\&M}$ is the annual operation and maintenance primary energy consumption of the PV system (in MJ oil-equivalent), it may vary from year to year and a lifetime average is recommended, maintenance requirements tend to be higher at the end of life of the system.

Two important clarifications should be emphasized in this definition. Firstly, the amount of energy produced by the PV system is a net delivery to the grid (or to AC or DC consumption in off-grid systems), and therefore, the self-consumption of energy required for operation and maintenance of the system must be subtracted in the denominator of Eq. 6.10. Secondly, the produced energy is a mean energy, which must be calculated as an average of the system production throughout its lifetime, including degradation losses. The recommended energy unit is primary energy in MJ oil-equivalent since it covers all kinds of energy generated or used; the conversion efficiency strongly affects the results; in some cases kWh can be used (it facilitates electricity calculations) but inverse conversion factors for primary energy contributions should also be applied in this case.

The cumulative (primary) energy demand was presented in Chap. 3 for general life cycle inventory cases, and it is described here more specifically for a PV system

$$\mathrm{CED} = E_{\mathrm{mat}} + E_{\mathrm{manuf}} + E_{\mathrm{trans}} + E_{\mathrm{inst}} + E_{\mathrm{EOL}}, \tag{6.11}$$

where (all primary energy in MJ oil-equivalent):

- E_{mat} is the primary energy demand to produce the materials embedded in the PV system;
- E_{manuf} is the primary energy demand to manufacture the PV system;
- E_{trans} is the primary energy demand to transport all materials and components of the PV system during its life cycle;
- E_{inst} is the primary energy demand to install the PV system;
- E_{EOL} is the primary energy demand for end-of-life management.

The EPBT calculation is a concept used to quantify the beneficial effects of producing energy from a renewable source. In early years of PV development, it was used to demonstrate that PV systems indeed produced a net environmental benefit, since the calculated EPBT was shorter than the system lifetimes. The focus was (and still is) the displacement of non-renewable energy generation by photovoltaic energy generation. But the most common approach is to include in the calculation all contributions to the energy mix of the grid electricity that is being displaced, that is, renewable and non-renewable sources of energy production, whose relative contribution depends on the mix (it may contain a large amount of hydro or wind contribution). In this case, the displacement concept behind EPBT approach is not strictly speaking a substitutional one. The annual electricity generation ($E_{G,a}$ in Eq. 6.10) is converted into its equivalent primary energy with a calculation based on the efficiency of electricity conversion at the demand side, which uses the current average or average non-renewable (in attributional LCAs) or the long term marginal (in decisional/consequential LCAs) grid mix where the PV plant is being installed [18]. Since the electricity mix is changing in time with increasing renewable contribution every year in many countries around the world, this conceptual approach to EPBT may lead to misleading interpretations, especially in the consequential LCA approach with uncertainties in future energy mix. The EPBT of a PV technology may change significantly without the technology having changed at all because of the impacts of variations of grid mix and efficiency.

To avoid this ambiguity, another conceptual approach can be used: the **non-renewable energy payback time (NREPBT)** [18]. It considers the photovoltaic electricity as a replacement for only the non-renewable energy sources included in the energy mix. In the NREPBT calculation, the renewable primary energy is not accounted for, neither on the demand side (during manufacturing), nor during the operation phase. It provides information about the time required by the PV system to generate the electricity equivalent to only the non-renewable contribution of the CED of the system; the conversion factors from primary energy to electricity should be modified accordingly in Eq. 6.11. The newly calculated CED is not the total energy embedded in the PV system, but the non-renewable energy embedded in the system, a NRCED, and strongly depends on the characteristics of the energy mix of the country were materials and modules were manufactured. Also, the annual generated electricity is converted to primary energy considering only the non-renewable contribution to the mix (it considers only the non-renewable primary energy to electricity conversion efficiency).

Nevertheless, both the EPBT and the NREPBT depend on the electricity mix, the first one is commonly used to provide larger-scale utility grid replacement information, while the second one is more locally sensitive on the particular conditions of the mix where the PV systems were fabricated or installed for its operational phase. While the renewable contribution to the electricity mix is low, both EPBT and NREPBT deliver similar results, but if the share of renewable sources becomes important, as it is envisaged in the future, the values will diverge. The geographical scale and time of the mix and its conversion efficiency must be clearly indicated in any LCA study which informs about payback times [34, 36].

6.3.2 Technology Dependence of the Energy Payback Time

In Table 6.6, a comparison of EPBT for different PV technologies is presented; the environmental parameters are the same in all cases (insolation of 1700 kWh/m^2 per year, average performance ratio to account for all losses, including temperature losses, of PR = 0.75 and a lifetime of 25 years). Technical parameters with impact on EPBT (power conversion efficiency, lifetime and degradation rate) are best for each technology at the time of the cited reference.

The EPBT of all PV technologies is ranging from 4 years or less for crystalline silicon technologies (reduced from 6 to 7 reported in early 90s) to less than one year for thin film CdTe and some organic and hybrid emerging technologies. In some extreme cases, even EPBT of a few days have been reported as a possiblity for the "factories of the future" [15]. The values for crystalline and thin film inorganic technologies have been stable since some years ago, with some promising results in experimental silicon heterojunction technologies which are about to reach 1 year EPBT. In the case of emerging organic and inorganic technologies, the embedded energy in the devices (contributing to CED) is strongly dependant on the electrodes to be used, with special focus on transparent conductive oxides (ITO or FTO), which contribute to more than 70% of embedded energy, and thus there is a great potential to reduce EPBT in organic and hybrid technologies by replacing ITO or FTO by other electrode alternatives described in Sect. 7.4 which have lower CED [12].

The hybrid perovskite/silicon tandem is a promising option which has reached a very high power conversion efficiency with small size research cells (29.5%) and thus have a great potential to reduce EPBT. In this case, the calculation requires taking into consideration the different lifetimes of a crystalline silicon (>25 years) and perovskite (around 1 year) cells; the perovskite top cell will reach its end of life much sooner and it is important that once it ceases operation, it does not diminishes the power conversion efficiency of the silicon cell that will be operating many more years. For its calculation, Monteiro et al. considered two scenarios, one in which the perovskite cell is transparent at the end of life (1 year), the other one in which it is opaque [33]; the difference between scenarios reduces the EPBT between –11.7% and –18.7% depending on the electrode and architecture of the tandem and in all cases the EPBT is lower than any crystalline silicon technology single junction cell.

Table 6.6 Energy Payback time of PV technologies for the same environmental parameters and best technical parameters at the time of cited references

Technology	EPBT (years)	References
c-Si	4.1	Bhandari et al. [5][a]
c-Si	2.3	de Wild-Scholten [39]
mc-Si	3.1	Bhandari et al. [5]
Si heterojunction	1.5	[29]
a-Si	1.4	de Wild-Scholten [39]
a-Si	2.3	Bhandari et al. [5]
CdTe	1.2	Bhandari et al. [5]
CdTe	0.6	Leccisi et al. [28]
CIGS	1.7	Bhandari et al. [5]
CIGS	1.1	Leccisi et al. [28]
OPV	1.3	Espinosa et al. [13]
OPV	0.6	Espinosa et al. [12]
OPV	1 day	Espinosa et al. [15]
Perovskite Ag	1.1	Espinosa et al. [16]
Perovskite Au	1.1	Serrano-Luján et al. [38]
Peroskite Au[b]	1.1–1.5	Celik et al. [7]
Perovskite Al	0.9	You et al. [41]
Perovskite/Si (HIT) tandem Ag[c]	1.5–1.7	Monteiro Lunardi et al. [33]
Perovskite/Si (HIT) tandem Au	1.5–1.7	Monteiro Lunardi et al. [33]
Perovskite/Si (p-n) tandem Al	1.3–1.6	Monteiro Lunardi et al. [33]

[a]Data from reference [5] are mean values from different sources harmonized to the environmental parameters under consideration.
[b]The range of values reported in reference [7] depend on the use of solution process (low CED), vacuum process (high CED) and HTL free architecture (lowest CED).
[c]The low and high values represent two scenarios where the perovskite cell becomes opaque or transparent after 1 year lifetime, as described in more detail in reference [33]

6.3.3 Geographical Dependence of the Energy Payback Time

The geographical dependence of the EPBT is related on the one side with the environmental parameters (mainly irradiance and temperature) and on the other side with technical parameters (mainly local grid efficiency) that depend on the location where the PV system is built and operated. As an example of the range of variation of EPBT values depending on environmental parameters and grid efficiency, Table 6.7 includes data for c-Si (Cz PERC, PCE = 19.6% modules with 60 cells) rooftop PV systems, in different geographical locations (with different environmental parameters and grid efficiency as indicated in the table) and for modules fabricated in the European Union (EU) and China with BoS components fabricated in the EU in both cases.

Table 6.7 Comparison of EPBT (EU/China) for c-Si (Cz PERC, PCE = 19.6% modules with 60 cells) roof-top systems in different locations. Data from Fraunhofer ISE 2021 [17]

Location	Houston	Arica	Ottawa	Brussels	Catania	C. Town	Cairo	Jaipur	Lanzhou	Perth
Country	USA	Chile	Canada	Belgium	Italy	S.Africa	Egypt	India	China	Australia
EPBT EU	0.86	0.86	1.28	1.15	0.97	0.48	0.61	0.40	0.89	0.69
EPBT China	0.95	0.93	1.46	1.26	1.05	0.52	0.66	0.44	0.94	0.74
Irradiation[a]	1913	2279	1566	1249	2048	2163	2416	2242	1799	2166
Grid efficiency[b]	9.2%	11.0%	11.9%	8.5%	11.2%	6.1%	8.2%	4.9%	8.9%	8.4%

[a]Irradiation is measured at module level (= Global Tilted Irradiation GTI) in kWh/m^2/year
[b]Grid efficiency: Electric to primary energy conversion ratio in percent as kWh$_{Grid}$/MJ$_{eq}$

The share of EPBT corresponding to each part of the PV system is very similar for all locations and for the c-Si modules considered for Table 6.7 is: poly-Silicon production (22.5%), ingot/wafering (15.5%), cell (5.4%), module (17.8%) and BoS (35.7%) manufacture and finally a contribution from transport, which can be more geographically dependant, around 3%.

As expected and as can be deduced from the case studies presented above (and many others in the scientific literature), the Energy Payback Time of PV systems is strongly dependent on the geographical location where the system is built and operated: for example, PV systems in North Europe need around 1.5 years to balance their embedded energy, while PV systems in South Europe equal their embedded energy after 1 year and less, depending on the technology installed and the grid efficiency. The reduction in EPBT has been constant since the late 90s when typical EPBT was around 3 years for a crystalline silicon roof-top PV system installed in a location with irradiation of 1700 kWh/m^2 per year (typical of southern Europe) to less than one year in 2020. The variation of the EPBT today ranges from 0.4 to 1.5 years for all commercial technologies in 2020; this implies that a typical roof-top PV system produce net clean electricity for about 97% of their lifetime, assuming a life span of 30 years or more [17].

6.3.4 Energy Return on (Energy) Investment

The Energy Return On (Energy) Investment (EROI) is defined as the ratio of energy delivered by a system to the energy required to deliver that energy. For an energy production system to provide a positive net energy "return" to the end user, the gross energy return must be larger than the total energy "invested" in the chain of energy harvesting and transformation processes that make up the system itself. Only if EROI is larger than 1, there is a net energy return. The EROI is defined for the whole lifetime of the energy system, and a direct relationship between EPBT and EROI can be established if the lifetime (LT) of the PV system is known (or assumed)

$$\mathrm{EROI} = \frac{\mathrm{LT}}{\mathrm{EPBT}}. \tag{6.12}$$

The EROI is an indicator more strongly focussed on returns for society and if a direct link between CED, EPBT and EROI is established (for example by combining Eqs. 6.10, 6.11 and 6.12), the results of EROI calculations can be misleading and special care should be put to CED contributions (for example, excluding some flows or fine tuning of weighting) when it is going to be used for an EROI calculation. This is due to CED contributions that are not a measure of diverted energy that would have been "useful" to the society as the EROI purpose requires. A few examples are the loss of heating value of coal when transported or stored due to fugitive dust emissions from coal stockpiles, or methane emissions from oil wells, or mischaracterisation of solar power as qualitatively equivalent to chemical energy content of combusted fuels; any of them have an impact on society benefits, but it is difficult to extract them from the CED calculations. EROI can be affected by the system boundary chosen for the study, which defines the stage of the energy supply chain at which an energy carrier is identified as the system's output; the energy carrier to which each calculated EROI applies must be clearly defined and comparison of EROI calculations with different energy carriers should be avoided (for example, liquid fuels versus electricity) [36]. Furthermore, comparison of EROI values across different energy technologies must be carried out carefully and with full transparency about the considered parameters [4].

Hundreds of results are published in scientific journals that report on CED, EPBT and EROI values for PV technologies, and a careful revision of methodological consistency indicates that only around 20% deliver values that allow a consistent harmonized comparison of different parameters throughout the different technologies; the need for harmonization of parameters considered for the calculations is important to reduce the large range of reported values [5]. Without harmonization, the range of values is very large even within the same technology, since assumptions (or measurements) for lifetime, system design and performance ratio strongly affect the results; in Table 6.8 the lower, higher and mean harmonized EROI measured values for commercial technologies are presented.

Table 6.8 Energy return on (energy) investment, EROI of commercial PV technologies, including modules and BoS components. The reference included in the table is for the minimum values, the maximum values are obtained from reference [39] and the mean harmonized values are obtained from reference [5]

EROI:	Min.	Max.	Mean	References
c-Si	2.2–3.3[a]	15.3	8.73	[6, 20]
mc-Si	3.4	24.2	11.6	[6]
a-Si	7.7	21.7	14.5	[1]
CIGS	7.1	29.7	19.9	[35]
CdTe	13.3	44.1	34.2	[35]

[a]These two values for c-Si refer to stand-alone (2.2) and grid connected systems (3.3)

The best energy return on energy investment is obtained for CdTe technology, which is the most successful EROI alternative to crystalline silicon within the group of thin film solar cells. Nevertheless, the market share of PV capacity installation per year is dominated by crystalline silicon, and the share of thin film technologies have been reduced in recent years despite its better EROI values.

References

1. Alsema EA (2000) Energy pay-back time and CO_2 emissions of PV systems. Progress Photovolt: Res Appl 8(1):17–25
2. Alsema EA, Nieuwlaar E (2000) Energy viability of photovoltaic systems. Energy Policy 28(14):999–1010
3. Anctil A, Babbitt CW, Raffaelle RP, Landi BJ (2013) Cumulative energy demand for small molecule and polymer photovoltaics. Progress Photovolt: Res Appl 21(7):1541–1554
4. Arvesen A, Hertwich EG (2015) More caution is needed when using life cycle assessment to determine energy return on investment (EROI). Energy Policy 76:1–6
5. Bhandari KP, Collier JM, Ellingson RJ, Apul DS (2015) Energy payback time (EPBT) and energy return on energy invested (EROI) of solar photovoltaic systems: A systematic review and meta-analysis. Renew Sustain Energy Rev 47:133–141
6. Bizzarri G, Morini G (2007) A life cycle analysis of roof integrated photovoltaic systems. Int J Environ Technol Manag 7(1–2):134–146
7. Celik I, Song Z, Cimaroli AJ, Yan Y, Heben MJ, Apul D (2016) Life Cycle Assessment (LCA) of perovskite PV cells projected from lab to fab. Solar Energy Mater Solar Cells 156:157–169
8. Celik I, Phillips AB, Song Z, Yan Y, Ellingson RJ, Heben MJ, Apul D (2017) Environmental analysis of perovskites and other relevant solar cell technologies in a tandem configuration. Energy Environ Sci 10(9):1874–1884
9. Chatzisideris MD, Espinosa N, Laurent A, Krebs FC (2016) Ecodesign perspectives of thin-film photovoltaic technologies: A review of life cycle assessment studies. Life Cycle Environ Ecol Impact Anal Sol Technol 156:2–10
10. Collier J, Wu S, Apul D (2014) Life cycle environmental impacts from CZTS (copper zinc tin sulfide) and Zn3P2 (zinc phosphide) thin film PV (photovoltaic) cells. Energy 74:314–321
11. Dodd N, Espinosa N (2021) Solar photovoltaic modules, inverters and systems: options and feasibility of EU Ecolabel and Green Public Procurement criteria. Tech. Rep. KJ-NA-30474-EN-N, EU-JRC, European Commission-Joint Research Centre. https://www.doi.org/10.2760/29743, iSBN 978-92-76-26819-2 ISSN 1831-9424, https://doi.org/10.2760/29743
12. Emmott CJM, Urbina A, Nelson J (2012) Environmental and economic assessment of ITO-free electrodes for organic solar cells. Solar Energy Mater Solar Cells 97:14–21
13. Espinosa N, García-Valverde R, Urbina A, Krebs FC (2011) A life cycle analysis of polymer solar cell modules prepared using roll-to-roll methods under ambient conditions. Solar Energy Mater Solar Cells 95(5):1293–1302
14. Espinosa N, García-Valverde R, Urbina A, Lenzmann F, Manceau M, Angmo D, Krebs FC (2012) Life cycle assessment of ITO-free flexible polymer solar cells prepared by roll-to-roll coating and printing. Solar Energy Mater Solar Cells 97:3–13
15. Espinosa N, Hösel M, Angmo D, Krebs FC (2012) Solar cells with one-day energy payback for the factories of the future. Energy Environ Sci 5(1):5117–5132
16. Espinosa N, Serrano-Luján L, Urbina A, Krebs FC (2015) Solution and vapour deposited lead perovskite solar cells: ecotoxicity from a life cycle assessment perspective. Solar Energy Mater Solar Cells 137:303–310
17. Fraunhofer-ISE (2021) Photovoltaics Report 2021. Tech. rep., Fraunhofer Institute for Solar Energy Systems, ISE, Germany. https://www.ise.fraunhofer.de/content/dam/ise/de/documents/publications/studies/Photovoltaics-Report.pdf

18. Frischknecht R, Stolz P, Heath G, Raugei M, Sinha P, de Wild-Scholten MJ (2020) Methodology Guidelines on Life Cycle Assessment of Photovoltaic. Tech. rep., International Energy Agency, PVPS Task 12: PV Sustainability, iSBN: 978-3-906042-99-2
19. Fthenakis VM, Kim HC, Alsema E (2008) Emissions from photovoltaic life cycles. Environ Sci Technol 42(6):2168–2174
20. García-Valverde R, Miguel C, Martí nez Bejar R, Urbina A (2009) Life cycle assessment study of a 4.2 kW(p) stand-alone photovoltaic system. Solar Energy 83(9):1434–1445. https://doi.org/10.1016/j.solener.2009.03.012. Go to ISI://WOS:000269289200002, type: Journal Article
21. García-Valverde R, Cherni JA, Urbina A (2010) Life cycle analysis of organic photovoltaic technologies. Progress Photovolt 18(7):535–558
22. Gong J, Darling SB, You F (2015) Perovskite photovoltaics: life-cycle assessment of energy and environmental impacts. Energy Environ Sci 8(7):1953–1968
23. Huld T, Amillo AMG (2015) Estimating PV module performance over large geographical regions: the role of irradiance, air temperature wind speed solar spectrum. Energies 8(6):5159–5181. https://doi.org/10.3390/en8065159
24. Huld T, Friesen G, Skoczek A, Kenny RP, Sample T, Field M, Dunlop ED (2011) A power-rating model for crystalline silicon PV modules. Solar Energy Mater Solar Cells 95(12):3359–3369
25. International Electrotechnical Commission (1998) IEC 61724. Photovoltaic system performance monitoring guidelines for measurement, data exchange, and analysis
26. Kato K, Hibino T, Komoto K, Ihara S, Yamamoto S, Fujihara H (2001) A life-cycle analysis on thin-film CdS/CdTe PV modules. PVSEC 11 - PART III 67(1):279–287. https://doi.org/10.1016/S0927-0248(00)00293-2, https://www.sciencedirect.com/science/article/pii/S0927024800002932
27. Kim H, Fthenakis V (2011) Comparative life-cycle energy payback analysis of multi-junction a-SiGe and nanocrystalline/a-Si modules. Progress Photovolt: Res Appl 19(2):228–239
28. Leccisi E, Raugei M, Fthenakis V (2016) The energy and environmental performance of ground-mounted photovoltaic systems-a timely update. Energies 9(8). https://doi.org/10.3390/en9080622
29. Louwen A, van Sark W, Schropp R, Turkenburg W, Faaij A (2015) Life-cycle greenhouse gas emissions and energy payback time of current and prospective silicon heterojunction solar cell designs. Progress Photovolt: Res Appl 23(10):1406–1428
30. Meijer A, Huijbregts MAJ, Schermer JJ, Reijnders L (2003) Life-cycle assessment of photovoltaic modules: Comparison of mc-Si, InGaP and InGaP/mc-Si solar modules. Progress Photovolt: Res Appl 11(4):275–287
31. Mohr NJ, Schermer JJ, Huijbregts MAJ, Meijer A, Reijnders L (2007) Life cycle assessment of thin-film GaAs and GaInP/GaAs solar modules. Progress Photovolt: Res Appl 15(2):163–179
32. Mohr NJ, Meijer A, Huijbregts MAJ, Reijnders L (2013) Environmental life cycle assessment of roof-integrated flexible amorphous silicon/nanocrystalline silicon solar cell laminate. Progress Photovolt: Res Appl 21(4):802–815
33. Monteiro Lunardi M, Wing Yi Ho-Baillie A, Alvarez-Gaitan JP, Moore S, Corkish R (2017) A life cycle assessment of perovskite/silicon tandem solar cells. Progress Photovolt: Res Appl 25(8):679–695
34. Raugei M (2013) Energy pay-back time: methodological caveats and future scenarios. Progress Photovolt: Res Appl 21(4):797–801
35. Raugei M, Bargigli S, Ulgiati S (2007) Life cycle assessment and energy pay-back time of advanced photovoltaic modules: CdTe and CIS compared to poly-Si. Energy 32(8):1310–1318
36. Raugei M, Frischknecht R, Olson C, Sinha P, Heath G (2016) Methodological Guidelines on Net Energy Analysis of Photovoltaic Electricity. Tech. Rep. Report T12- 07: 2016, International Energy Agency, PVPS Task 12: Subtask 2.0, LCA, iSBN 978-3-906042-39-8
37. Sengül H, Theis TL (2011) An environmental impact assessment of quantum dot photovoltaics (QDPV) from raw material acquisition through use. J Clean Prod 19(1):21–31
38. Serrano-Luján L, Espinosa N, Larsen-Olsen TT, Abad J, Urbina A, Krebs FC (2015) Tin- and lead-based perovskite solar cells under scrutiny: an environmental perspective. Adv Energy Mater 5(20):1501119

39. de Wild-Scholten M (2013) Energy payback time and carbon footprint of commercial photovoltaic systems. Thin-Film Photovolt Solar Cells 119:296–305
40. Yasuda K, Morita K, Okabe TH (2014) Processes for production of solar-grade silicon using hydrogen reduction and/or thermal decomposition. Energy Technol 2(2):141–154
41. You J, Hong Z, Yang YM, Chen Q, Cai M, Song TB, Chen CC, Lu S, Liu Y, Zhou H, Yang Y (2014) Low-temperature solution-processed perovskite solar cells with high efficiency and flexibility. ACS Nano 8(2):1674–1680

Chapter 7
Impacts of Solar Electricity

The impact assessment of solar electricity is the next stage of the Life Cycle Assessment. Once the inventory or materials and energy have been established, a specific method has to be applied to evaluate the impacts generated in different categories by all the elements of the inventory. Each LCA study will depend on the specific technology under evaluation and on the selected functional unit. If the whole life cycle is considered, for example, for a functional unit referred to the generated electricity averaged over the lifetime of the PV system, also the operating conditions influence the final result. Several studies have been published about LCA on different PV technologies since the late 90s. Most of them have been focussed on crystalline silicon and thin film technologies, both already in the market. More recently, many LCA studies of emerging technologies have been published and the Task 12 group of the IEA-PVPS has been working to compile and systematize them [19, 20]. This chapter will focus first on crystalline and thin film impact assessment results in different impact categories, and compare them with the impacts of other energy technologies; then a specific section is devoted to emerging technologies (organic and hybrid) and a final section focuses on land occupancy requirements and agrivoltaics, a more recent issue which is still not fully integrated in LCA methodology (although there is an impact category, "land occupancy" that is already assessed by some LCIA methods).

The main drivers to reduce impacts per functional unit of generated electricity is the technological improvement of PV modules reflected in power conversion efficiency (PCE) that is the parameter which more strongly influences results based on "service" functional unit. Particularly, in thin film technologies, the main objective to reduce LCA impacts is first the improvement of PCE and then the module dematerialization (reducing material content in each layer of the cell, a task difficult to achieve for thin film, and already very challenging even for crystalline silicon beyond 100μm thickness), this combined challenge was already pointed out in the first LCA studies of thin film technologies (CdTe and CIGS mainly) carried out in the 90s but has been kept as the main challenge since then in more recent studies [10, 27].

A. Urbina, *Sustainable Solar Electricity*, Green Energy and Technology,
https://doi.org/10.1007/978-3-030-91771-5_7

Then, research at laboratory scale must be translated into manufacturing lines, but the new developments only become profitable for manufacturers if the production can be up-scaled; this was a requisite clearly identified for thin film technologies which were very promising in the past decade despite its lower PCE compared to crystalline silicon [35]. But since then, its production stalled and the prediction was fulfilled but in a regressive way because CdTe and especially CIGS technologies could not accomplish the task of up-scaling production and have lost market share since then.

A detailed and updated Life Cycle Inventory for PV technologies was carried out by Andreas Wade et al in 2017 with the objective of including the PV module manufacturing routes in the Product Environmental Footprint (PEF) category rules [46]. The functional unit used for the impact assessment is 1 kWh (DC) of electricity produced with PV modules of different technologies, the reference flow is the photovoltaic module, expressed in the maximum power output measured in kW_p (kilowatt peak) under standard conditions and the product system of the electricity production considered three stages: manufacturing, use and end of life. This FU is not exactly the one recommended by the IEA-PVPS-Task 12 working group in the "Methodology guidelines on LCA of Photovoltaic", where the 1kWh AC electricity is prioritized, but has the advantage that the contribution of inverters and AC cabling is not considered and therefore the LCA provides an impact evaluation more focussed on PV technology [18].

Nevertheless, since the FU is the produced electricity, some assumptions for the operational stage of the PV modules have been considered. The following use phase scenario was considered: optimally oriented PV modules mounted on a slanted roof in Europe with an average annual yield of 1090 kWh/kW_p (excluding degradation), which corresponds to 975 kWh/kW_p including an average 10.5% production loss during 30 years or 0.7% per year due to degradation.

The life cycle impact assessment categories and indicators used to compare the PV technologies are shown in the graphs of the following sections and summarized in Table 7.1. They have been considered relevant for the "Product Environmental Footprint" [46]; several different methodologies exist as was explained in Chap. 3, and up to 27 categories have been developed, but for the purpose of comparing PV technologies, the 15 categories considered below are enough to provide a clear overview of impacts for mono-Si, multi-Si, a-Si/micro-Si, CdTe and CIGS PV modules.

The contribution to each category of the different PV module phases is presented in Fig. 7.1: raw material supply and module production; installation and mounting; module operation; dismantling and recycling and finally, a future potential environmental benefits that result from recycling the PV modules (50% of the net potential benefits are allocated to the PV system delivering the goods and 50% are allocated to the product system reusing the recycled goods in the future). The calculation was carried out for an "average" module in Europe within a 3 kW_p PV system installed on a slanted roof.

Table 7.1 Impact categories and indicators considered relevant for the "Product Environmental Footprint" and that are used to compare PV technologies in the figures of Sects. 7.1, 7.2 and 7.3

Impact category	Indicator	Units
Climate change	Radiative forcing as Global Warming Potential (GWP100)	kg CO_{2eq}
Ozone depletion	Ozone Depletion Potential (ODP)	kg CFC-11_{eq}
Human toxicity, cancer effects	Comparative Toxic Unit for humans	CTUh, c
Human toxicity, non-cancer effects	Comparative Toxic Unit for humans	CTUh, n-c
Particulate matter/respiratory effects	Intake fraction for fine particles	kg $PM2.5_{eq}$
Ionizing radiation, human health	Human exposure efficiency relative to U235	kBq $U235_{eq}$
Photochemical ozone formation	Tropospheric ozone concentration increase	kg $NMVOC_{eq}$
Acidification	Accumulated Exceedance (AE)	mol H^{+}_{eq}
Eutrophication, terrestrial	Accumulated Exceedance (AE)	mol N_{eq}
Eutrophication, freshwater	Fraction of nutrients reaching freshwater end compartment (P)	kg P_{eq}
Eutrophication, marine	Fraction of nutrients reaching marine end compartment (N)	kg N_{eq}
Ecotoxicity, freshwater	Comparative Toxic Unit for ecosystems	CTUe
Land use	Soil Organic Matter	kg C deficit
Resource depletion, water	Water abstraction related to local scarcity of water	m^3 $water_{eq}$
Resource depletion, mineral, fossil	Scarcity	kg Sb_{eq}

The values obtained for the indicator units in the impact categories are meaningful for the expert LCA developer, but a comparison with other alternative sources of electricity is required to put its value into context and compare with the impacts of different technologies that supply electricity. In this case, the full balance of system contribution (inverters included) has to be included in the impact of the functional unit (service electricity, in AC). The United Nations Development Programme carried out a detailed study to evaluate the impacts of electricity generation in different scenarios which is used to compare impacts of solar electricity with other renewable or fossil fuel electricity sources [43].

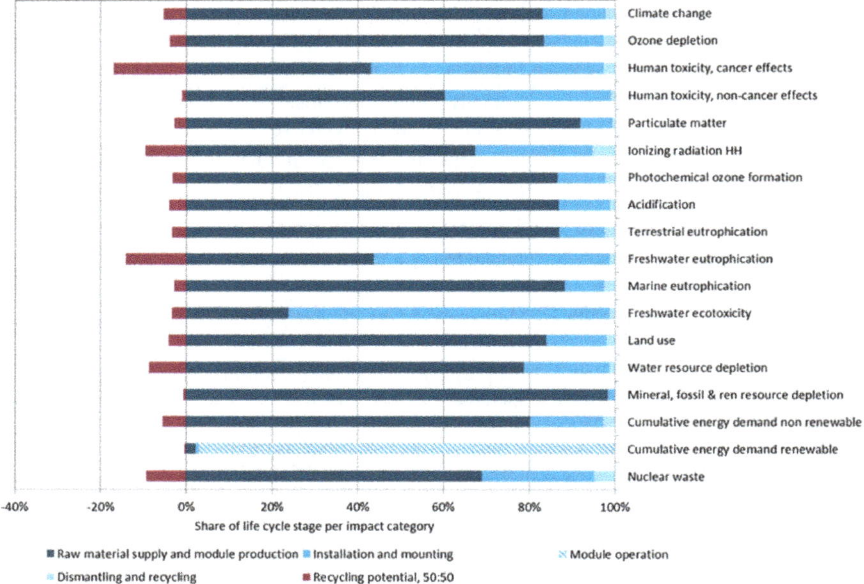

Fig. 7.1 Environmental impact results (characterized, indexed to 100%) of 1 kWh of DC electricity produced with a residential scale (3 kW_p) PV system with average PV panels mounted on a slanted roof (Reproduced with permission from [46])

7.1 Human Health Impacts

Single crystal Si modules have the largest impacts and CdTe the lowest in all categories related to human health. This may seem contradictory since the impact of modules containing cadmium may be expected higher than other technologies which do not contain toxic materials. But this is an example of how LCA can provide insightful evaluation of risks. The contribution to impact categories arises from any process related to material processing or module manufacture and wafer-based silicon technologies are high consumers of energy, with the largest primary embedded energy, which, due to the energy mix of the electricity consumed, create higher health risks (the grid efficiency of about 35% to convert from primary energy to electricity is usually considered in most LCA studies) (Fig. 7.2).

All PV technologies have impacts in the same order of magnitude, when an endpoint impact category is evaluated using different methodologies as explained in Chap. 3, a single score can be obtained for "human health" impact. In this case, the comparison to other renewable and non-renewable technologies for the production of the same functional unit of electricity is more clear. As an example of this kind of comparative results, a summary graph of a study carried out by the United Nations Environmental Programme is shown in Fig. 7.3. The study was carried out using the ReCiPe(H) impact assessment method to evaluate an aggregated human health

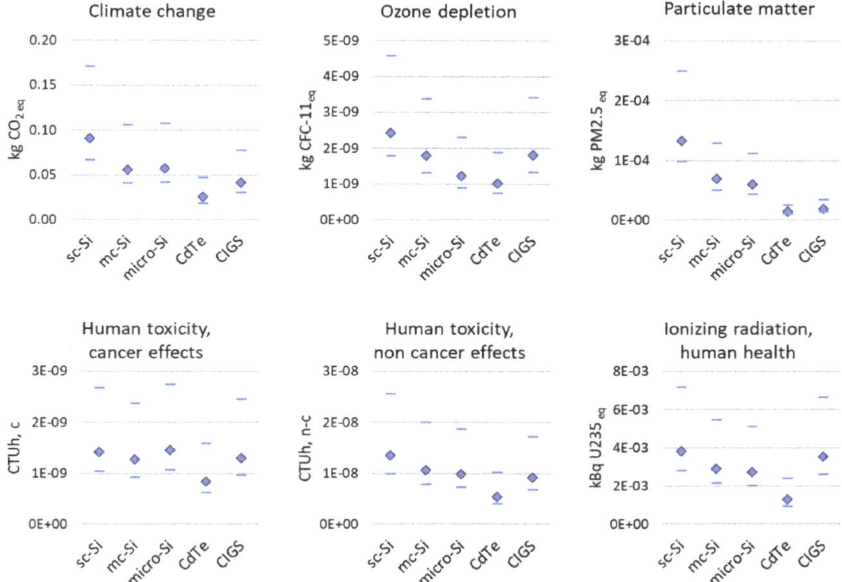

Fig. 7.2 Average (diamond), maximum and minimum (bars) climate change and human health impacts of 1 kWh DC electricity produced with mounted mono-Si, multi-Si, a-Si/micro-Si, CdTe and CIS/CIGS PV modules. *Data source* [46] based on Ecoinvent data v2.2+

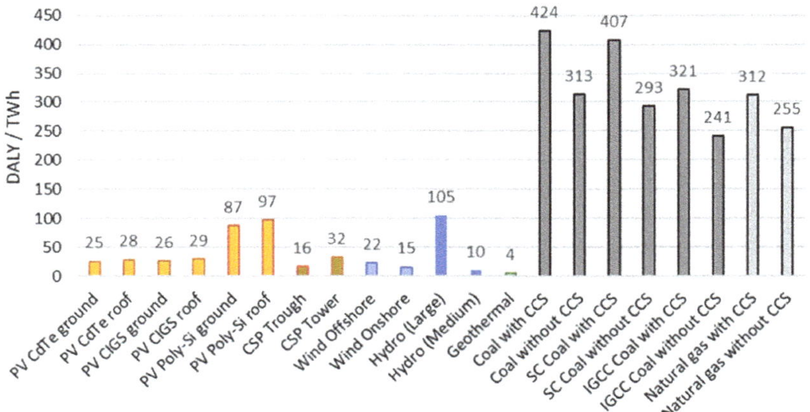

Fig. 7.3 Comparative human health impact of electricity production, in aggregated endpoint category measured in units of disability adjusted life years (DALY) per TWh of electricity generated following different damage pathways according to the ReCiPe (H) impact assessment methods (CCS: CO_2 capture and storage, IGCC: integrated gasification combined cycle, SC: supercritical). *Data source* [43, 46]

impact in units of disability adjusted life years (DALY) per TWh of produced electricity; in these units of DALY/TWh, the aggregated impact on human health for thin film technologies are for multi-crystalline silicon 87 DALY/TWh and 97 DALY/TWh for ground or roof-top mounted systems respectively; for CdTe, 25 DALY/TWh and 28 DALY/TWh for ground or roof-top mounted systems, respectively, and for CIGS 26 DALY/TWh and 29 DALY/TWh for ground or roof-top mounted systems, respectively. Again it is confirmed that wafer-based crystalline silicon PV technologies have three times more human health impacts than thin film technologies, although this group of PV technologies have around five times less impact than any fossil fuel technology.

7.2 Environmental Impacts

When environmental impacts, including climate change, are considered, a similar trend can be observed (Fig. 7.4). All crystalline wafer-based silicon technologies have larger impacts than thin film technologies, with CdTe being the one with slightly lower imacts in all categories and always of the same order of magnitude.

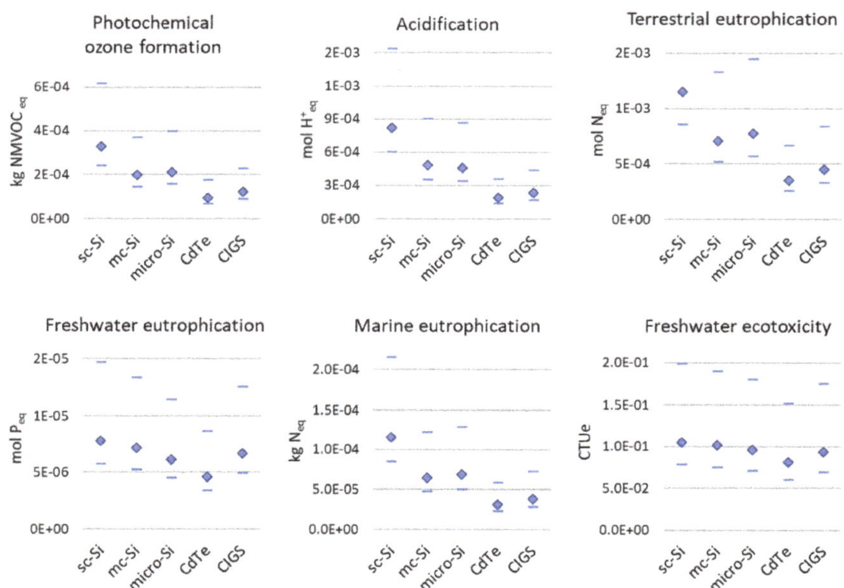

Fig. 7.4 Average (diamond), maximum and minimum (bars) environmental impacts of 1 kWh DC electricity produced with mounted mono-Si, multi-Si, a-Si/micro-Si, CdTe and CIS/CIGS PV modules. *Data source* [46] based on Ecoinvent data v2.2+

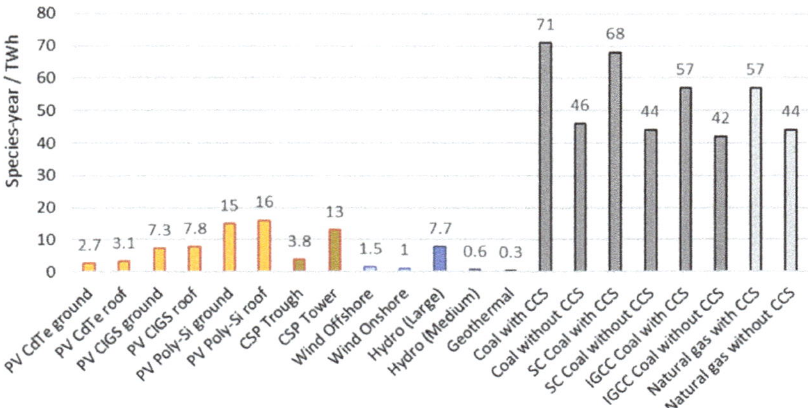

Fig. 7.5 Ecosystem impacts of electricity production measured in species-year affected per TWh of electricity following different damage pathways according to the ReCiPe (H) impact assessment method (CCS: CO_2 capture and storage, IGCC: integrated gasification combined cycle, SC: supercritical). *Data source* [43, 46]

Again, when an endpoint aggregated calculation is carried out, the comparison to other electricity production technologies is more clear; in this case, the aggregated ecosystem impact measured in species-year affected per TWh of electricity following different damage pathways according to the ReCiPe (H) impact assessment method is shown in Fig. 7.5 and the climate change impact measured in Green House Gases (GHG) emissions in grams of CO_{2eq} per kWh of produced electricity throughout the lifetime of the energy system is shown in Fig. 7.6, note the logarithmic scale in this graph illustrating the large difference in life cycle emissions of PV technologies compared to any fossil fuel technology. For the PV technologies, the emissions, taking the upper limits, range from 14 gCO_{2eq}/kWh for CdTe technology or 17 gCO_{2eq}/kWh for CIGS technology (typical of thin film) to 58 gCO_{2eq}/kWh for polycrystalline silicon technology, in all cases considering ground mounted systems. The range varies depending on the geographical location of the module factory because the grid mix of the country where the module is manufactured have an influence on the embedded emissions that are used for the calculation; also the location where the system is installed has a strong influence on the electricity produced during its lifetime (the performance ratio depends on irradiance and temperature). In some cases, the climate change impacts are evaluated in terms of "avoided" emissions, a parameter that depends even more strongly on the geographical location where the PV system is installed because the solar electricity is (ideally) replacing electricity that otherwise would have been consumed from the grid, via the electricity grid mix of the country under consideration; this substitutional approach is also used to evaluate "avoided" NO_x and SO_x emissions, but it is not considered as part of a LCA study and will be discussed in Sect. 12.4.

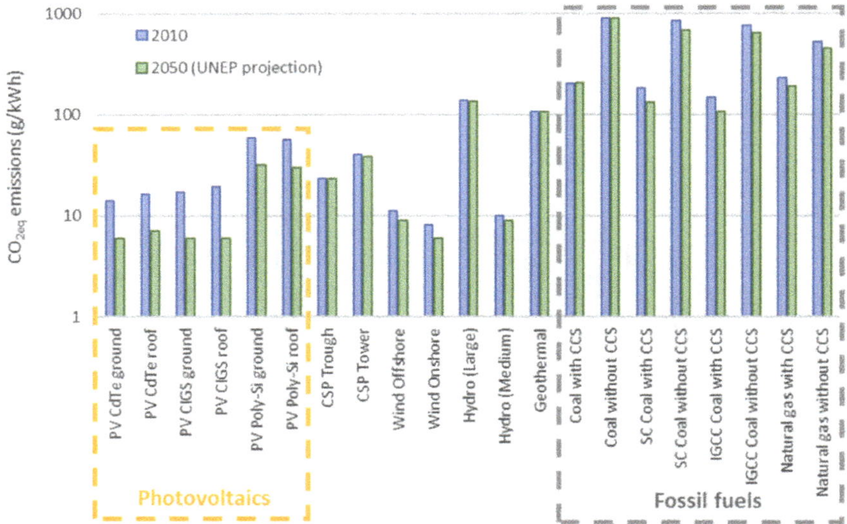

Fig. 7.6 Life cycle carbon emissions of different electricity supply technologies, modelled for 1 kWh produced in Europe (CCS: CO_2 capture and storage, IGCC: integrated gasification combined cycle, SC: supercritical). *Data source* [43, 46]

7.3 Land use, Water, Mineral, Fossil and Renewable Depletion Impacts

In this case, the trend observed in all other impact categories is altered for mineral, fossil and renewable resource depletion because CIGS technology has much stronger impacts than the other technologies as shown in Fig. 7.7, where a different axis has been included in the graph, emphasizing that CIGS has impacts that are an order

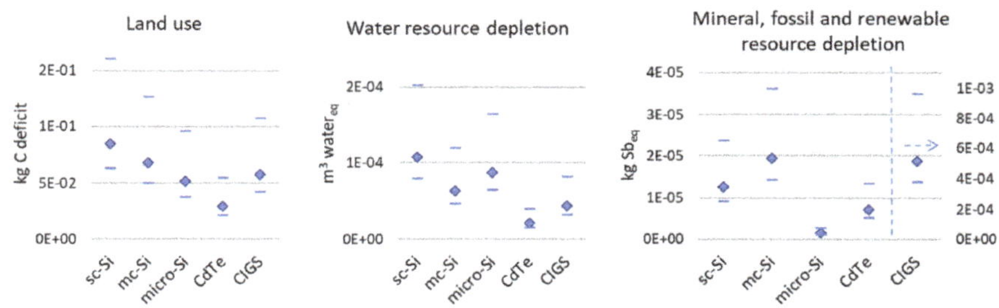

Fig. 7.7 Average (diamond), maximum and minimum (bars) environmental impacts of 1 kWh DC electricity produced with mounted mono-Si, multi-Si, a-Si/micro-Si, CdTe and CIS/CIGS PV modules; note the different axis scale for CIGS in the resource depletion graph. *Data source* [46] based on Ecoinvent data v2.2+

of magnitude higher than the other technologies. The impacts on mineral resource depletion is strongly linked to materials demand presented in Chap. 5, where tellurium and selenium were the two elements posing higher risks for supply; when LCA is carried out, the impact is much higher on CIGS technology, putting both gallium and selenium as the main contributor to impacts.

Detailed LCA studies which project energy scenarios to the future point that the demand of copper per unit of generated solar electricity, mainly embedded in BoS components, is larger than other fossil fuel alternatives and may require between 11 and 40 times more copper in a scenario with 39% electricity generation from renewable sources in 2050 (the BLUE scenario proposed by the IEA in 2010); the more ambitious NZE2050 proposes 33% for PV generation, a lower value that is compared to a baseline scenario, but this apparently striking result is smoothed by the consideration that the copper demand for all low carbon technologies projected for 2050 can be supplied with only two years of current copper world supply, although future supply may be a concern [26].

Water use should be considered in PV LCA studies. Water is consumed as an input during the operational phase for cleaning purposes (maintenance work) and it is also used during fabrication and end-of-life phases of the PV module lifetime in the associated industrial processes. Water should be included in the inventory, and its impact on several categories must be taken into account, especially when water scarcity is becoming a pressing issue in many regions in the world, some of them with high irradiance levels, and therefore, locations where PV systems are likely to be installed. Either by direct use or indirectly via sub-processes, the use of water in the life cycle of PV modules must be addressed. For the water depletion category, thin film amorphous and micro-Si technology consumes slightly more water than multi-crystalline silicon; water consumption is produced during manufacturing phase (quantities are well defined by processes and depend on the technology) and during the operation phase when water consumption strongly depends on the cleaning strategies and may vary strongly between different locations; unfortunately, regions with higher insolation often coincide with water scarcity and atmospheric conditions where dust particles in suspension are common, leading to a higher water demand for module cleaning. Rain is only effective for larger particles (pollen >50 μm, dust >20 μm) and may lead to problems of dew formation which reduces module efficiency, and glass self-cleaning surface treatment for self-cleaning purposes are being developed although the extra cost added to module price is not yet competitive [6, 39].

When land occupation is considered, the comparison to other energy technologies is unfavourable for solar electricity, but two issues are evident in Fig. 7.8 and must be emphasized: roof-top versus ground PV systems and the high land impacts of coal-based technologies. The first issue refers to the strong difference between roof-top and ground mounted PV systems, roof-top systems do not have an impact on land occupation during operational phase because the land was already occupied (the same for building integrated systems), while ground PV plants have a much stronger impact on land occupation and also on landscape visual impact (although

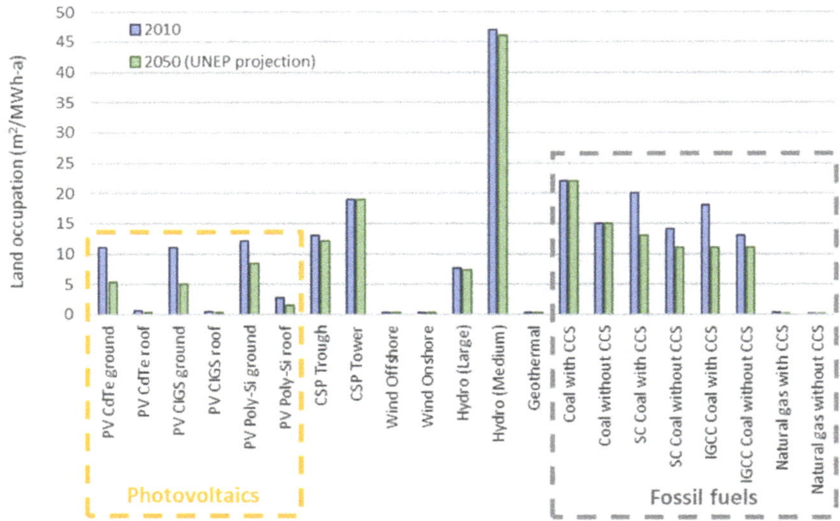

Fig. 7.8 Land occupation is required for the production of electricity. For coal energy, the dark green bar represents open pit mines (land use largely associated with the mine itself) and the total size of the bar reflects the land use associated with coal from underground mines (CCS: CO_2 capture and storage, IGCC: integrated gasification combined cycle, SC: supercritical). *Data source* [43, 46]

this later "category" is not included in LCA methodologies). The occupation of very large swathes of land by MW scale PV plants also poses an additional risk on biodiversity which is still not included on the other environmental categories due to the lack of clear attributional routes to this categories via land occupation [25, 45]. Furthermore, the construction of very large plants (>100 MW_p) which occupies hundreds of hectares is raising social concern and competition with agricultural exploitation of land. A more deep discussion of the possibilities of combination of PV systems with agricultural use of land is presented in Sect. 7.5.

The second issue is the much larger land impact of coal-based technologies. All of them, including integrated gasification combined cycle (IGCC) with or without carbon capture and storage technology have larger land impacts than all renewables (with the exception of large reservoir hydropower); this land requirement occurs in the initial stages of coal extraction, due to land occupation of open pit mines or the use of hardwood as structural material in underground mines; during the burning of coal in plants, the land required is lower (as expected), but the full life cycle impact of coal on land is very large, larger than any ground mounted PV plant.

7.4 The Rapidly Evolving Impacts of Emerging PV Technologies

In the previous sections, the commercial PV technologies were analysed in more detail since its massive deployment is already occurring (especially for crystalline silicon technologies) and experimental data from systems deployed worldwide are available and although some dispersion in the impact assessment is still found, the results are robust and can be used to characterize the impacts of a given technology.

For the case of emerging technologies, the challenge is larger since the variety of materials, processing methods, lifetimes and end-of-life options makes difficult to assign impacts to a given representative technology at industrial level; on the contrary, the published results are case studies aimed at comparing the selection of materials or methods within a technological family. Additionally, there is still a lack of long term experiments where PV systems of emerging technologies are deployed in a variety of locations (with different environmental parameters and climate conditions) and during a few years to provide operational case study data. Some round-robin experiments were carried out with initial focus on measuring power conversion efficiency in outdoor conditions, and analyse the rapid degradation of organic and hybrid modules with the aim to propose a consensus for measurement protocols which allows fair comparison of results [29, 32, 36]. On the other hand, there is still no commercial deployment of organic or hybrid emerging PV technologies, beyond a few small systems for demonstration purposes (BIPV systems, stand-alone mini systems), and therefore, its real impacts on LCA categories are still lacking enough empirical results measured in operational systems; also the future factories that may produce PV modules at MW scale are still under development or working as pilot-plants and no established production line can be considered as representative of the technology. A common task for all of them is extending the lifetime of devices, a parameter that needs to be significantly improved if emerging technologies want to compete against inorganic technologies and gain a significant market share.

Nevertheless, many LCA studies for organic and hybrid technologies have been published in recent years and the information they provide is useful for decision-making with regard to materials selection, device architectures and processing routes selected with the aim to minimize impacts and at the same time keeping good performance parameters of the manufactured modules (mainly power conversion efficiency and lifetime).

The first LCA studies of organic technologies already pointed to the high relative contribution to impacts that the evaporated transparent conducting oxide (TCO), up to 86% of energy embedded in the materials or 74% of total cumulative energy demand of organic module processing (with variations depending on the printing technologies used for the other layers) [16, 21]. Similar important contribution of TCO on the substrate is found on other impact categories, with a share of more than 50% in all categories analyzed in the previous section for inorganic technologies (see Table 7.1). These findings lead to the proposal of several technological variations for the electrodes of organic solar cells, ranging from silver nanowires, carbon nanotubes, PEDOT:PSS with silver grid, but all of them keeping a high share of total embedded energy in the modules, which only reduced to a range between 36% and 50%, lower than ITO, but still high [14]. Only highly conductive PEDOT:PSS used as a transparent conducting electrode reduced significantly its contribution to 6%. After TCOs, the substrate is the second contributor to impacts, either glass or plastic-based (mostly PET). Active layers for different combinations of organic donor and acceptors always had a much lower impact in all categories of life cycle impacts than the electrodes plus substrate block, and therefore, the main recommendation to reduce impacts for this technology is improving alternatives for metal-based electrodes [4]. Detailed LCA studies based on the benchmark P3HT:PCBM blend for organic solar cells on FTO/PET substrate and roll-to-roll production showed great potential to reduce environmental impacts more than 90% when compared with crystalline silicon for the same functional unit of $1W_p$ of peak power; small variations are produced by different routes of PCBM synthesis [41]. Power conversion efficiencies of organic solar cells are improving steadily and the more challenging problem remains the extension of lifetimes to become a really competitive technology that can reach the market; if a significant share of solar electricity in future scenarios such as NZE2050 is achieved in the next few years by organic technologies, the aggregated environmental impacts of solar electricity will be strongly reduced for any functional unit under consideration in the corresponding LCA study.

Perovskite solar cells are a promising technology which has already reached PCE above 20% and lifetimes of more than 10,000 hours. The rapid evolution of both PCE and lifetime indicates that perovskite technology is a strong candidate for a commercial penetration in the coming years; probably the first step towards this commercialization is the perovskite/Si tandem approach. The balance between PCE, lifetime and environmental impacts strongly depends on the selection of materials for the solar cell structure (active layer, HTL, ETL and encapsulant material). The main concern about the sustainability of perovskite solar cell technology is the presence of lead in the active layer of the most efficient alternatives; nevertheless the lead content has a lower impact than other components of the cell, for example, in the Human Toxicity or Freshwater Ecotoxicity categories: evaporation of transparent conducting oxides for electrodes is the main contributor with around one third of the total, followed by active layer with 28%, spiro-OMeTAD (HTL) (20%) and titanium dioxide (17%) (results are similar in studies with different impact assessment methods such as ReCiPe or TRACI and percentages are an average of several studies [8, 24, 44]. When evaporation is avoided, the contribution of active layer becomes the most important, but in this case, studies are for spin-coated layers, a method which will not be up-scaled to industrial production. Several studies are coincident in the identification of complex chemical processing routes for spiro-OMeTAD as the main contributor to impacts in perovskite technology, and therefore, from an LCA perspective, the research for good HTL replacements is important: polymers, fullerene derivatives or small molecules have been tested with good results, thus reducing impacts while keeping good PCE and extended lifetimes [3, 17]. The studies which focus on the future industrial production of perovskite or perovskite/Si tandem cells deliver promising results built upon the experience of roll-to-roll processing or organic solar cells and providing recommendations for better encapsulation in order to extend lifetimes; early results with pre-industrial slot-die coating methods have been recently improved to provide good efficiency and reproducibility with low environmental impacts [2, 5, 37].

7.5 Size Dependant Impacts of PV Systems: Land Occupancy and Agrivoltaics

The size of photovoltaic systems, measured in installed capacity or square meters of module surface, can be roughly divided into two main categories: roof-top and ground mounted systems, with several subcategories. Roof-top PV systems, in the scale of a few kW_p to a few hundreds of kW_p of installed capacity, do not generate an extra demand of land occupancy since they are installed in already occupied land, similarly for building integrated (BIPV) systems; roof-top and BIPV systems contribute to distributed generation injected into grids, usually within a self-consumption framework (with or without net balance). The monetary economic return of roof-top systems is slightly lower than ground mounted ones due to limited use of economy of scale advantages, but the socioeconomic return must be evaluated more carefully since roof-top systems require extra engineering design time and extra installation time per unit of installed power. If the economic return includes job creation, the economic savings of end-users that benefit from self-consumption, and the reduction of demand (since users are more aware of its consumption), the distributed roof-top systems appear as the most beneficial scheme. For example, in the United Kingdom, a detailed study on 302 households with roof-top PV systems that participated in smart grid demonstration project found a 24% reduction of the average UK household's annual electricity demand, which implies savings of 138 GBP per year per household (assuming a cost of electricity of 0.15 GBP/kWh) [34]. When a broad approach for socioeconomic impacts is included in the studies, the apparently obvious better monetary return of larger plants is modulated and smaller systems provide more global socioeconomic benefits; a multicriteria decision support system analysis led to the following recommendation for policymakers: "the smaller the better" to maximize community scale benefits with PV system installations [23]. And with regards land occupancy, if the small PV systems are roof-top or building integrated, then it is a much better solution.

But the economy of scale of investments in large plants provides better monetary return, with break-even points at shorter times. This is moving the momentum generated by policies oriented to promote roof-top systems to an economically driven push for large plants in the scale of a few to hundreds of MW_p of installed capacity. In this case, the competition for land occupancy is becoming a problem. Many locations optimal for PV system installation are already occupied by small farms or larger agro-industrial facilities for food production, thus generating a competition for land use. The displacement of food production by solar electricity production is already happening in several countries, and an important concern is growing at two levels: from a global point of view and in the context of climate change, the reduction of land availability for crops may generate a shortage of food production in certain areas already stressed by extreme weather events and increasing droughts; and at the local level, the displacement of traditional cultivation areas because the economic returns of solar electricity per square meter of occupied land is larger than most crops have already lead to tensions and even protests from farmers. The quantitative measurement of this impact of land occupancy is not included in the conventional category of "land use" of many LCIA methods, but should be taken into consideration in a sustainability evaluation of large PV plants.

A technical report from the International Electrotechnical Commission about land usage of photovoltaic plants (IEC TR 63149:2018) describes the mathematical models and provides examples for the calculation of land requirement by large PV plants, minimizing the distance between arrays to avoid shading and reasonably reducing the land usage of PV large plants. It can be calculated for different tracking systems where the ground requirement ratio (GRR), which is the ratio between the land occupancy of the PV system and the surface of the PV generator (all the modules), varies between 2 and 6; for fixed systems is around 2, with uniaxial tracking is around 4 and with biaxial tracking is around 6, in all cases with shadow losses below 2%. For large plants of several MW_p of installed capacity, the land requirement rises to several hectares, and in some cases, to hundreds of hectares. Furthermore, the change in land use induced by the massive deployment of PV systems fostered by investment in the energy transition may lead to an unwanted increment of emissions induced by the reduction of carbon sinks that result from the change in land coverage: a calculation by Dirk-Jan van de Ven et al. for European Union, Japan, India and South Korea analyses several scenarios up to 2050 and indicates that penetration of solar electricity generation between 25% to 80% into the electricity mix of those countries may lead to a net release up to 50 gCO_{2eq}/kWh and a land occupancy between 0.5% and 5% depending on the country, PV technology and land management [45]. This is an important result which alerts about the need to focus on global analysis of sustainability that go beyond LCA to inform stakeholders and governments in order to regulate land use for photovoltaic large plants.

The integration of activities that require rural land occupancy is possible: photovoltaic technology and agriculture have now been combined in the concept of "agrivoltaics" with the simultaneous use of land for solar electricity production and farming (agriculture or livestock). It was first proposed at an early stage of PV deployment, in the 80s [22]; and during the past ten years, it has been broadly applied and demonstrated as a good alternative to minimize land occupancy impacts of PV systems and maximize the benefits of the food-energy-water nexus, especially in drylands [9]. The use of photovoltaic modules integrated into greenhouses is a growing area of research, with PV greenhouses providing a good example of shared use of land for agriculture and photovoltaic applications [7, 12, 15]. But the more clear realization of the concept of agrivoltaics is the large photovoltaic system mounted on special supports adapted to be compatible with land cultivation.

The economic assessment of agrivoltaic systems compared to equivalent nominal capacity ground or roof-top systems indicates that the capital expenditure is about 33% higher in agrivoltaic systems, which is mainly due to increased cost of balance of system components (requiring support structures compatible with agricultural work) and higher land costs, with larger GRR to avoid too much shadowing in the crops. This extra investment increases the LCOE of solar electricity produced with agrivoltaics systems between 5 and 15 euros per MWh with respect ground mounted systems (90 euro per MWh in the cited study); and EPBT to around 9 years, higher than other PV systems [1]. But this lower economic performance of energy production of agrivoltaic systems must be balanced by the benefits of sharing land occupation with crop production and a combined economic analysis is more adequate to evaluate the returns of land use.

This is a complex task. Detailed monitorization of microclimatic conditions to compare a vegetated ecosystem, a traditional PV system and an agrivoltaic system, where solar panels share land with crops under the panels have delivered a clear conclusion: the photosynthetically active regions of the solar spectrum are reduced for the plants, but this could be beneficial in dry locations to boost crop resilience to droughts and at the same time improving PV performance due to the reduced temperature of the panels (compared to the case where there is no vegetation under the modules); the agrivoltaic system prevents photosynthesis depression due to heat and light stress and this allows higher carbon uptake for growth and reproduction of the plants while the transpirational cooling from the crops induces a reduction in PV module temperature that are 1.2 ± 0.3 °C lower compared with a traditional PV setting and water losses are reduced since soil moisture remained between 5% and 15% higher in the agrivoltaic system depending on the irrigation patterns [9]. These promising results build upon other studies focussed on specific crops which evaluated the balance between crop yield reduction and PV module coverage (in greenhouses or in open field agrivoltaic systems), although crop yield was reduced, the combined yield of crop and PV was improved with respect to the separated systems [13, 28, 33]. The impacts on agro-labourer's health working under solar panels have not been studied in detail, but it may be envisaged that the additional risks arising from the presence of the PV system (cabling, structures) can be compensated by the better conditions arising from lower sun exposure and reduced temperatures. The mounting evidence of the aggregated benefits of agrivoltaics is boosting the investment in this kind of PV systems although they require a higher capital expenditure. Agrivoltaics, therefore, creates a synergy which reduces competition for land exclusively in terms of monetary returns and may contribute simultaneously to food and energy security, especially in drylands.

Two further considerations about the impact of large PV plants must be addressed, in both cases with substantial research being published in recent years. The first one is the impact on local biodiversity of the installation of a large PV plant, with a special focus on vegetation patterns that are modified by the presence of the solar modules, with clear differences in areas under the modules and between the arrays due to different irradiation patterns. This vegetation can be managed to obtain ecosystem benefits and reduce soil erosion, for example, with species for the soil between the photovoltaic panels that include perennial grasses (*Lolium perenne, Dactylis glomerata*) and perennial herbs (*Pastinaca sativa, Trifolium repens, Silene latifolia, Galium album*), while species recommended for the soil under the photovoltaic panels are particularly perennial herbs (*Achillea millefolium, Potentilla anserina, Plantago major*); a few extra maintenance work will enable these plants and other native species to create functional ecosystems, including local fauna recovery and reducing erosion or desertification in locations with high insolation and low precipitation and humidity [42].

The second consideration is about landscape protection. Landscape integrity is difficult to evaluate because it combines ecosystem preservation with visual and cultural concerns, and therefore, requires a multidisciplinary approach that must be added to the technical agrivoltaic design and production yield calculations. The inclusive multidisciplinary approach will deliver results which are closer to a global sustainability concept than the exclusively technical approach, but it is more difficult to evaluate quantitatively. Sconamiglio and Toledo have recently proposed a new methodology based on considering the agrivoltaic system as a three- dimensional landscape pattern (land surface topology plus the height of PV panels), it includes the definition of a multidimensional performance matrix with several parameters that can be calculated for the assessment and design of agrivoltaic systems (greenhouses and open field crops with agrivoltaic patterns). The matrix also evaluates the trade-offs between the different yields involved in the agrivoltaic system (agriculture, electricity, ecological services, cultural values, community services) and it is useful for stakeholders and policymakers to make decisions related to the regulation of land use in the context of an energy transition that will require large swathes of land for photovoltaic systems [38, 40].

7.6 Impacts of Module Transportation During Manufacture, Installation and End of Life

Transportation is present in any LCA study, and play an important part in life cycle inventories of energy demand in three main steps: initial stages when raw material from mining or recycling sites are transported to the location where purification of material is carried out and then to locations where cells and/or modules are manufactured; intermediate step when the manufactured module will have to be transported to the PV system construction site and a final step where decommissioned modules are transported to recycling or land-filling sites.

These initial steps, that is, the transport means required in the supply chain for module manufacture, has to be included in the cradle to gate LCA. The standard databases include complete and updated inventories of different kinds of transport (ship, lorry, train, plane, etc…of different sizes and consuming different fuels); the transport contribution for a LCA study of a PV technology will require a geographical analysis of supply chains and a calculation of distances. This effort has been accomplished by several research groups which found a contribution of transport to several impact categories ranging between 5% and 30%, interestingly with more contribution from transport to recycling or landfilling sites at end of life and recommendation to limit this final phase transportation to a range lower than 100 km [30, 31]. The economic cost of transportation only increases marginally with distance according to a USA case study that could be extrapolated to other regions; transport cost starts with a baseline quantity ranging between 2 and 2.5 USD per square meter of c-Si modules and per 1000km of transportation, slightly more for other technologies (depending on the weight of modules), and adding only a marginal contribution even for module distribution (or collection) when the distance of 1000 km is doubled [11]. The problem with this kind of study is that they rely on strong assumptions about the routes (distance and time) and transport means, mainly a combination of ships and lorries of different size. Case studies provide robust results but it is difficult to obtain general conclusions applicable to other cases. The databases (for example Ecoinvent) provide an easy way to include transport impacts in the processes under evaluation, but a separated contribution analysis is often useful to provide recommendations to improve logistics in three main phases of the module life cycle: first for import/export transport requirements of materials in the supply chain for the manufacture of modules, then its transport to the location where the PV system will be finally built, and finally, end-of-life routes to collection, recycling or landfilling sites.

References

1. Agostini A, Colauzzi M, Amaducci S (2021) Innovative agrivoltaic systems to produce sustainable energy: an economic and environmental assessment. Appl Energy 281:116102
2. Alberola-Borrás JA, Baker J, De Rossi F, Vidal R, Beynon D, Hooper K, Watson T, Mora-Seró I (2018a) Perovskite photovoltaic modules: life cycle assessment of pre-industrial production process. iScience 9. https://doi.org/10.1016/j.isci.2018.10.020
3. Alberola-Borrás JA, Vidal R, Mora-Seró I (2018) Evaluation of multiple cation/anion perovskite solar cells through life cycle assessment. Sustain Energy Fuels 2(7):1600–1609
4. Anctil A, Babbitt CW, Raffaelle RP, Landi BJ (2013) Cumulative energy demand for small molecule and polymer photovoltaics. Progress Photovolt: Res Appl 21(7):1541–1554
5. Angmo D, DeLuca G, Scully AD, Chesman AS, Seeber A, Zuo C, Vak D, Bach U, Gao M (2021) A lab-to-fab study toward roll-to-roll fabrication of reproducible perovskite solar cells under ambient room conditions. Cell Rep Phys Sci 2(1):100293
6. Appels R, Lefevre B, Herteleer B, Goverde H, Beerten A, Paesen R, De Medts K, Driesen J, Poortmans J (2013) Effect of soiling on photovoltaic modules. Solar Energy 96:283–291
7. Aroca-Delgado R, Pérez-Alonso J, Callejón-Ferre AJ, Velázquez-MartíB (2018) Compatibility between crops and solar panels: an overview from shading systems. Sustainability 10(3). https://doi.org/10.3390/su10030743
8. Babayigit A, Ethirajan A, Muller M, Conings B (2016) Toxicity of organometal halide perovskite solar cells. Nat Mater 15:247–251. https://doi.org/10.1038/nmat4572
9. Barron-Gafford GA, Pavao-Zuckerman MA, Minor RL, Sutter LF, Barnett-Moreno I, Blackett DT, Thompson M, Dimond K, Gerlak AK, Nabhan GP, Macknick JE (2019) Agrivoltaics provide mutual benefits across the food-energy-water nexus in drylands. Nat Sustain 2(9):848–855
10. Bergesen JD, Heath GA, Gibon T, Suh S (2014) Thin-film photovoltaic power generation offers decreasing greenhouse gas emissions and increasing environmental co-benefits in the long term. Environ Sci Technol 48(16):9834–9843

11. Celik I, Lunardi M, Frederickson A, Corkish R (2020) sustainable end of life management of crystalline silicon and thin film solar photovoltaic waste: the impact of transportation. Appl Sci 10(16). https://doi.org/10.3390/app10165465
12. Cossu M, Yano A, Solinas S, Deligios PA, Tiloca MT, Cossu A, Ledda L (2020) Agricultural sustainability estimation of the European photovoltaic greenhouses. Euro J Agron 118:126074
13. Dinesh H, Pearce JM (2016) The potential of agrivoltaic systems. Renew Sustain Energy Rev 54:299–308
14. Emmott CJM, Urbina A, Nelson J (2012) Environmental and economic assessment of ITO-free electrodes for organic solar cells. Solar Energy Mater Solar Cells 97:14–21
15. Emmott CJM, Roehr JA, Campoy-Quiles M, Kirchartz T, Urbina A, Ekins-Daukes NJ, Nelson J (2015) Organic photovoltaic greenhouses: a unique application for semi-transparent PV? Energy Environ Sci 8(4):1317–1328
16. Espinosa N, García-Valverde R, Urbina A, Krebs FC (2011) A life cycle analysis of polymer solar cell modules prepared using roll-to-roll methods under ambient conditions. Solar Energy Mater Solar Cells 95(5):1293–1302
17. Espinosa N, Serrano-Luján L, Urbina A, Krebs FC (2015) Solution and vapour deposited lead perovskite solar cells: ecotoxicity from a life cycle assessment perspective. Solar Energy Mater Solar Cells 137:303–310
18. Frischknecht R, Stolz P, Heath G, Raugei M, Sinha P, de Wild-Scholten MJ (2020) Methodology Guidelines on Life Cycle Assessment of Photovoltaic. Tech. rep., International Energy Agency, PVPS Task 12: PV Sustainability, iSBN: 978-3-906042-99-2
19. Fthenakis V, Kim H (2011) Photovoltaics: life-cycle analyses. Progress Solar Energy 185(8):1609–1628. https://doi.org/10.1016/j.solener.2009.10.002, https://www.sciencedirect.com/science/article/pii/S0038092X09002345
20. Fthenakis V, Kim HC, Frischknecht R, Raugei M, Sinha P, Stucki M (2011) Life Cycle Inventories and Life Cycle Assessments of Photovoltaic Systems. Tech. Rep. IEA-PVPS T12-02:2011, International Energy Agency - Photovoltaic Power Systems Programme - Task12
21. García-Valverde R, Cherni JA, Urbina A (2010) Life cycle analysis of organic photovoltaic technologies. Progress Photovolt 18(7):535–558
22. Goetzberger A, Zastrow A (1982) On the coexistence of solar-energy conversion and plant cultivation. Int J Solar Energy 1(1):55–69
23. Guerrero-Liquet GC, Oviedo-Casado S, Sanchez-Lozano JM, Socorro García-Cascales M, Prior J, Urbina A (2018) Determination of the optimal size of photovoltaic systems by using multi-criteria decision-making methods. Sustainability 10(12). https://doi.org/10.3390/su10124594, Go to ISI://WOS:000455338100260, type: Journal Article
24. Hauck M, Ligthart T, Schaap M, Boukris E, Brouwer D (2017) Environmental benefits of reduced electricity use exceed impacts from lead use for perovskite based tandem solar cell. Renew Energy 111:906–913
25. Hernández R, Easter S, Murphy-Mariscal M, Maestre F, Tavassoli M, Allen E, Barrows C, Belnap J, Ochoa-Hueso R, Ravi S, Allen M (2014) Environmental impacts of utility-scale solar energy. Renew Sustain Energy Rev 29:766–779
26. Hertwich EG, Gibon T, Bouman EA, Arvesen A, Suh S, Heath GA, Bergesen JD, Ramirez A, Vega MI, Shi L (2015) Integrated life-cycle assessment of electricity-supply scenarios confirms global environmental benefit of low-carbon technologies. Proc Natl Acad Sci 112(20):6277
27. Kato K, Hibino T, Komoto K, Ihara S, Yamamoto S, Fujihara H (2001) A life-cycle analysis on thin-film CdS/CdTe PV modules. PVSEC 11—Part III 67(1):279–287. https://doi.org/10.1016/S0927-0248(00)00293-2, https://www.sciencedirect.com/science/article/pii/S0927024800002932
28. Kavga, Angeliki, Tripanagnostopoulos, Georgios, Zervoudakis, George, Trypanagnostopoulos Yiannis (2018) Growth and Physiological Characteristics of Lettuce (Lactuca sativa L.) and Rocket (Eruca sativa Mill.) Plants Cultivated under Photovoltaic Panels. Notulae Botanicae Horti Agrobotanici Cluj-Napoca 46(1). https://doi.org/10.15835/nbha46110846, https://www.notulaebotanicae.ro/index.php/nbha/article/view/10846. Section: Research Articles

29. Khenkin MV, Katz EA, Abate A, Bardizza G, Berry JJ, Brabec C, Brunetti F, Bulović V, Burlingame Q, Di Carlo A, Cheacharoen R, Cheng YB, Colsmann A, Cros S, Domanski K, Dusza M, Fell CJ, Forrest SR, Galagan Y, Di Girolamo D, Grätzel M, Hagfeldt A, von Hauff E, Hoppe H, Kettle J, Köbler H, Leite MS, Liu SF, Loo YL, Luther JM, Ma CQ, Madsen M, Manceau M, Matheron M, McGehee M, Meitzner R, Nazeeruddin MK, Nogueira AF, Odabaşl C, Osherov A, Park NG, Reese MO, De Rossi F, Saliba M, Schubert US, Snaith HJ, Stranks SD, Tress W, Troshin PA, Turkovic V, Veenstra S, Visoly-Fisher I, Walsh A, Watson T, Xie H, Yildirim R, Zakeeruddin SM, Zhu K, Lira-Cantú M (2020) Consensus statement for stability assessment and reporting for perovskite photovoltaics based on ISOS procedures. Nat Energy 5(1):35–49
30. Latunussa CE, Ardente F, Blengini GA, Mancini L (2016) Life cycle assessment of an innovative recycling process for crystalline silicon photovoltaic panels. Life Cycle, Environ, Ecol Impact Anal Sol Technol 156:101–111
31. Lunardi MM, Alvarez-Gaitan JP, Bilbao JI, Corkish R (2018) Comparative life cycle assessment of end-of-life silicon solar photovoltaic modules. Appl Sci 8(8). https://doi.org/10.3390/app8081396
32. Madsen MV, Gevorgyan SA, Pacios R, Ajuria J, Etxebarria I, Kettle J, Bristow ND, Neophytou M, Choulis SA, Roman LS, Yohannes T, Cester A, Cheng P, Zhan X, Wu J, Xie Z, Tu WC, He JH, Fell CJ, Anderson K, Hermenau M, Bartesaghi D, Koster LJA, Machui F, Gonzalez-Valls I, Lira-Cantu M, Khlyabich PP, Thompson BC, Gupta R, Shanmugam K, Kulkarni GU, Galagan Y, Urbina A, Abad J, Roesch R, Hoppe H, Morvillo P, Bobeico E, Panaitescu E, Menon L, Luo Q, Wu Z, Ma C, Hambarian A, Melikyan V, Hambsch M, Burn PL, Meredith P, Rath T, Dunst S, Trimmel G, Bardizza G, Muellejans H, Goryachev AE, Misra RK, Katz EA, Takagi K, Magaino S, Saito H, Aoki D, Sommeling PM, Kroon JM, Vangerven T, Manca J, Kesters J, Maes W, Bobkova OD, Trukhanov VA, Paraschuk DY, Castro FA, Blakesley J, Tuladhar SM, Roehr JA, Nelson J, Xia J, Parlak EA, Tumay TA, Egelhaaf HJ, Tanenbaum DM, Ferguson GM, Carpenter R, Chen H, Zimmermann B, Hirsch L, Wantz G, Sun Z, Singh P, Bapat C, Offermans T, Krebs FC (2014) Worldwide outdoor round robin study of organic photovoltaic devices and modules. Solar Energy Mater Solar Cells 130:281–290
33. Marrou H, Dufour L, Wery J (2013) How does a shelter of solar panels influence water flows in a soil-crop system? Euro J Agron 50:38–51
34. McKenna E, Pless J, Darby SJ (2018) Solar photovoltaic self-consumption in the UK residential sector: new estimates from a smart grid demonstration project. Energy Policy 118:482–491
35. Raugei M, Bargigli S, Ulgiati S (2007) Life cycle assessment and energy pay-back time of advanced photovoltaic modules: CdTe and CIS compared to poly-Si. Energy 32(8):1310–1318
36. Reese MO, Gevorgyan SA, Jøorgensen M, Bundgaard E, Kurtz SR, Ginley DS, Olson DC, Lloyd MT, Morvillo P, Katz EA, Elschner A, Haillant O, Currier TR, Shrotriya V, Hermenau M, Riede M, Kirov KR, Trimmel G, Rath T, Inganäs O, Zhang F, Andersson M, Tvingstedt K, Lira-Cantú M, Laird D, McGuiness C, Gowrisanker SJ, Pannone M, Xiao M, Hauch J, Steim R, DeLongchamp DM, Rösch R, Hoppe H, Espinosa N, Urbina A, Yaman-Uzunoglu G, Bonekamp JB, Breemen AJJMv, Girotto C, Voroshazi E, Krebs FC (2011) Consensus stability testing protocols for organic photovoltaic materials and devices. Solar Energy Mater Solar Cells 95(5):1253–1267
37. Schmidt TM, Larsen-Olsen TT, Carlé JE, Angmo D, Krebs FC (2015) Upscaling of perovskite solar cells: fully ambient roll processing of flexible perovskite solar cells with printed back electrodes. Adv Energy Mater 5(15):1500569
38. Scognamiglio A (2016) Photovoltaic landscapes: design and assessment. A critical review for a new transdisciplinary design vision. Renew Sustain Energy Rev 55:629–661
39. Simsek E, Williams MJ, Pilon L (2021) Effect of dew and rain on photovoltaic solar cell performances. Solar Energy Mater Solar Cells 222:110908
40. Toledo C, Scognamiglio A (2021) Agrivoltaic Systems design and assessment: a critical review, and a descriptive model towards a sustainable landscape vision (Three-Dimensional Agrivoltaic Patterns). Sustainability 13(12). https://doi.org/10.3390/su13126871

41. Tsang MP, Sonnemann GW, Bassani DM (2016) A comparative human health, ecotoxicity, and product environmental assessment on the production of organic and silicon solar cells. Progress Photovolt: Res Appl 24(5):645–655
42. Uldrijan D, Kováčiková M, Jakimiuk A, Vaverková MD, Winkler J, (2021) Ecological effects of preferential vegetation composition developed on sites with photovoltaic power plants. Ecol Eng 168:106274
43. UNEP (2016) Green Energy Choices: The benefits, risks and trade-offs of low-carbon technologies for electricity production. Tech. rep., United Nations Environment Programme. https://wedocs.unep.org/handle/20.500.11822/7694, iSBN number: 978-92-807-3490-4
44. Urbina A (2020) The balance between efficiency, stability and environmental impacts in perovskite solar cells: a review. J Phys: Energy 2(2):022001
45. van de Ven DJ, In Capellán-Peréz, In Arto, Cazcarro I, de Castro C, Patel P, Gonzalez-Eguino M (2021) The potential land requirements and related land use change emissions of solar energy. Sci Rep 11(1):2907
46. Wade A, Stolz P, Frischknecht R, Heath G, Sinha P (2018) The Product Environmental Footprint (PEF) of photovoltaic modules-Lessons learned from the environmental footprint pilot phase on the way to a single market for green products in the European union. Progress Photovolt: Res Appl 26(8):553–564

Chapter 8
Recycling and End of Life of PV Technologies

Photovoltaic technology is an example of the application of **reduce, reuse and recycle** strategies to increase the sustainability and diminish the environmental burden of the final product. The "3R" approach is still in its initial stage for PV systems, but the technology is ready to be applied and has progressed very fast in the past few years. "Reducing" has been already achieved with regards to the reduced amount of material that is required for the manufacture of modules with improved power conversion efficiency; the mass of material required per unit of peak power (kg/kW_p) in the manufacture of modules, with special focus on silicon use in crystalline silicon technology and critical raw materials in thin film technologies have been analyzed in detail in Chap. 5. In this chapter, both "reusing" and "recycling" strategies will be presented and discussed. After a few methodological considerations about recommendations for LCA methodology with regard to recycling, the following sections will be devoted, first to describe reusing strategies, then to present recycling state-of-the-art technology and prospects and finally to discuss end-of-life strategies and how the production lines for PV modules could be modified to take into account requirements for higher recyclability of the modules.

The Life Cycle Assessment of the end-of-life phase of a PV system should include at least the following aspects [11]:

- Deconstruction, dismantling;
- Transports;
- Waste processing;
- Recycling and reuse;
- Disposal.

All of these LCA studies are required in any of the "3R" strategies. Reuse is strongly affected by deconstruction, dismantling and transport of used modules (it will require further activities to be evaluated like new tests for relabelling and the logistics of secondary markets); recycling require, on top of the previous ones, a

A. Urbina, *Sustainable Solar Electricity*, Green Energy and Technology,
https://doi.org/10.1007/978-3-030-91771-5_8

detailed study of all the physical and chemical processes involved in the recycling stage. If neither reuse or recycle is applied (or only partially applied), then the module will be disposed off, and landfilling or other disposal strategies should be used. Ideally, this final stage should be avoided if a 100% reuse and/or recycling of modules is achieved in the future.

Methodologically, care should be put in LCA studies considering reused and recycled material to avoid double accountability of inputs in the process flows. The recycled material could originate from any industrial activity which recovers and recycles material or the PV industry itself, which are increasingly recovering and recycling components and materials from modules. In all cases, recommendations from ISO14040 standard in clause 4.3.4.3 should be applied for the allocation procedures for recycling [13, 14]. In particular, the following points should be taken into account: (i) the possible change in the properties of recycled or reused materials compared to the original ones, (ii) the required recovery processes between the original and the subsequent product systems and (iii) the possibility that the inputs and outputs associated with unit processes for extraction and processing of raw materials and final disposal of products are shared by more than one product. All these points may lead to a redefinition of the system boundary and the product systems or the allocation procedures are often classified as open or closed loops according to the criteria summarized in Fig. 8.1. Combinations of open and close allocation and product systems can be considered in a LCA study but the choices must be explained and assumptions clearly stated in the final report.

When a PV module is fabricated using some amount of recycled material, it is recommended to perform several analyses on material recycling using the recycled content (cut-off) allocation approach as default and the end-of-life (avoided burden) recycling approach in a sensitivity analysis. If the analysis is focussed on the recycled material embedded in the PV module (or substance required for processing of the module), it serves to the strong sustainability concept since it contributes to reduce the amount of primary materials use and increase the recycled content. On the contrary, an end-of-life recycling approach results in a higher eco-efficiency of primary raw materials production compared to secondary production, and the loss

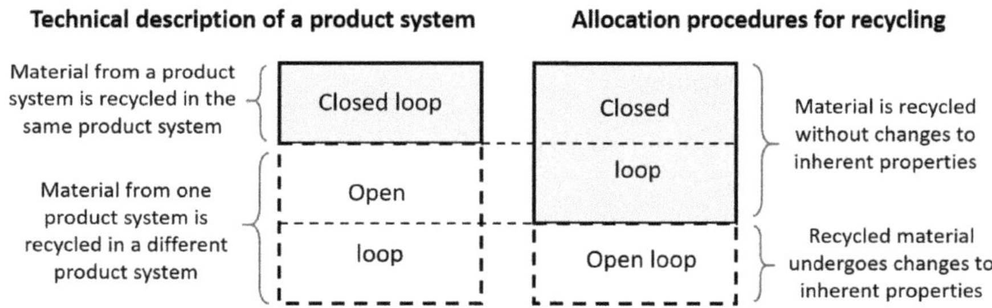

Fig. 8.1 Distinction between a technical description of a product system and allocation procedures for recycling according to standards [13, 14]

in natural capital is compensated by human-made capital, it, therefore, serves better to the weak sustainability concept [10, 26]. For a consequential LCA study, where impacts are projected to the future and depend on choices and scenarios, allocation of recycling and multi-output processes are based on system boundary expansion to reduce consequential uncertainty [7]; in the case of PV electricity production, this uncertainty is inherently introduced when the future displaced energy production of alternative energy sources and their "avoided" impacts are calculated. A detailed definition of future scenarios is required and a sensitivity analysis should be carried out.

The LCI and LCA reports with recycled material must be very clear on the applied recycling approach, since different stakeholders may opt for one or another depending on their interests and they will benefit from a sensitivity analysis of the results before making the final decisions about investment in factories (industrialist) or regulatory frameworks (policymakers).

When sustainability is considered beyond standardized LCA methods, the importance of recycling materials and reusing or recycling PV modules is enhanced. Vasilis Fthenakis proposed a sustainability metrics where environmental impacts, system costs and availability of resources appear linked to each other in a prospective LCA approach and lead to the important conclusion that the concerns related to critical materials availability (especially indium and tellurium) can be reduced with enhanced recovery of materials during module production and recycling of spent modules, together with the conventional approaches of PCE increment and layer thickness reduction, but with the advantage that recycling contributes to the three dimensions proposed for the sustainability metrics [12].

8.1 Reusing PV Modules

Many of the modules that are dismantled in PV systems after its T_{80} lifetime have been reached (between 25 and 30 years) are still functional modules delivering power and its operational lifetime may be extended in many cases another 10–20 years. The dismantling of useful PV modules is increasing due to revamping and re-powering of large PV plants, where some underperforming modules are replaced by new modules after a few years of operation (with or without rearrangement of strings), the dismantling of PV modules and associated BoS components (cables, regulators, inverters) require skilled workforce. Also, modules which failures can be repaired; in many cases the failure is just a problem of cabling or junction box which can be repaired in an elementary workshop, in other cases, failures of sealing or framing require more skilled workers but can still be easily repaired, finally, reparation of damaged cells will require specialized facilities. In all cases, secondary markets for modules and for components are created.

In recent years, an important stockpile of PV modules that can have a second life without requiring a recycling process is growing. Only a good logistic approach and the regulation of a secondary market is required, standards for reused modules should be kept equal to standards for new ones; a new power rating (ideally in standard test conditions) should be carried out and a relabelling of peak values and new warranties must be included in the reused module as a proof of its real functionality. There is still no regulation in this regard, and an international coordination should be carried out both for the relabelling requirements and for the secondary market regulations. In Europe, directives for waste hierarchy are applied (within the framework of waste from electrical and electronic equipment directives), but the regulation is lost when PV modules are exported for reuse.

The market for reuse of PV modules is strongly driven by logistic costs: efficient collection and transport networks will be the main contributors to added cost, storage and export duties can have a relative impact on final costs; in all cases assuming that the price of dismantled modules will be very low. Often, the secondary market is linked to rural electrification projects in developing countries, Non-Governmental Organizations (NGOs) for development promote energy projects for rural livelihoods that may benefit of this secondary market, but also a wide network of local small companies with technical capacity to design and build small PV systems with reused modules can be an economic drive for local development with economic support from local banks or institutional loans and, therefore, a growing cross-border secondary market for PV modules is expected.

8.2 Recycling PV Modules: Recovery of Components and Materials

The International Renewable Energy Agency published an estimation of the amount of waste projected from now to 2050 considering a lifetime for modules of 30 years and it emphasizes that recycling or reusing of PV modules can provide a stock of up to 78 million tonnes of raw materials and other valuable components (of which, China accounts for 20 million tonnes, USA 10 million, India and Japan 7.5 and Germany 4.5 million as the main contributors); this "waste", may provide an economic value of the recovered material that could exceed 50 billion USD [29].

The amount of waste arising from PV systems at its end of life has been estimated by different researchers and institutions. Based on IRENA's REmap2030 scenario and country data, Karsten Wambach and Knut Sander presented a model assuming a constant annual addition of PV capacities from 2014 to 2029 in a country multiplied by an annual growth factor of 1.083 and a reduction in the mass required per installed capacity due to technological improvement and module (conversion factor of 79 kg/kW installed was used), the modules were to be dismantled after 40 years and partial earlier replacement was accounted for in the maximum waste scenario; they found the total expected waste per country that is presented in Table 8.1 [27].

The values that they calculated for 2020 have demonstrated a good accuracy for the amount of waste really produced in 2020 and are similar in its 2030 projection to the values provided by IRENA-IEA PVPS in its 2018 report: cumulative waste

Table 8.1 Cumulative PV module waste (in kilotonnes) in two scenarios (minimum and maximum) as described in [27]

Cumulative kt	2020		2030	
	Min	Max	Min	Max
Europe	30	417	744	2350
Asia	23	225	465	2690
Africa	1	5.6	1.4	130
North America	11	79	154	739
Latin America	0.3	2.8	6.4	52
Oceania	1.8	14.6	27	115
Middle East	1.3	10	24	219
ROW	4	63	125	815
Total:	72.4	817	1546.8	7110

ranging between a minimum of 1.7 million tonnes and 8 million tonnes in 2030 and between 60 and 78 million tonnes in 2050 for regular loss and early loss scenarios commented above [17, 18, 29].

PV modules' end-of-life treatment has been included in the European directive on waste electrical and electronic equipment (WEEE) [9], requiring 80/85% of recovery/recycling rate for PV modules (from 2018 onwards) and making module installers accountable for their electronic waste and requiring solar producers to recycle. Discussions are ongoing with the purpose to approve at the end of 2021 a new ecolabel regulation which will require that PV modules shall be designed to allow for easy disassembly for recycling by a specialist firm using ordinary tools [6, 8]. In Japan, there is no specific regulations for end-of-life PV panels, which, therefore, must be treated under the general regulatory framework for waste management (the Waste Management and Public Cleansing Act), but project developers and owners have to contribute to a decommissioning fund and are liable for waste treatment. There is currently no regulatory framework for PV recycling in the United States at federal level, only Washington and North Carolina states have some regulatory framework, California, where most PV is currently installed consider PV modules as "universal waste" and Arizona, Florida and Texas as "common waste"; this is in stark contrast with the initiative by the NSF/ANSI to propose the 457 Sustainability Leadership Standard for Photovoltaic Modules [21]. China currently has no specific regulations for end-of-life PV modules, although the 13th Five Year Plan (FYP) for 2016–2020 already pointed to create regulations and accelerate the management of PV modules end of life, the recently approved 14th FYP (2021–2025) concedes a leading role to PV technology and high capacity (>100 MW/year) recycling demonstration lines and on-site recycling for the large Chinese plants is planned. Nevertheless, a very large regulatory work is still required worldwide, and quotas for recycling should be imposed and fomented with incentives.

Dismantling of PV modules in a PV system is a task that can be carried out easily requiring elementary technical skills in small systems, but some additional expertise is required in large PV plants were dismantling modules may involve working with high voltage strings. Additionally, the logistics of module transport to storage points and then to recycling plants require regulations about the management of electronic waste which are often local regulations or at most country regulations. The role of local authorities and waste management service providers is critical for a proper recovery of PV waste and more international coordination of regulations is strongly required. In many cases, the PV waste to be recovered is dispersed in small to medium size installations, either in urban areas (roof-top, BIPV systems) or for rural electrification; this fact requires end users' and stakeholders' involvement in the process of recovery, the owners of the systems or the technicians that built the system should collaborate in an extended liaison to guarantee this recovery service. In the case of large PV plants, the tasks are highly concentrated and the owner of the plant (usually a large company) is responsible for proper PV module recovery and transport at end of life.

Once the modules have been dismounted and transported to the recycling site, the recovery process of materials is accomplished. Until very recently, the principal sites for crystalline silicon PV module recycling were glass recovery plants were recycling of the laminated glass used in the modules was carried out by low-cost processing, mainly involving mechanical (automated but also manual in many cases) dismantling of frame and recovery of the glass which were subsequently ground. In terms of mass, since glass accounts for more than 75% and frame around 7–10% of total panel weight, it is an efficient recycling process. But the recovery of other materials like the PV cells or the metals of electrodes is not carried out by these elementary processes. The early stages of recycling of thin film technology modules have been developed by a first solar plant in Germany for CdTe modules using a combination of chemical and mechanical processes that achieved a recovery rate of 90% for glass and nearly 95% for semiconductor materials [17].

Several projects involving companies and research centres have demonstrated pilot plants for a more advanced recycling industrial processes for all PV technologies. The IEA-PVPS Task 12 workgroup published recently some important reports: first, a survey on "Life Cycle Inventory of Current Photovoltaic Module Recycling Processes in Europe" carried out in operational recycling pilot plants in 2017 by contacting 16 recyclers throughout Europe, although only 5 provided life cycle inventory data [28]; then two reports in 2018, devoted to "End-of-Life Management of Photovoltaic Panels: Trends in PV Module Recycling Technologies" [17] and "Life Cycle Assessment of Current Photovoltaic Module Recycling" [26], which analysed the processes for recycling crystalline silicon (c-Si) and thin film modules and its environmental impacts.

The general flow of the end of life for any PV technology can be described in three main steps carried out in three different locations: (i) the initial dismounting of failing or damaged PV modules or decommissioning of underperforming modules (in revamping operations or dismantling of the system) at the PV plant site, (ii) recyclers or intermediate processors disassemble frames, junction boxes, glass, in a process that separate valuable materials (glass, plastics, metals, compound semiconductors) and final disposal (landfilling or burning) of non-recoverable materials and (iii) in a specialized material processing plant, some of the outputs of the recycler's site are further processed for the purification or refining of materials, which can be sold in the market, additional waste for final disposal (landfilling) can be generated at this stage.

An initial life cycle inventory (LCI) compilation for PV recycling routes was carried out using as functional unit the "processing of one metric tonne of crystalline silicon PV modules" and as input data all the information collected with the survey on five of the existing pilot plants on a batch to batch scale, but extrapolating the energy consumption to an annual scaled-up average, with plant capacity estimated around 200,000 tonnes per year [28]. The bulk materials: glass, aluminium, copper were recovered by using a combination of mechanical crushing, shredding, milling, grinding and ulterior sorting and separation methods [9]. The recycled cullet feedstock glass can be used for foam or fiberglass production and the obtained metals can be further purified in downstream processes in smelters or metal recycling industrial sites. Other foils, including residues of metals and cells were incinerated or landfilled. Incineration can be carried out in the recycling plant or sold to a waste-for-energy standard plant if halogenated content of plastics is less than 1% measured as chlorine; if it is higher, it can only be incinerated in a specialized hazardous-waste facility according to European regulations. Only in one plant, the diluted ash resulting from in-situ incineration was further processed for recovery of silicon using a leaching process and for recovery of metals by ulterior electrolysis of the leachate. Regarding electricity consumption, all plants ranged in the order of 50–100 kWh per tonne of processed module for the mechanical process and an extra 494 kWh/tonne for the metal processing. The yield of the recycled glass varied between 59% and 75% and for non ferrous metals between 13.5% and 21.8%.

An important conclusion of the report was the recommendation to improve the processing of the "polymer fraction" (the foil) which is the mixture of different plastic compounds (ethylene-vinyl-acetate, EVA; polyethylene-terephthalate, PET; polyester and Tedlar®) resulting from the mechanical processes, mixed with silicon and metal components. Efforts to separate particles using thermo-mechanical processing can be carried out at the cost of increased energy inputs. Recovery metals, and especially valuable silver, together with optimization of the industrial processes to reduce energy consumption and increase yields could make the PV recycling cost competitive. If similar methods are applied to thin film (CIGS, CdTe, III-V) technologies, valuable and scarce materials could be recovered, representing a small fraction in weight but a large fraction in price of recycled products.

In Fig. 8.2, a schematic representation of all possible steps involved in c-Si module recycling is presented. The processes can be classified into those that eliminate the encapsulant (mostly EVA) from the laminated structure of the module (using thermal, chemical or mechanical methods), a difficult step in the recycling process after disassembly of frame and glass; and those that recover the metals from the Si cells, involving additional chemical steps to complete the cycle of a full recycling of the module, and sometimes involving the use of metallurgical techniques in a metal

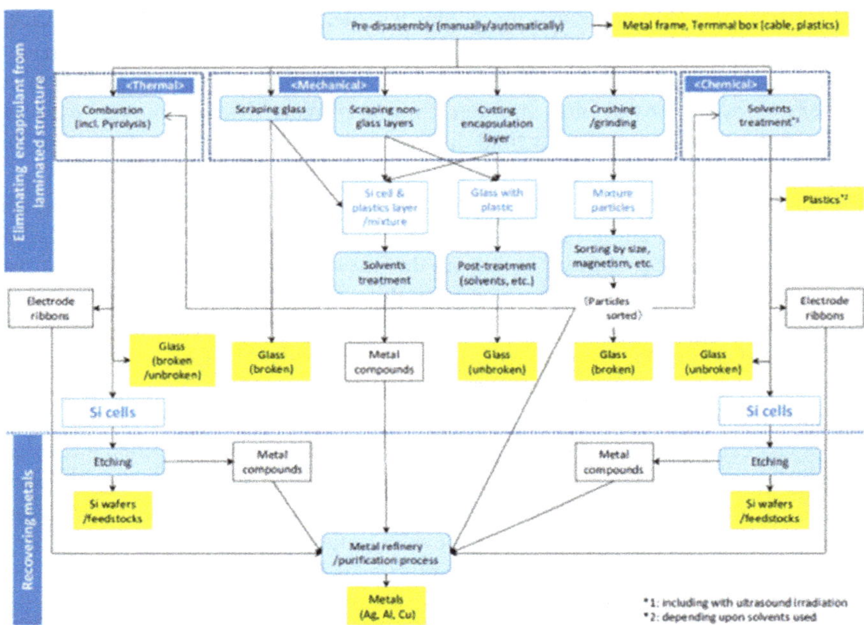

Fig. 8.2 Possible processes for crystalline Si PV module recycling, from Report IEA-PVPS T12-10:2018 (Reproduced with permission from [17])

refining industry, always struggling to reduce costs of the recovered materials in comparison to the purified material obtained from raw mineral mining. The chemical processes, such as the use of solvent treatments to eliminate the encapsulant from the laminated structures will enable the recovery of Si cells and other semiconductors and metals, but it requires liquid waste treatment facilities for halogenated solvents, strong acids (such as hydrofluoric, nitric or sulphuric) and alkali hydroxides at a very large scale to become competitive [4]. Recovery rates achieved in research laboratories where Si wafers were recovered and treated are very high 80, 79 and 90% for Si, Cu and Ag, respectively (and 93% removal rate for Pb), which are very promising results, but this method is still not applied at industrial scale, where wafer recovery has never been successful [15].

Similarly, the processes for recycling thin film modules are classified in those that eliminate the encapsulant from the laminated structures and those that recover the metals and substrate glass from the modules as indicated in Fig. 8.3. In this case, the recovery rate for metals and semiconductors is higher, and may reach up to 95%. For CdTe technology, hazardous waste of cadmium and cadmium hydroxide is produced during the use of strong acids (mostly sulphuric) for stripe-off the metals from the glass; for CIGS, small amounts of gallium, selenium, selenium hydroxide can be released and are not usually recycled, only recovery of indium is applied in this case. The main difference between recycling methods for c-Si and thin film technologies

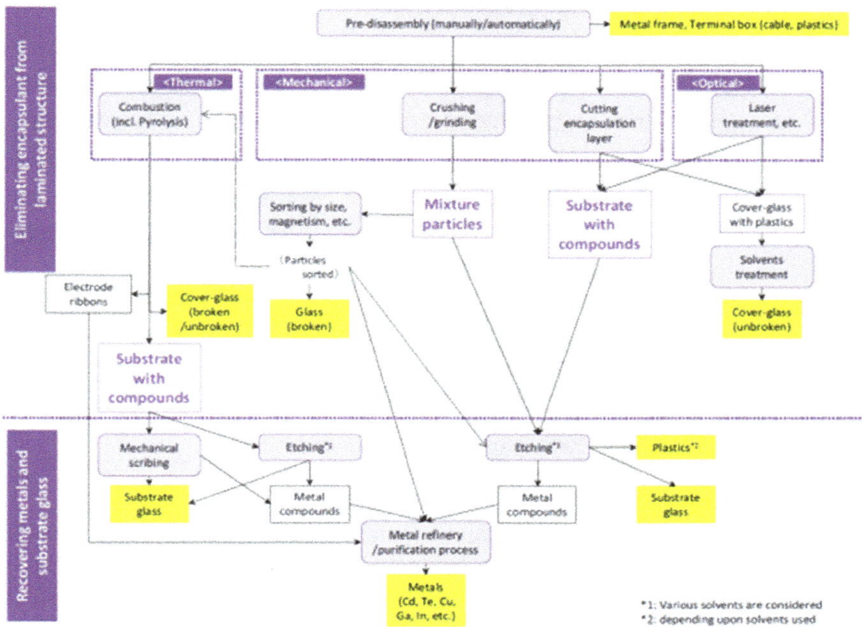

Fig. 8.3 Possible processes for thin film module recycling, from Report IEA-PVPS T12-10:2018 (Reproduced with permission from [17])

is the objective of the elimination of the encapsulant material: for c-Si it focuses on the separation of front glass, and then recovery of cells and metals is left for another stage, while in thin film, the objective is to recover both front glass and substrate glass which contains the semiconducting material layers. In both cases, the final recycled materials are grouped in glass (broken, ground or unbroken), plastics, metals from frame (almost always aluminium) and elements from cells, especially metals, and semiconductors in thin films, since Si cells or material is still not recycled at end of life (although Si is often recovered in the initial module manufacture steps of crystalline or multicrystalline ingots or bricks sawing for manufacture of wafers).

The IEA-PVPS Task12 workgroup carried out a Life Cycle Assessment on recycling processes for c-Si and CdTe technologies with the inventories collected from previous studies. The functional unit was the recycling of 1 kg of used framed c-Si and unframed CdTe PV modules at the place of installation and the applied methodology for impact assessment indicators was ILCD-Midpoint (2011). The study included the impacts of dismantled modules transport to the recycling site. The results, shown in Fig. 8.4, emphasize the importance of recovered materials: for c-Si modules, the potential benefits due to recovered copper have the highest impact in the indicators mineral, fossil and renewable resource depletion and human toxicity non-cancer effects, while the avoided burdens of aluminium recovery have a high contribution to cancer effects in humans, which is mainly due to chromium (VI) emissions to

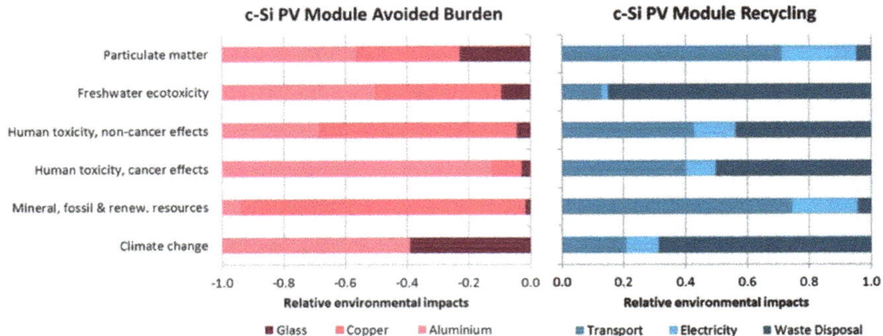

Fig. 8.4 Relative contributions of recovered materials to the potential benefits (left) and relative contributions of the recycling processes to the environmental burdens (right) of first generation c-Si PV module recycling based on data from four European recyclers and presented in 2018 by the IEA-PVPS Task12 workgroup (Reproduced with permission from [26])

water in the production of primary aluminium. Waste disposal is also responsible for the major part of the freshwater ecotoxicity, human toxicity and climate change impacts. Transportation and electricity supply (for the recycling process) have significant contributions in climate change and human toxicity (by consumption of fossil fuels) [26].

For CdTe recycling, the LCA carried out with inventory data from the First Solar plant in Germany, delivered the results shown in Fig. 8.5. The largest benefits are obtained (as may be expected) in the mineral, fossil and renewable resource depletion impact categories, which are 750 times higher than the impacts caused by the recycling of CdTe PV modules. In other indicators, the beneficial balance is lower and even negative for the case of human toxicity cancer effects, where the use of hydrogen peroxide in the recycling process of metals and semiconducting materials strongly increases health risks. On the other hand, glass recovery dominates the avoided burdens in climate change, particulate matter, human toxicity cancer effects and freshwater ecotoxicity impact categories. Again, transport and electricity supply has a relatively high impact in particulate matter, human toxicity non-cancer effects and mineral, fossil and renewable resource depletion. In this case, waste disposal impacts are much lower due to the lower weight of material for landfilling or burning.

Keiichi Komoto et al. projected the above presented methods into a recycling industrial line and calculated its environmental impacts using LCA methodology and LIME2 category indicators. In particular, they assumed a thermal approach that could by applied to crystalline and amorphous silicon and CIGS thin film technology with the following steps: aluminium frame dismantling, back sheet removal by a milling process and disposed as industrial waste, then the EVA encapsulant is thermally decomposed in a muffle furnace at temperature reaching 500 °C in several steps (the decomposition gas is sucked and burned out) and the heat generated from the combustion of the EVA resin is thermally recycled to the furnace. The recovered

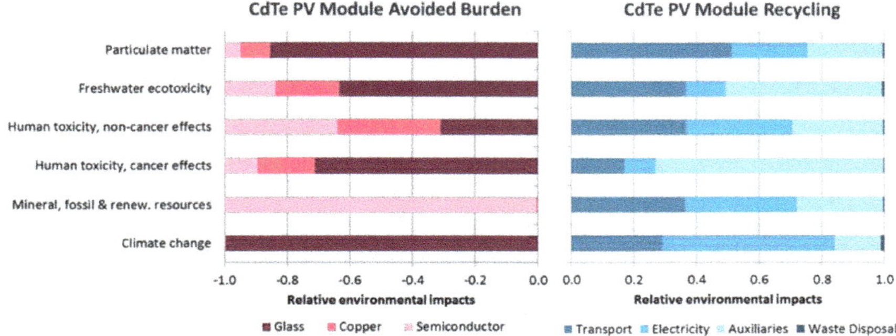

Fig. 8.5 Relative contributions of recovered materials to the potential benefits (left) and relative contributions of the recycling processes to the environmental burdens (right) of first generation CdTe PV module recycling based on data for the First Solar recycling facility in Germany and presented in 2018 by the IEA-PVPS Task12 workgroup (Reproduced with permission from [26])

glass is recycled to float glass and metals are recovered from the cells, silver in the case of Si and the CIGS metals are recovered by a cyclone collector and further processed in a metal refining industry. The expected secondary material recovery rates are about 77% of crystalline Si module, 96% of thin film Si module and 85% for CIGS modules. The conclusion of the LCA study is a net environmental benefit in four analysed indicators (Global warming potential; Acidification potential; Human toxicity potential (HTP); and Abiotic resource depletion potential), while on the other hand, the avoided environmental burdens by recovered materials don't outweigh photochemical ozone creation potential caused by the recycling efforts, and impact arising from the transport of modules to the recycling site, thus emphasizing the importance of logistics in the recycling strategy [17].

Other recycling methods have been demonstrated experimentally at laboratory scale, although not implemented at industrial level for thin film technologies; for both CdTe and CIGS, it has been proved that the environmental impacts associated with wet mechanical methods that avoid the use of chemicals are lower and have similar efficiency in recovery of indium and tellurium than conventional methods [1]; another group has shown innovative methods based on crushing, sulfuric acid leaching, precipitation and filtration, as common processes for thin film module recycling and then extraction with surfactant (in toluene solution) and electrodeposition for CIGS or decantation and electrowinning for CdTe; it was found that the process for CdTe has lower impacts compared to CIGS process, mainly due to the enviromental "credits" that the recovery of materials provided to the LCA study [23].

The recovery of glass is still a dominant part of the recycling process of PV modules of any technology. Regulations by the European Committee for Electrotechnical Standardization (CENELEC) stablish purity requirements that must be accomplished by the recycled glass, where the content of hazardous substances in output glass fractions shall not exceed the following defined limit values (from standards EN50625-1 and EN50625-2-4:

- 1 mg/kg (dry matter) cadmium (Si-based PV)/10 mg/kg (dry matter) cadmium (non-Si-based PV)
- 1 mg/kg (dry matter) selenium (Si-based PV)/10 mg/kg (dry matter) selenium (non-Si-based PV)
- 100 mg/kg (dry matter) lead

Despite the importance of future PV recycling needs, in terms of the required annual weight processing capacity once the PV plants now in operation reach its end of life, still few facilities are operating worldwide, mostly in Europe (due to strict regulatory framework for PV waste). In contrast, a very large research and development effort is being carried out by several research centres and universities, often in collaboration with companies in order to improve recycling methods and to reduce its energy consumption and environmental impacts. The IEA-PVPS carried out an analysis of R&D activities related to recycling PV modules and found that the numbers of effective patents directly related to PV recycling technology are 128 for c-Si and 44 for thin film modules (survey from 1976 to 2016), with the first patents found in mid 90s), very few until 2011 (mostly in Europe) and an increasing number since then (mostly in China, Korea and Japan). Regarding c-Si technology, patents targeting the removal of encapsulant (mainly EVA) account for 45% of total, thus emphasizing that it is the most challenging step; it is followed by frames (30%), solar cells (24%), and Cu ribbons (1%) recovery strategies, classifying by methods, 40% of patents are filled for mechanical processes, 25% combined methods, 19% chemical and 15% thermal [17]. For thin film technology, the number of patents is still dominated by the USA, followed by Asia, with a stronger focus on semiconductor recovery and a domination of combined methods (64%) indicating the higher technological complexity. This trend indicates that the amount of patents is still reduced, that the geographical location of filing is moving from Europe to Asia and that the knowledge progress is still dominated by mechanical, relatively low-tech processes focussed on the most challenging process, which is the separation of the encapsulant from the cells.

But so far, silicon cells cannot be recovered as Si wafers at an industrial scale although some laboratory scale success has been achieved [20]. The recycled glass is a low grade product and the recovery of scarce or expensive metals are still minimal with very low yields. Higher purity levels for the recovery of glass and silicon are required in order to approach quality and value of the original material. R&D has progressed steadily in the past ten years, yet commercialization of recycling procedures, especially recovery of semiconductors and metals is still at an early stage. There are plenty of room for improvement in this research field, especially moving from proof-of-concept already demonstrated experimentally to a large scale industrial facilities devoted to recycling of PV modules, which is strongly required to close the loop for a really circular economy for photovoltaics.

Several demonstration projects have been carried out with the purpose to recover crystalline silicon cells without the need to broke them. The approaches are based on incineration of EVA encapsulant materials, but in most cases, the solar cell is accidentally broken during the process. A research into the mechanisms that lead to this unwanted destruction of cells pointed to the combined effect of the formation of bubbles between EVA and the glass and the deformation of EVA by thermal stress putting too much pressure on the cell; a simple pretreatment of glass cracking and EVA patterning allowed researchers to recover unbroken cells which underwent acid treatment to obtain second generation silicon wafers that then were used as input in a standard module fabrication process delivering modules with same efficiency as first generation modules (18.5% compared to original 18.7%). It is a remarkable result which opened the door to a new path for PV module recycling [20].

The LCA study of a complete recovery and recycling process combining several methods (mechanical, thermal and chemical with acid treatment and electrolysis) was carried out by Cynthia Latunussa et al; they were careful to exclude environmental benefits arising from the use of secondary materials, and therefore, the results are a proper LCA evaluation of the recycling process of 1 tonne of crystalline silicon waste modules; the conclusions indicate that the major contributor to impacts is the incineration of the encapsulation layers illustrated as an example in the global warming potential impact category and being the major contributor in Human Toxicity (cancer, 50%) and Freshwater Ecotoxicity (75%), but similar in others; recovery of silicon, silver, copper and aluminium comes second in impacts and transport is also an important contributor in all categories, varying between 10% for freshwater eutrophization to 80% in the abiotic depletion potential for minerals, and therefore, a clear recommendation arises: it would be convenient to explore decentralized schemes for PV recycling industrial sites of smaller capacity, but strategically distributed; also fluorinated plastics in the module components are major contributors to impacts and eco-design approaches should point to reduction of the use of this materials and simplify the disassembly process with clever designs for aluminium frames; a final recommendation of the study is to encourage manufacturers to use recycled glass for the production of new modules [19]. The potential reduction of economic cost of using high purity recycled silicon or intact silicon wafers in a circular manufacturing process can be as high as 20% and provide long term environmental and economic benefits [5].

Futhermore, recent studies indicate that recovery of key minerals (copper, palladium, gold and silver) from printed circuit boards could require as little as 5% of the energy as compared to primary supply from mining [24, 25]. This is a promising result that can be extended to PV module recycling (and BoS electronic components) to emphasize the large potential to reduce environmental impacts by recycling metallic components.

Recycling at end of life of perovskite technology has been demonstrated recently, including examples of recovery of materials at end of life [2, 16]; lead was recovered up to 99.8% by using deep eutectic solvents in recovered cells [22]. Furthermore, in another example, Chen et al fabricated perovskite solar cells with lead recycled from lead-acid batteries achieving the same efficiency as the control cells [3]. The impacts of the processing routes used for the recycling have not been studied so far and should be included in any future global assessment including second generation cells or use of recovered or recycled materials.

8.3 Recovery and Reuse of Substances Required for PV Module Manufacture

Several materials and substances are required during the manufacturing of solar cells, these materials are not embedded in the final cells, but they could be considered as a limiting factor for PV module production and also contribute to environmental impacts that are evaluated by the LCA methodology. If part of the substances used during the manufacture are recovered and reused in the factory or recovered and sent to another production centre for further processing, the overall LCA impacts can be reduced. This approach is already implemented in the two standard recycling routes used for crystalline silicon and thin film technologies, as described in the previous section) [17, 29].

It is more complex to recover and reuse the solvents used in organic and hybrid technology manufacture. Either chlorinated or nor chlorinated solvents pose important risks to human health and environment, and therefore, reducing its consumption is paramount to reduce impacts of the overall production process. Regulations in many countries require avoiding any spill of substances out of the factory, and the storage of used solvents needs to comply with strict rules (see Chap. 11). Then, another cycle for recovery of used solvents is initiated, but in this case it is usually left out of the scope of the LCA studies of PV cell and module manufacture.

References

1. Berger W, Simon FG, Weimann K, Alsema EA (2010) A novel approach for the recycling of thin film photovoltaic modules. Resour Conserv Recycl 54(10):711–718. https://doi.org/10.1016/j.resconrec.2009.12.001, https://www.sciencedirect.com/science/article/pii/S0921344909002808
2. Binek A, Petrus ML, Huber N, Bristow H, Hu Y, Bein T, Docampo P (2016) Recycling Perovskite Solar Cells To Avoid Lead Waste. ACS Appl Mater Interfaces 8(20):12881–12886. https://doi.org/10.1021/acsami.6b03767
3. Chen PY, Qi J, Klug MT, Dang X, Hammond PT, Belcher AM (2014) Environmentally responsible fabrication of efficient perovskite solar cells from recycled car batteries. Energy Environ Sci 7(11):3659–3665. https://doi.org/10.1039/C4EE00965G
4. Deng R, Chang NL, Ouyang Z, Chong CM (2019) A techno-economic review of silicon photovoltaic module recycling. Renew Sustain Energy Rev 109:532–550. https://doi.org/10.1016/j.rser.2019.04.020, https://www.sciencedirect.com/science/article/pii/S1364032119302321
5. Deng R, Chang N, Lunardi MM, Dias P, Bilbao J, Ji J, Chong CM (2020) Remanufacturing end-of-life silicon photovoltaics: feasibility and viability analysis. Progress Photovolt: Res Appl n/a(n/a). https://doi.org/10.1002/pip.3376. Publisher: John Wiley & Sons, Ltd
6. Dodd N, Espinosa N (2021) Solar photovoltaic modules, inverters and systems: options and feasibility of EU Ecolabel and Green Public Procurement criteria. Tech. Rep. KJ-NA-30474-EN-N, EU-JRC, European Commission-Joint Research Centre. https://doi.org/10.2760/29743, iSBN 978-92-76-26819-2 ISSN 1831-9424
7. Ekvall T, Weidema BP (2004) System boundaries and input data in consequential life cycle inventory analysis. Int J Life Cycle Assess 9(3):161–171. https://doi.org/10.1007/BF02994190
8. European Commission-Joint Research Centre (2021) Discussion paper on potential Ecodesign requirements and Energy Labelling scheme(s) for photovoltaic modules,

inverters and systems. Tech. rep., European EU-JRC, European Commission-Joint Research Centre. https://susproc.jrc.ec.europa.eu/product-bureau/sites/default/files/2021-04/Discussion%20paper%20Ecodesign%20Photovoltaic%20Products.pdf
9. European Parlament and European Council (2012) Directive 2012/19/EU on waste electrical and electronic equipment (WEEE). https://eur-lex.europa.eu/legal-content/EN/TXT/?uri=CELEX:02012L0019-20180704
10. Frischknecht R (2010) LCI modelling approaches applied on recycling of materials in view of environmental sustainability, risk perception and eco-efficiency. Int J Life Cycle Assess 15(7):666–671. https://doi.org/10.1007/s11367-010-0201-6
11. Frischknecht R, Stolz P, Heath G, Raugei M, Sinha P, de Wild-Scholten MJ (2020) Methodology Guidelines on Life Cycle Assessment of Photovoltaic. Tech. rep., International Energy Agency, PVPS Task 12: PV Sustainability, iSBN: 978-3-906042-99-2
12. Fthenakis V (2012) Sustainability metrics for extending thin-film photovoltaics to terawatt levels. MRS Bull 37(4):425–430. https://doi.org/10.1557/mrs.2012.50, https://www.cambridge.org. Edition: 2012/04/09 Publisher: Cambridge University Press
13. International Organization for Standardization (2006a) ISO 14040:2006 Environmental management - Life cycle assessment - Principles and framework. https://www.iso.org/standard/37456.html, technical Committee : ISO/TC 207/SC 5 Life cycle assessment ICS : 13.020.10 Environmental management 13.020.60 Product life-cycles
14. International Organization for Standardization (2006b) ISO 14044:2006 Environmental management - Life cycle assessment - Requirements and guidelines. https://www.iso.org/standard/38498.html, technical Committee : ISO/TC 207/SC 5 Life cycle assessment ICS : 13.020.10 Environmental management 13.020.60 Product life-cycles
15. Jung B, Park J, Seo D, Park N (2016) Sustainable system for raw-metal recovery from crystalline silicon solar panels: from noble-metal extraction to lead removal. ACS Sustain Chem Eng 4(8):4079–4083. https://doi.org/10.1021/acssuschemeng.6b00894. Publisher: American Chemical Society
16. Kadro JM, Pellet N, Giordano F, Ulianov A, Müntener O, Maier J, Grätzel M, Hagfeldt A (2016) Proof-of-concept for facile perovskite solar cell recycling. Energy Environ Sci 9(10):3172–3179. https://doi.org/10.1039/C6EE02013E
17. Komoto K, Lee JS, Zhang J, Ravikumar D, Sinha P, Wade A, Heath G (2018a) End-of-Life Management of Photovoltaic Panels: Trends in PV Module Recycling Technologies. Tech. Rep. Report IEA-PVPS T12-10:2018, IEA PVPS Task12, Subtask 1, Recycling, iSBN 978-3-906042-61-9
18. Komoto K, Oyama S, Sato T, Uchida H (2018b) Recycling of PV modules and its environmental impacts. In: 2018 IEEE 7th World Conference on Photovoltaic Energy Conversion (WCPEC) (A Joint Conference of 45th IEEE PVSC, 28th PVSEC & 34th EU PVSEC), pp 2590–2593. https://doi.org/10.1109/PVSC.2018.8547691, journal Abbreviation: 2018 IEEE 7th World Conference on Photovoltaic Energy Conversion (WCPEC) (A Joint Conference of 45th IEEE PVSC, 28th PVSEC & 34th EU PVSEC)
19. Latunussa CE, Ardente F, Blengini GA, Mancini L (2016) Life Cycle Assessment of an innovative recycling process for crystalline silicon photovoltaic panels. Life Cycle, Environ, Ecol Impact Anal Sol Technol 156:101–111. https://doi.org/10.1016/j.solmat.2016.03.020, https://www.sciencedirect.com/science/article/pii/S0927024816001227
20. Lee JK, Lee JS, Ahn YS, Kang GH, Song HE, Kang MG, Kim YH, Cho CH (2018) Simple pretreatment processes for successful reclamation and remanufacturing of crystalline silicon solar cells. Progress Photovolt: Res Appl 26(3):179–187. https://doi.org/10.1002/pip.2963. Publisher: John Wiley & Sons Ltd
21. NSF (2017) NSF 457 Sustainability Leadership Standard for Photovoltaic Modules. Tech. Rep. 457i1r1, NSF International. https://standards.nsf.org/apps/group_public/download.php/36153/JC%20Memo%20and%20Ballot%20457i1r1.pdf
22. Poll CG, Nelson GW, Pickup DM, Chadwick AV, Riley DJ, Payne DJ (2016) Electrochemical recycling of lead from hybrid organic-inorganic perovskites using deep eutectic solvents. Green Chem 18(10):2946–2955. https://doi.org/10.1039/C5GC02734A

23. Rocchetti L, Beolchini F (2015) Recovery of valuable materials from end-of-life thin-film photovoltaic panels: environmental impact assessment of different management options. J Clean Prod 89:59–64. https://doi.org/10.1016/j.jclepro.2014.11.009, https://www.sciencedirect.com/science/article/pii/S0959652614011809
24. Seabra D, Caldeira-Pires A (2020) Destruction mitigation of thermodynamic rarity by metal recycling. Ecol Indic 119. https://doi.org/10.1016/j.ecolind.2020.106824, https://www.sciencedirect.com/science/article/pii/S1470160X20307627
25. Spooren J, Binnemans K, Björkmalm J, Breemersch K, Dams Y, Folens K, González-Moya M, Horckmans L, Komnitsas K, Kurylak W, López M, Mäkinen J, Onisei S, Oorts K, Peys A, Pietek G, Pontikes Y, Snellings R, Tripiana M, Varia J, Willquist K, Yurramendi L, Kinnunen P (2020) Near-zero-waste processing of low-grade, complex primary ores and secondary raw materials in Europe: technology development trends. Resour Conserv Recycl 160:104919. https://doi.org/10.1016/j.resconrec.2020.104919, https://www.sciencedirect.com/science/article/pii/S0921344920302378
26. Stolz P, Frischknecht R (2018) Life Cycle Assessment of Current Photovoltaic Module Recycling. Tech. Rep. Report IEA-PVPS T12-13:2018, International Energy Agency, PVPS Task 12: Subtask 2.0, LCA, iSBN 978-3-906042-69-5
27. Wambach K, Sander K (2015) Perspectives on management of end-of-life photovoltaic modules. In: Proceedings of the 31st European photovoltaic solar energy conference and exhibition. Hamburg, Germany, 10.4229/EUPVSEC20152015-7EO.2.5. https://www.eupvsec-proceedings.com/proceedings?paper=33471
28. Wambach K, Heath G, Libby C (2017) Life Cycle Inventory of Current Photovoltaic Module Recycling Processes in Europe. Tech. Rep. Report IEA-PVPS T12-12:2017, IEA PVPS Task12, Subtask 2, LCA, iSBN 978-3-906042-67-1
29. Weckend S, Wade A, Heath G (2016) End-of-life management: Solar Photovoltaic Panels. Tech. Rep. Report Number: T12-06:2016, IRENA in collaboration with IEA-PVPS Task 12. https://www.irena.org/publications/2016/Jun/End-of-life-management-Solar-Photovoltaic-Panels, iSBN: 978-92-95111-99-8

Chapter 9
Balance of System (BoS) and Storage

When a life cycle assessment study of a complete photovoltaic system is carried out, an important contribution comes from the balance of system (BoS) components. From a life cycle assessment perspective, BoS is becoming an important contributor to impacts, both environmental and economic, with an increasing share of impacts compared to the contribution of modules. In particular, the Joint Research Centre (European Commission) Methodology Guidelines on Life Cycle Assessment of Photovoltaic recommends to include at least the following items in a LCA study [15]: Manufacture of the mounting system; manufacture of the cabling; manufacture of the inverters and manufacture of all further components needed to produce electricity and supply it to the grid (e.g. transformers for utility-scale PV). To this recommended list, the manufacture of batteries should be added for a LCA study of stand-alone systems which requires energy storage.

Together with the PV modules, the electronic system for power management is the component with most contributions to environmental impacts (and also to economic cost). Basically, there are two stages in the power management of solar electricity generated with a photovoltaic system: DC/DC regulators and DC/AC inverters.

Regulators are used for adapting DC power to different loads and they include a maximum power point tracker device to optimize the PV module operation: individual modules or strings of interconnected modules must always work at its maximum power point for a given irradiance and temperature, independently of the load that is connected to them. Originally, regulators were designed to be used with one or a few modules in stand-alone systems that supply power to a DC load (for example, solar home systems for rural electrification); they manage the energy generated by the PV system, the energy stored in the battery and the energy consumed by the loads with the objective to provide a good service (often using priority load control algorithms) and at the same time protecting the battery against excessive discharging or overcharging and thus extending its lifetime. Nowadays, most DC/DC converters are designed to work at a single module level or at string level with current optimization

A. Urbina, *Sustainable Solar Electricity*, Green Energy and Technology,
https://doi.org/10.1007/978-3-030-91771-5_9

in PV systems connected to the grid and without any storage device. In this case, in order to supply the generated power to the grid, the regulator requires an additional connection to a DC/AC inverter and all the electronic equipment is integrated into a single device. The final combined apparatus including mppt tracker, DC/DC power management and DC/AC inversion is commonly called **inverter**.

9.1 Life Cycle Assessment of BoS Electronic Components

The electronic equipments required for power management are grouped into three "inverter" typologies which include in the same equipment the DC/DC regulator with maximum power point tracking systems, DC/AC inverter, monitorization and communication devices, safety switches both at DC input and the AC output lines (either monophase or three-phase output) and ground/earth connections; its market share is presented in Table 9.1. They are: string inverters, the most commonly used so far; central inverters, for large size PV plants, and micro-inverters, designed to work at single module level and which require additional electronic equipment to connect more than one module to the grid or AC application. Since all stages of the power management and the required components are actually integrated into a single electronic block, when considering LCA studies of electronic components, their contribution is often reduced to the "inverter". There is one exception to this general rule: rural electrification in isolated livelihoods where there is no grid and the electricity supply by the PV generator is DC and no inverter is required in the system; in this case, only regulators (with internal mppt tracker) are considered in the LCA study, optional small DC/AC inverters for specific AC loads may be included and the contribution of batteries to LCA becomes important; this case will be presented in Sect 9.4.

The evolution of inverter design and nominal power has been fast and strongly relying on regulations for PV feed-in tariffs or other subsidy policies (for example, the limit of 100 kW_p for eligibility for a subsidy scheme was a driver for a strong development of this size of inverter). All designs have been optimized and now work with efficiencies >98%, with the exception of micro-inverters, a more recent

Table 9.1 Typology of inverters, depending on its size. Best efficiencies and market share are provided for 2019 typical products. Data from Fraunhofer ISE 2021 PV report [14]

Inverter/Converter	Power	Efficiency (%)	Market share (%)
String inverters	Up to 150 kWp	Up to 98	61.6
Central inverters	More than 80 kWp	Up to 98.5	36.7
Micro-inverters	Module power range	90–97	1.7
DC/DC converters[a]	Module power range	Up to 99.5	5.1

[a]DC/DC Converters still require a DC/AC inverter in order to be connected to the grid, that is why total market share is larger than 100% in column four

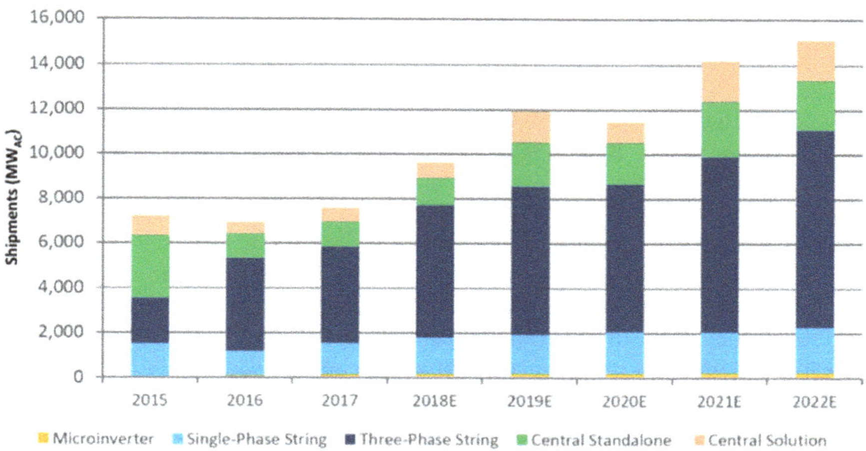

Fig. 9.1 Inverter shipments to the European Union (AC-rated MW power per year; E = Estimate). Reproduced with permission from [11]

development, which still has room for improvement from actual efficiency around 90% with some manufacturers already approaching efficiencies of large size inverters. The evolution of the global market is illustrated by the shipments to Europe during the past years. Europe was the strongest market a few years ago, and although it is not the first one any more, it is the better example to illustrate the evolution of PV size, and therefore, inverter size. In Fig. 9.1, the evolution of shipments of inverters to Europe is presented, the market now is dominated by three-phase string inverters (75%), followed by centralized inverters (26%) that are used as stand-alone systems or in combination with transformers, single phase string inverters (20%) and more recently micro-inverters (1% at module or a few modules level) complete the technological portfolio [11]. Average lifetimes of inverters are about 15 years, and recent developments with improved metal-oxide-semiconductor field-effect transistors (MOSFET, with wider bandgap, they keep high efficiency at higher temperatures) have extended the lifetime to 20 years; improved design for better maintenance and for easy replacement of some parts more prone to failure (electronic control subsystems, for example) will extend this lifetime further with the consequent big impact on LCA contribution of inverters during the lifetime of PV systems (currently at an average of 30 years).

There are two main approaches to evaluate the human health and environmental impacts of the inverters (in the broad meaning of the integrated electronic component described above). One is focussed on the calculation of the cumulative energy demand (CED) required to manufacture the inverter, which comprises energy embedded in the materials and process energy. This energy is added to the CED of other PV system components and it is used for energy payback time calculations, which was the main parameter to compare the "sustainability" of different energy systems or different PV technologies when solar photovoltaic energy is considered. The inverter is an important contributor for the BoS total embedded energy; the CED strongly depends on the nominal power of the equipment. When the CED data is reported as primary energy per nominal power (MJ/kW), the expected scaling down is observed, but with an attenuated power law decay indicating that very large inverters will meet a lower limit that is very difficult to surpass unless a breakthrough in inverter design or materials is achieved (it is not expected in the coming years). The second approach is a full LCA study of a PV system where the inverter is included and it contributes to impacts in all categories. In this LCA approach, the CED is often included as a part of the life cycle inventory and its global contribution is in general low, about 1–1.5% adding all impact categories. The advantage of the LCA approach is that the contribution of inverters to other impact categories can be analyzed in detail and targeted recommendations for manufacturers can be provided with the aim to reduce its impacts.

The cumulative energy demand of inverters was calculated by Erik Alsema in pioneering work in late 90s, delivering an early value for primary energy embedded in small inverters (3 kW) of about 1000 MJ/kW [2, 3]. Fifteen years later, several detailed LCA studies of small and medium inverters had reduced this value; for example, the study carried out by Laura Tschümperlin (Treeze Ltd., commissioned by the Swiss Federal Office of Energy SFOE, reference [51]) is now widely used as a reference and had found that the cumulative energy demand for manufacture has been reduced by a factor of four in comparison to "old" inverters, with a minimal impact contribution of 1.5% to total impacts. The minerals, fossil and renewable resource depletion has been reduced by a factor of two, with tantalum being the individual element most affecting this category (65%). On the other hand, impacts on human health are in general higher: for human toxicity cancer effects results are similar, while for human toxicity non-cancer effects and particulate matter they had risen around 30% compared to the "old" inverters due mainly to the higher complexity of electronic control printed boards and microelectronic circuits which makes them the main contribution to these categories (55% and 58%, respectively). Similarly, impacts on ecosystems, illustrated, for example, by freshwater ecotoxicity category, has also risen around 30% with respect to "old" inverters, again mainly due to the contribution of the electronic printed board and the metals included in the new inverters (impacting in metal depletion category). A graphical summary of results is presented in Fig. 9.2 for new inverters of different sizes where the data are normalized as impact per nominal AC power unit of the inverter; an economy of scale in impacts is observed with the best fit provided by a power law, having a faster reduction of impacts when increasing the size of the inverter from small (around 5 kW) to medium (around 20 kW) and then a lower rate of reduction that is extrapolated to inverters of 100 kW and is indicated in the figure by a dotted red line.

It is important to emphasize that the standard lifetime of inverters is 15 years, and therefore, during the lifetime of a PV system (either roof-top or large plant) will require at least one replacement of inverters, a fact that should be taken into account in LCA studies comprising a whole lifetime of systems. Uncertainties in data related to inverters are still large, but new results are constantly emerging and databases are regularly updated.

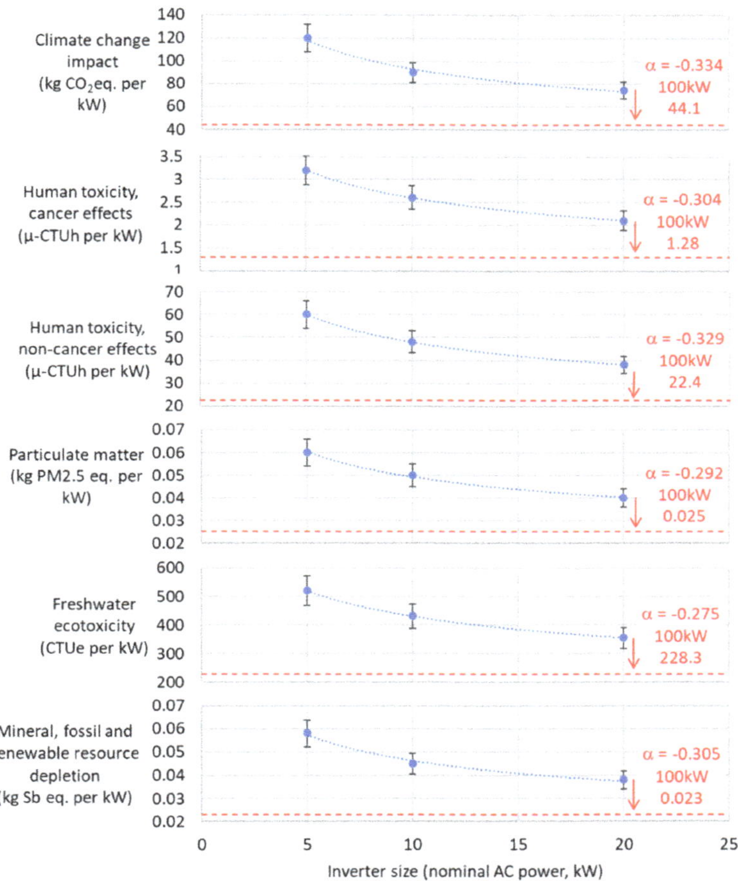

Fig. 9.2 Impacts of manufacture and disposal of solar inverters on six impact categories, the error bars indicate a 10% error, the dotted red line indicates an extrapolation to 100 kW AC nominal power inverters; fits are power laws $\propto x^{\alpha}$ with exponent α shown in each graph. Source of data used in the graphs: [51, 52]

In 2021, the market share of string inverters is estimated to be 52%; that is, half of power conversion electronics that is currently installed is still used in residential, small and medium commercial applications in PV systems up to 150 kW_p. The market share of central inverters, with applications mostly in large commercial and utility-scale systems, is about 44%, and the trend seems to indicate that this market share for large inverters will grow in the coming years since large PV plants are expected to represent the largest share of future PV system construction. A small part of the market (about 1%) is represented by micro-inverters (used on the module level and connected to electronic modules which optimize the self-consumption or the injection to the grid of the AC electricity generated by each module working with independent maximum point trackers per module). The market for DC/DC converters, also called "power optimizers", is estimated to be in the same range [14]. The observed trends for future developments of inverters (and power management in general) are: digitalization, repowering, new features for grid stabilization and optimization of self-consumption; storage; utilization of innovative semiconductors (SiC or GaN) in electronic components of the inverters which allow very high efficiencies and more compact designs; increase to 1500 V_{DC} maximum for string voltage and the corresponding improvement on safety and ground connections of frames and rack structures. It is desirable that this innovations will also lead to a reduction in its LCA impacts, which has not been the case for the transition from "old" to "new" generation of inverters.

9.2 Life Cycle Assessment of BoS Structural and Mechanical Components

The total cumulative primary energy demand for total Balance of System manufacture and installation is presented in Fig. 9.3; several studies reported values for CED measured or calculated in different conditions, hence the large dispersion of published data around a mean value with high standard deviation (as can be observed in the small lines). Roof-top systems BOS has a mean CED of 623 MJ/m^2 and ground systems BOS has a mean CED of 923 MJ/m^2, the conversion to embedded energy per nominal power (installed capacity), MJ/kW$_p$, will depend strongly on the efficiency of the PV modules, and therefore, in this case, the units of CED per square meter are more meaningful to be used for LCA studies of any system despite that the final CED for the whole system is commonly reported as MJ/kW$_p$. In Fig. 9.3, a statistical analysis of data reported in several publications is presented and the average value is a CED for total BoS components of 623.24 MJ/m^2 and 932.26 MJ/m^2 for roof-top and ground mounted systems, respectively. In this case, the functional unit of 1 m^2 is the one which allows a better comparison and a meaningful average, because typology and efficiencies of the modules to be installed can be very different, but with low impact on the CED of the BoS components. Only as an indicative calculation, for Si modules with PCE = 18%, the values of BoS contribution to CED would be 3.46 MJ/W$_p$ and 5.18 MJ/W$_p$, respectively.

The main source of uncertainty in LCA calculations related to BOS components other than inverters is the amount of concrete and steel to be used in the mounting racks, which strongly differ between ground mounted or roof-top systems. On the other hand, there is no expected technological innovation on concrete and steel production that could change their manufacturing process contribution to LCA categories. A recent report by the Joint Research Centre of the European Commission provides estimations for the use of concrete (60.7 tonnes/MW) and steel (67.9 tonnes/MW) required for photovoltaic systems, future scenarios considered in this report keep these figures without significant changes in all but the most optimistic

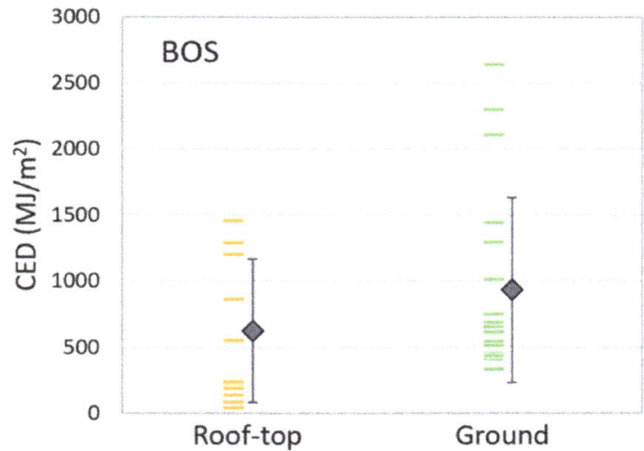

Fig. 9.3 Cumulative energy demand for the balance of system (BoS) of different PV system configuration (roof-top *vs* ground mounted) showing a mean value (diamond) with standard deviation error bars and the large dispersion of reported data in the literature for different scenarios and systems (bars); with data from [5, 16, 30] and references therein

scenarios where a small reduction in materials intensity use is predicted for 2050 (concrete: 48.6 tonnes/MW, steel: 54.3 tonnes/MW) [7].

9.3 Introduction to Electricity Storage for PV Systems

All future scenarios of major supranational organizations (such as IEA, IRENA, OCDE) as well as specific plans at the country level point to an acceleration of the energy transition that is already occurring. This transition will require a high penetration of renewable energy in the electricity sector and other sectors such as transport. The management of power generated by renewable sources, especially in electricity generation and consumption, already demands the use of energy storage in some applications, but a much larger storage capacity will be required in the coming years [26]. In the NZE2050 scenario of the International Energy Agency, demand for batteries for transport reaches around 14 TWh in 2050, 90 times more than in 2020 [22]. The variability of photovoltaic electricity generation and the mismatch in real time between generation and consumption requires energy storage at different scales [1]: ranging from small scale in local isolated systems to medium or large scale, for example, in electricity grids with distributed renewable generation in order to stabilize the grid and manage generation and consumption.

9.3.1 Electricity Storage Technologies

There are several available technologies for electricity storage. At very large scale, pumped hydro energy storage is at present time the only available technology for economic storage of energy; it is not an efficient method, but it is competitive when the alternative is to stop generating and disconnecting from the grid the photovoltaic (or wind) production sites. This storage method requires specific infrastructure and reversible pumps usually associated with hydroelectric dams [9]; however, due to the location restrictions of the hydro pumped storage plants, the expansion of this technology is limited. At smaller scales, the technologies for storage of electricity can be organized into three main groups: superconducting magnetic energy storage [37, 45], electric double layer capacitor [44] and the most reliable nowadays: systems based on electrochemical energy storage usually grouped with the common name of batteries.

In a different group, the production of chemical fuels using renewable electricity should also be mentioned. The most extended is the production of hydrogen using electrolyzers whose electricity supply comes from a renewable source; the photovoltaic systems can be coupled to the electrolyzers either by using DC-DC power management electronics or with direct connection using clever design rules [18, 19]. The so-called "green" hydrogen (it is green only if produced with renewable electricity) can be used as an energy vector in transport or heating sectors (by burning) or electricity generation (by closing the cycle with fuel cells). The combination of electrolyzers and fuel cells are usually included as a single block for electricity storage covering a large energy range and the possibility of physical separation between the electrolyzer and the fuel cell makes it a reliable system, although unfortunately with very low round trip global efficiency (35–45%) and much higher capital cost (17 USD/W) as compared to batteries (2–5 USD/W) makes this combination unreliable for competing in the market for small and medium applications [1, 42].

9.3.2 Battery Technologies

For solar electricity, the coupling of PV generators to batteries has been since the early development of photovoltaics the most common storage mean for isolated small systems where there is no grid, or more recently batteries have been used as a buffer to optimize self-consumption in systems with grid connection. The rapid development of electric vehicles that have batteries can be used as a complement for solar electricity storage, either to be used during driving or as a battery system managing electricity supply and demand in local systems (with or without photovoltaic production on site). Batteries are also considered as a means of storage in large PV plants, with a much larger storage capacity in order to moderate the intermittency of generation before injecting into the grid; these large storage systems can also be used to store electricity from the grid at some specific time slots depending on real-time electricity prices and global supply-demand equilibrium required by electricity grid managers.

Therefore, batteries will be required at very different scales: from very small energy supply for portable gadgets to medium and large systems in combination with photovoltaic electricity generation either in isolated locations or for grid-connected locations. Grid stabilization will require both power quality and flexibility of the energy management from short time scales (ms) to hourly variability (typical of PV systems); storage will be unavoidable when renewable electricity penetration reaches more than 80% [21, 26]. And batteries will be also required especially for the massive penetration of electric transport using small or medium size batteries.

Battery technology is constantly developing, with many alternatives already in the market and many others very close to commercialization. Lead-based, alkaline or lithium-ion-based batteries are the two most broadly used, with Li-ion gaining market steadily. In fact, for photovoltaic applications, Li-ion batteries represent currently 90% of the market [24]. Applications of batteries are very different, therefore, demanding different characteristics both on size and cyclability. A summary of operational parameters for different battery technologies is presented in Table 9.2. Two broad families must be taken into account: primary batteries designed for a single use (not rechargeable) and rechargeable batteries. Advanced lead–acid, alkaline (sodium–sulfur and sodium–sulfur–chlorine), Nickle–Cadmium and Li-ion batteries are considered cyclable and well established technologies, with round trip efficiencies around 85% and operating cycles above 3000 (Li-ion above 7000).

Besides those already established kinds of battery, emerging technologies are being developed, such as redox flow batteries (Fe–Cr, V–Br or V–O_2 cells) which are a promising alternative in which the redox processes are carried out in the battery but the fluids are stored in external containers and delivered on demand [54]. They are scalable and flexible, have a high round trip efficiency (85%) and low environmental impacts, although the only one which has reached the market is the VRB option which uses vanadium (IV-V)/vanadium (II-III) dissolved in aqueous sulfuric acid, but it is difficult to scale down to small systems and to use for electric vehicles [1]. Besides all-vanadium options, there are other technologies combining different metals: vanadium/bromine; iron/chromium; Fe-EDTA/bromine, Zinc/Cerium,

Table 9.2 Typical parameters of some current and emerging battery technologies

Battery	Top power MW	Top energy MWh	Energy density Wh/kg	Round-trip efficiency (%)	Cycles ($\times 10^3$)	References
Lead acid	10–40	01–10	25–50	78–80	3	[48]
Na-S	34	10	150–120	85–90	4.3–6	[40]
Na-Ni-Cl	1	6	95–120	85	3—4	[39, 40]
Ni-Cd			40–60	75	>15	[42, 49]
Redox flow	2–100	6–120	9–50	85	>13	[1]
Li-ion	16	20	80–250	95	4–8	[20, 28, 42, 56]
Zn-ion			250	80	1	[27]
Li/Air			600	66		[23, 39, 56]
Zn/Air	Power density		350–500	60	1–2	[10, 31, 32, 47]
	200 mW/cm^2		776 prim., 378 s.			

etc…with potential for the redox couple in the range of 1.2–3.4 V [54, 55]; recent advances in redox flow batteries using low cost carbon polymer composites and graphene-based nanoparticles have extended their lifetime [8, 33], and accelerated degradation charge–discharge studies have shown that bench-scale vanadium redox flow batteries (VRFB) can be adequate for storage of solar photovoltaic electricity and wind electricity [34, 36].

Other systems, such as Metal/Air batteries, have attracted much attention recently as rechargeable batteries due to the high capacity and energy densities. Nowadays, Zn/Air primary is the only Metal/Air battery with a real commercial single use application, but no rechargeable Metal/Air batteries have been sufficiently improved to reach a commercial level. Li/Air, Na/Air, Al/Air and Zn/Air batteries are the main systems that are under investigation and can achieve an energy density theoretically ten times higher than current Li-ion batteries [17, 46, 56]. Furthermore, from a sustainability point of view, those batteries which focus on the use of Zn as negative electrode have many advantages, such as its low cost, abundance of Zn in the natural medium or the availability of use aqueous-based electrolytes. Finally, aqueous Zn-ion batteries have also been proposed recently as a cheap alternative (see discussion on capital costs of batteries in Sect. 10.3), although they are still far from commercialization [27].

9.4 Overview of Life Cycle Assessment Applied to Batteries

For the fabrication of any kind of battery, a large amount of raw material and energy are consumed during the process; waste and end-of-life recycling or disposal also generates an important environmental impact [10]. All the solutions for renewable energy storage provided by batteries should ideally be sustainable from an environmental and an economical point of view. The result of its sustainability evaluation depends mainly on the processing routes for battery fabrication, the efficiency and cyclability of the battery during the operational phase and the decommission of the battery including recycling and/or land-filling at the end of life [6]. The Life Cycle Assessment methodology can be applied to evaluate the impacts of battery production; the large variety of technologies makes compulsory using the same functional unit to quantify the impacts and define clearly the scope of the LCA study.

9.4.1 Phases in LCA for Batteries

In order to define a functional unit based on the service that the battery provides to the end-user, in this case, the amount of electricity stored and delivered in the battery throughout its lifetime, it is necessary to clarify the limits for depth of discharge

(DOD) that each kind of battery considers for safe operation (usually around 80%) and its nominal lifespan. The lifespan is the number of cycles for which cell capacity does not fall below the specified limit, that can be the DOD or a lower value for optimal cycling (for example, 60% of nominal storage capacity, expressed in Ah). An additional difficulty for a functional unit based on service for batteries is that this limit used to define the lifespan decreases with time. A complete LCA study for batteries should comprise at least three phases: production phase, where raw materials or materials from recycling input should be considered; use phase for the different applications (such as electric vehicles or energy storage for photovoltaic systems) including maintenance and end-of-life phase where final collection, disposal or recycling of the used battery is carried out. In each stage, besides material input (which generates a material inventory), inputs of energy and gas emissions must be taken into consideration, as well as other emissions treated as waste (that may or may not be recycled). Each of the main phases may be subdivided into stages depending on the scope of the LCA. Usually, the selected functional unit of the LCA is product-based, when LCA focuses on production phase, or service-based, when LCA includes also the use phase.

9.4.2 Phases in LCA Including Second Life of Batteries

Different use phase of batteries poses different requirements for their peak power, maximum energy storage and cyclability. It is possible to use a battery for a second application once their lifetime for their primary application has been exhausted. In particular, batteries designed for their use in electric vehicles can be used for solar applications after they are discarded, since the requirements that the battery needs to meet for its second application are less demanding; nevertheless, a reconditioning phase is often required to adapt the old battery for its second life. This processing changes the scope of the LCA and provides different figures for the overall use of any kind of battery at the end of its two possible lives (see Fig. 9.4). The industry for battery reconditioning for second life will develop if the economical return is good; initial studies are promising for Li-ion batteries designed for electric vehicles and with a second life for energy storage in photovoltaic systems: cost of battery range between 150 and 250 USD/kWh for the new battery and after reconditioning, the refurbished battery may be back into market costing between 44 and 180 USD/kWh depending on model; since the cost of refurbishment is around 25–50 USD/kWh, the market possibilities are open although the marketing model is still uncertain [29].

From a LCA point of view two roads are open regarding recycling of battery components and reconditioning as indicated in Fig. 9.4: first, fabrication of the initial battery that may or may not use recycled materials or components is an initial

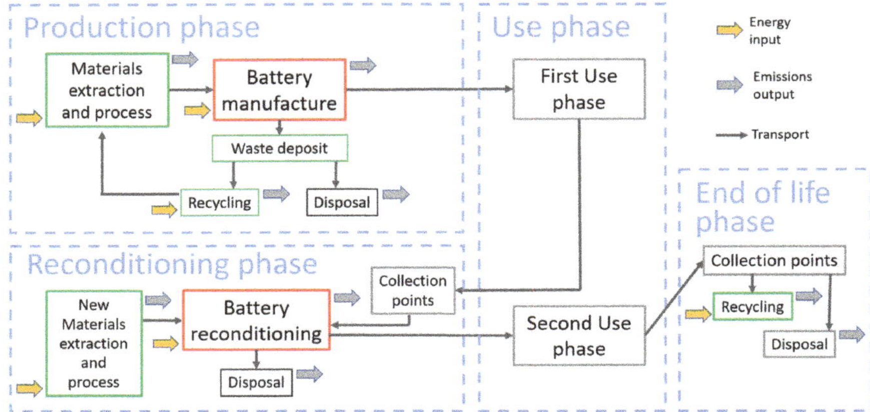

Fig. 9.4 Scope and phases of Life Cycle Assessment for batteries with second life, including a reconditioning phase and recycling processes during manufacture and at end of life

production phase, followed by the first life use phase; then an intermediate production phase that is called reconditioning phase is included leading to a second use phase. The difficulty will arise with the selection of a single functional unit for the full LCA, since it should include services which may be very different depending on the application during the two lives of the battery: electric vehicle often use km of service, while photovoltaic applications use energy (kWh_{AC}) delivered throughout its lifetime. At least two collection points are needed for a cycle of batteries for two lives, the first one as an intermediate point between first life and reconditioning; this stage is important to recover the maximum parts or whole battery after the end of first life, disposal of some parts may be required, eventual recycling of materials and pieces at this stage are included in the reconditioning process. The second collection point at the end of second life leads to recycling of materials and disposal.

9.4.3 Results of LCA for Batteries

Production of raw materials by the mining industry creates important environmental impacts. The review carried out by Dehghani-Sanij et al. in reference [10] indicates that 85% of lead production worldwide is used in the fabrication of lead–acid batteries according to the International Lead Association (ILA) "Lead Uses—Statistics and Lead Facts" web site[1] and to [38]. Similarly, already in 2017 around 45% of Li production was devoted to the fabrication of Li-ion batteries and 50% of cobalt and 10% of graphite production worldwide is used in battery electrodes. As mentioned

[1] International Lead Association Database: https://www.ila-lead.org/lead-facts/lead-uses–statistics.

in Chap. 5, graphite has been declared a strategic material by the European Union. Toxicity of lead, although it is efficiently recycled (more than 95%), probably will slowly reduce the production of lead–acid batteries, in spite of their low cost and good performance. The environmental burdens of manufacture of the Li-ion battery is dominated by the production of the negative and positive electrodes and the battery pack, while single cell, separator, lithium salt and solvent play a minor role [38]. The use of graphene has been proposed for several applications related to energy storage and could eventually reduce the environmental impact of those systems, especially those requiring metallic electrodes that could be replaced by graphene nanocomposite materials [25].

Recycling of components to recover Li or investment for the second life of Li-ion batteries could reduce environmental impacts up to 30% [56], but at present there is almost no industry dedicated to the recycling of lithium batteries since the economic return is very low and it will not happen unless there is an enforcement policy that makes recycling of Li compulsory [53].

Cumulative primary energy demand (CED) for current Li-ion battery production is from 350 to 650 MJ/kWh, which brings greenhouse gas (GHG) emissions to figures between 120 and 250 kg CO_{2eq}/kWh, although these calculations are strongly dependant on the country under consideration through the local electricity grid mix, these figures come from a study carried out in Sweden [42, 43]. For Li/Air battery, Zacrkisson et al. carried out a detailed LCA which calculated a climate change impact of 1100 kgCO_{2eq}/kWh of stored and delivered electricity, considering only the production phase of their study for the STABLE Li/Air battery prototype; the total impact including use and end-of-life phases is 1299 kgCO_{2eq}/kWh, thus showing that the higher impact comes from the production phase [56, 57]. The changing conditions with time for the use of any energy storage system must also be taken into account for the LCA which evaluates service-based functional units for a service extended in time, this approach was recently applied to Li-ion battery LCA [12].

The main impact categories are summarized in Table 9.3 for two kinds of Li-ion batteries and for a functional unit of 1 kg of battery. Conventional Li-ion batteries have values of 14.19 kg CO_{2eq}/1 kg or more recently reported impacts for redox flow batteries, of which the all-vanadium type has lower embedded emissions (2.86 kg CO_{2eq}/1 kg), both have lower values (even if the transport has been included in LCA), therefore, pointing to all- vanadium redox flow as the battery with lower emission impacts [13].

Often a second functional unit is used, in this case considering the electricity stored in the battery. For a FU of 1MJ energy storage capacity, the numbers for Li-ion battery for several impact categories are: climate change 17–27 kg CO_{2eq}; Human toxicity 3-5 kg 1,4-DB_{eq}; Metal depletion 28–44 kg Fe_{eq}, and Fossil depletion 2.2–3.4 kg Oil_{eq} [35]. Considering several studies, the average results for 1 kWh of energy storage capacity in Li-ion batteries for cumulative energy demand for production is 328 kWh and generates 110 kgCO_{2eq} of greenhouse gas emissions; when detailed information about cyclability and lifetime of the battery are available, the results for 1 kWh of electricity provided over the entire life cycle of a battery, the cumulative energy demand is reduced to 26 kWh and consequently, the GHG emissions are

Table 9.3 Main impact categories for LCA of Li-ion battery for a FU of 1kg of battery

Impact indicator	Unit	Lithium ion battery	
		NMP solvent[a]	Water solvent[b]
Cumulative energy demand	MJ/kg	90–97	88
Climate change	kg CO_{2eq}	12.5	4.4
Metal depletion	kg Fe_{eq}	20	20
Fossil fuel depletion	kg Oil_{eq}	1.6	1.5

[a][35, 49]
[b][35]

reduced to 74 gCO_{2eq} [41]. For emerging Zn/Air batteries, an initial LCA study based on laboratory scale delivered some cap values for environmental impacts that can be improved in an up-scaled factory: CED of 590.8 MJ/1 kg of fabricated battery (1780.3 MJ/1 kWh of stored energy), which translates into emissions amounting to 20.3 kg CO_{2eq}/kg of fabricated Zn/Air battery (61.2 kg CO_{2eq}/1 kWh of stored energy), those values (for FU 1 kWh) are estimated for a primary battery with a single cycle use; if cyclability is demonstrated with this kind of batteries, its impacts for FU of stored electricity can be reduced by at least two orders of magnitude [47].

These results emphasize the need to carry out a detailed LCA of the manufacturing process of any emerging battery technology, pointing to upscaling of production according to predicted demand for energy storage and not only analyzing cumulative energy demand or GHG emissions, but also toxicity, acidification and resources depletion [4, 42]. Also, in order to complete impact analysis, an integrated hybrid approach is recommended since economic impact may be important for LCA studies when functional units related to services are chosen and the entire background economy is affecting the service provided; the hybrid approach avoids truncation of the studies due to a limited scope and will also influence the recycling and disposal stages since these industrial activities will be strongly affected by economic incentives [50, 57].

References

1. Alotto P, Guarnieri M, Moro F (2014) Redox flow batteries for the storage of renewable energy: a review. Renew Sustain Energy Rev 29:325–335. https://doi.org/10.1016/j.rser.2013.08.001, http://www.sciencedirect.com/science/article/pii/S1364032113005418
2. Alsema EA (2000) Energy pay-back time and CO2 emissions of PV systems. Progress Photovolt: Res Appl 8(1):17–25. https://doi.org/10.1002/(SICI)1099-159X(200001/02)8:1<17::AID-PIP295>3.0.CO;2-C. Publisher: John Wiley & Sons, Ltd
3. Alsema EA, Nieuwlaar E (2000) Energy viability of photovoltaic systems. Energy Policy 28(14):999–1010. https://doi.org/10.1016/S0301-4215(00)00087-2, https://www.sciencedirect.com/science/article/pii/S0301421500000872
4. Armand M, Tarascon JM (2008) Building better batteries. Nature 451:652. https://doi.org/10.1038/451652a

5. Bhandari KP, Collier JM, Ellingson RJ, Apul DS (2015) Energy payback time (EPBT) and energy return on energy invested (EROI) of solar photovoltaic systems: a systematic review and meta-analysis. Renew Sustain Energy Rev 47:133–141. https://doi.org/10.1016/j.rser.2015.02.057, http://www.sciencedirect.com/science/article/pii/S136403211500146X
6. Bossche PVd, Vergels F, Mierlo JV, Matheys J, Autenboer WV (2006) SUBAT: an assessment of sustainable battery technology. J Power Sour 162(2):913–919. https://doi.org/10.1016/j.jpowsour.2005.07.039, http://www.sciencedirect.com/science/article/pii/S0378775305008761
7. Carrara S, Alves Días P, Plazzota B, Pavel C (2020) Raw materials demand for wind and solar PV technologies in the transition towards a decarbonised energy system. Tech. Rep. JRC119941, Joint Research Centre (European Commision). https://doi.org/10.2760/160859, https://publications.jrc.ec.europa.eu/repository/handle/JRC119941, iSBN: 978-92-76-16225-4 ISSN: 1831-9424
8. Chakrabarti M, Brandon N, Hajimolana S, Tariq F, Yufit V, Hashim M, Hussain M, Low C, Aravind P (2014) Application of carbon materials in redox flow batteries. J Power Sour 253:150–166. https://doi.org/10.1016/j.jpowsour.2013.12.038, http://www.sciencedirect.com/science/article/pii/S0378775313020065
9. Deane JP, Gallachóir BPO, McKeogh EJ (2010) Techno-economic review of existing and new pumped hydro energy storage plant. Renew Sustain Energy Rev 14(4):1293–1302. https://doi.org/10.1016/j.rser.2009.11.015, http://www.sciencedirect.com/science/article/pii/S1364032109002779
10. Dehghani-Sanij AR, Tharumalingam E, Dusseault MB, Fraser R (2019) Study of energy storage systems and environmental challenges of batteries. Renew Sustain Energy Rev 104:192–208. https://doi.org/10.1016/j.rser.2019.01.023, http://www.sciencedirect.com/science/article/pii/S1364032119300334
11. Dodd N, Espinosa N (2021) Solar photovoltaic modules, inverters and systems: options and feasibility of EU Ecolabel and Green Public Procurement criteria. Tech. Rep. KJ-NA-30474-EN-N, EU-JRC, European Commission-Joint Research Centre. https://doi.org/10.2760/29743, iSBN 978-92-76-26819-2 ISSN 1831-9424
12. Elzein H, Dandres T, Levasseur A, Samson R (2019) How can an optimized life cycle assessment method help evaluate the use phase of energy storage systems? J Clean Prod 209:1624–1636. https://doi.org/10.1016/j.jclepro.2018.11.076, http://www.sciencedirect.com/science/article/pii/S0959652618334796
13. Fernández-Marchante CM, Millán M, Medina-Santos JI, Lobato J (2019) Environmental and preliminary cost assessments of redox flow batteries for renewable energy storage. Energy Technol 1900914. https://doi.org/10.1002/ente.201900914
14. Fraunhofer-ISE (2021) Photovoltaics Report 2021. Tech. rep., Fraunhofer Institute for Solar Energy Systems, ISE, Germany. https://www.ise.fraunhofer.de/content/dam/ise/de/documents/publications/studies/Photovoltaics-Report.pdf
15. Frischknecht R, Stolz P, Heath G, Raugei M, Sinha P, de Wild-Scholten MJ (2020) Methodology Guidelines on Life Cycle Assessment of Photovoltaic. Tech. rep., International Energy Agency, PVPS Task 12: PV Sustainability, iSBN: 978-3-906042-99-2
16. Fthenakis V, Kim H (2011) Photovoltaics: life-cycle analyses. Progress Sol Energy 185(8):1609–1628. https://doi.org/10.1016/j.solener.2009.10.002, https://www.sciencedirect.com/science/article/pii/S0038092X09002345
17. Fu J, Cano ZP, Park MG, Yu A, Fowler M, Chen Z (2017) Electrically rechargeable zinc-air batteries: progress, challenges, and perspectives. Adv Mater 29(7):1604685. https://doi.org/10.1002/adma.201604685, https://onlinelibrary.wiley.com/doi/abs/10.1002/adma.201604685
18. García-Valverde R, Miguel C, Martínez-Béjar R, Urbina A (2008) Optimized photovoltaic generator-water electrolyser coupling through a controlled DC-DC converter. Int J Hydrogen Energy 33(20):5352–5362. https://doi.org/10.1016/j.ijhydene.2008.06.015, GotoISI://OS:000261009000008,type: Journal Article
19. García-Valverde R, Espinosa N, Urbina A (2011) Optimized method for photovoltaic-water electrolyser direct coupling. Int J Hydrogen Energy 36(17):10574–10586. https://doi.org/10.1016/j.ijhydene.2011.05.179, GotoISI://WOS:000295235200015, type: Journal Article

20. Goodenough JB, Park KS (2013) The Li-Ion rechargeable battery: a perspective. J Am Chem Soc 135(4):1167–1176. https://doi.org/10.1021/ja3091438
21. Hesse HC, Schimpe M, Kucevic D, Jossen A (2017) Lithium-Ion battery storage for the grid-A review of stationary battery storage system design tailored for applications in modern power grids. Energies 10(12):2107. https://doi.org/10.3390/en10122107 https://www.mdpi.com/1996-1073/10/12/2107
22. IEA (2021) Net Zero by 2050. A Roadmap for the Global Energy Sector. Tech. rep., International Energy Agency, net Zero by 2050 Interactive iea.li/nzeroadmap Net Zero by 2050 Data iea.li/nzedata
23. Imanishi N, Yamamoto O (2014) Rechargeable lithium-air batteries: characteristics and prospects. Mater Today 17(1):24–30. https://doi.org/10.1016/j.mattod.2013.12.004, http://www.sciencedirect.com/science/article/pii/S1369702113004586
24. IRENA (2021) Renewable Power Generation Costs in 2020. IEA-IRENA. https://www.irena.org/publications/2021/Jun/Renewable-Power-Costs-in-2020
25. Ji L, Meduri P, Agubra V, Xiao X, Alcoutlabi M (2016) Graphene-based nanocomposites for energy storage. Adv Energy Mater 6(16):1502159. https://doi.org/10.1002/aenm.201502159, https://onlinelibrary.wiley.com/doi/abs/10.1002/aenm.201502159
26. Kondoh J, Ishii I, Yamaguchi H, Murata A, Otani K, Sakuta K, Higuchi N, Sekine S, Kamimoto M (2000) Electrical energy storage systems for energy networks. Energy Convers Manag 41(17):1863–1874. https://doi.org/10.1016/S0196-8904(00)00028-5, http://www.sciencedirect.com/science/article/pii/S0196890400000285
27. Kundu D, Adams BD, Duffort V, Vajargah SH, Nazar LF (2016) A high-capacity and long-life aqueous rechargeable zinc battery using a metal oxide intercalation cathode. Nat Energy 1:16119. https://doi.org/10.1038/nenergy.2016.119
28. Landi BJ, Ganter MJ, Cress CD, DiLeo RA, Raffaelle RP (2009) Carbon nanotubes for lithium ion batteries. Energy Environ Sci 2(6):638–654. https://doi.org/10.1039/B904116H
29. Martí nez Laserna E, Gandiaga I, Sarasketa-Zabala E, Badeda J, Stroe DI, Swierczynski M, Goikoetxea A, (2018) Battery second life: Hype, hope or reality? a critical review of the state of the art. Renew Sustain Energy Rev 93:701–718. https://doi.org/10.1016/j.rser.2018.04.035, http://www.sciencedirect.com/science/article/pii/S1364032118302491
30. Leccisi E, Raugei M, Fthenakis V (2016) The energy and environmental performance of ground-mounted photovoltaic systems-a timely update. Energies 9(8). https://doi.org/10.3390/en9080622
31. Li Y, Dai H (2014) Recent advances in zinc-air batteries. Chem Soc Rev 43(15):5257–5275. https://doi.org/10.1039/C4CS00015C
32. Liu Q, Wang Y, Dai L, Yao J (2016) Scalable fabrication of nanoporous carbon fiber films as bifunctional catalytic electrodes for flexible Zn-Air batteries. Adva Mater 28(15):3000–3006. https://doi.org/10.1002/adma.201506112, https://onlinelibrary.wiley.com/doi/abs/10.1002/adma.201506112
33. Lobato J, Mena E, Millán M (2017) Improving a Redox Flow Battery Working under Realistic Conditions by Using of Graphene based Nanofluids. Chem Select 2(27):8446–8450. https://doi.org/10.1002/slct.201701042
34. López-Vizcaíno R, Mena E, Millán M, Rodrigo MA, Lobato J (2017) Performance of a vanadium redox flow battery for the storage of electricity produced in photovoltaic solar panels. Renew Energy 114:1123–1133. https://doi.org/10.1016/j.renene.2017.07.118, http://www.sciencedirect.com/science/article/pii/S0960148117307462
35. McManus MC (2012) Environmental consequences of the use of batteries in low carbon systems: the impact of battery production. Appl Energy 93:288–295. https://doi.org/10.1016/j.apenergy.2011.12.062, http://www.sciencedirect.com/science/article/pii/S0306261911008580
36. Mena E, López-Vizcaíno R, Millán M, Cañizares P, Lobato J, Rodrigo MA (2018) Vanadium redox flow batteries for the storage of electricity produced in wind turbines. Int J Energy Res 42(2):720–730. https://doi.org/10.1002/er.3858

37. Nomura S, Shintomi T, Akita S, Nitta T, Shimada R, Meguro S (2010) Technical and cost evaluation on SMES for electric power compensation. IEEE Trans Appl Superconduct 20(3):1373–1378. https://doi.org/10.1109/TASC.2009.2039745
38. Notter DA, Gauch M, Widmer R, Wäger P, Stamp A, Zah R, Althaus HJ (2010) Contribution of Li-Ion Batteries to the Environmental Impact of Electric Vehicles. Environ Sci Technol 44(17):6550–6556. https://doi.org/10.1021/es903729a
39. Ou Dongji (2017) State of the art of life cycle inventory data for electric vehicle batteries. Stockholm, Sweden
40. Pan H, Hu YS, Chen L (2013) Room-temperature stationary sodium-ion batteries for large-scale electric energy storage. Energy Environ Sci 6(8):2338–2360. https://doi.org/10.1039/C3EE40847G
41. Peters JF, Baumann M, Zimmermann B, Braun J, Weil M (2017) The environmental impact of Li-Ion batteries and the role of key parameters—a review. Renew Sustain Energy Rev 67:491–506. https://doi.org/10.1016/j.rser.2016.08.039, http://www.sciencedirect.com/science/article/pii/S1364032116304713
42. Posada JOG, Rennie AJR, Villar SP, Martins VL, Marinaccio J, Barnes A, Glover CF, Worsley DA, Hall PJ (2017) Aqueous batteries as grid scale energy storage solutions. Renew Sustain Energy Rev 68:1174–1182. https://doi.org/10.1016/j.rser.2016.02.024, http://www.sciencedirect.com/science/article/pii/S136403211600232X
43. Romare M, Dahllöf L (2017) A study with focus on current technology and batteries for light-duty vehicles. IVL Swedish Environmental Research Institute
44. Sahay K, Dwivedi B (2009) Supercapacitors energy storage system for power quality improvement: an overview. J Electr Syst 1:8
45. Sander M, Gehring R, Neumann H (2013) LIQHYSMES-A 48 GJ Toroidal MgB2-SMES for buffering minute and second fluctuations. IEEE Trans Appl Superconduct 23(3):5700505. https://doi.org/10.1109/TASC.2012.2234201
46. Santos F, Abad J, Vila M, Castro GR, Urbina A, Romero AJF (2018) In situ synchrotron x-ray diffraction study of Zn/Bi2O3 electrodes prior to and during discharge of Zn-air batteries: influence on ZnO deposition. Electrochimica Acta 281:133–141. https://doi.org/10.1016/j.electacta.2018.05.138, http://www.sciencedirect.com/science/article/pii/S001346861831185X
47. Santos F, Urbina A, Abad J, López R, Toledo C, Fernandez Romero AJ (2020) Environmental and economical assessment for a sustainable Zn/air battery. Chemosphere 250:126273. https://doi.org/10.1016/j.chemosphere.2020.126273, GotoISI://MEDLINE:32120147, type: Journal Article
48. Spanos C, Turney DE, Fthenakis V (2015) Life-cycle analysis of flow-assisted nickel zinc-, manganese dioxide-, and valve-regulated lead-acid batteries designed for demand-charge reduction. Renew Sustain Energy Rev 43:478–494. https://doi.org/10.1016/j.rser.2014.10.072, http://www.sciencedirect.com/science/article/pii/S1364032114008971
49. Sullivan JL, Gaines L (2012) Status of life cycle inventories for batteries. Energy Convers Manag 58:134–148. https://doi.org/10.1016/j.enconman.2012.01.001, http://www.sciencedirect.com/science/article/pii/S0196890412000179
50. Sun M, Wang Y, Hong J, Dai J, Wang R, Niu Z, Xin B (2016) Life cycle assessment of a bio-hydrometallurgical treatment of spent ZnMn batteries. J Clean Prod 129:350–358. https://doi.org/10.1016/j.jclepro.2016.04.058, http://www.sciencedirect.com/science/article/pii/S0959652616303262
51. Tschümperlin L, Stolz P, Frischknecht R (2016) Life cycle assessment of low power solar inverters (2.5 to 20 kW). Tech. rep., Treeze Ltd., Switzerland, commisioned by Swiss Federal Office of Energy SFOE. https://treeze.ch/fileadmin/user_upload/downloads/Publications/Case_Studies/Energy/174-Update_Inverter_IEA_PVPS_v1.1.pdf, version 174-Update Inverter_IEA PVPS_v1.1, 03.10.2016 15:37:00
52. Wade A, Stolz P, Frischknecht R, Heath G, Sinha P (2018) The Product Environmental Footprint (PEF) of photovoltaic modules-Lessons learned from the environmental footprint pilot phase on the way to a single market for green products in the European union. Progress Photovolt: Res Appl 26(8):553–564. https://doi.org/10.1002/pip.2956.Publisher: John Wiley & Sons Ltd

53. Wang X, Gaustad G, Babbitt CW, Bailey C, Ganter MJ, Landi BJ (2014) Economic and environmental characterization of an evolving Li-ion battery waste stream. J Environ Manag 135:126–134. https://doi.org/10.1016/j.jenvman.2014.01.021, http://www.sciencedirect.com/science/article/pii/S030147971400036X
54. Weber AZ, Mench MM, Meyers JP, Ross PN, Gostick JT, Liu Q (2011) Redox flow batteries: a review. J Appl Electrochem 41(10):1137. https://doi.org/10.1007/s10800-011-0348-2
55. Weber S, Peters JF, Baumann M, Weil M (2018) Life cycle assessment of a vanadium redox flow battery. Environ Sci Technol 52(18):10864–10873. https://doi.org/10.1021/acs.est.8b02073
56. Zackrisson M, Fransson K, Hildenbrand J, Lampic G, O'Dwyer C (2016) Life cycle assessment of lithium-air battery cells. J Clean Prod 135:299–311. https://doi.org/10.1016/j.jclepro.2016.06.104, http://www.sciencedirect.com/science/article/pii/S0959652616307818
57. Zhao S, You F (2019) Comparative life-cycle assessment of li-ion batteries through process-based and integrated hybrid approaches. ACS Sustain Chem Eng 7(5):5082–5094. https://doi.org/10.1021/acssuschemeng.8b05902

Part III
Beyond Life Cycle Assessment: Socioeconomics and Geopolitics of Solar Electricity

Part III goes beyond the standard approach to LCA and includes economic and social assessment of impacts. Economic evaluation of the economic cost of installed capacity and produced electricity is accomplished in this part. Comments on the geopolitics of photovoltaics provide the closing remarks of the whole book. In Chap. 10, the definition of economic parameters used to evaluate the impact of PV systems is provided. Those comprise the levelized cost of electricity, also with the modern definition of IEA, called the "value-adjusted" LCOE. Also, employment opportunities by sector and by country are analysed, including investigation on socioeconomic networks that range from Non-Governmental Organizations (NGOs) or other associations to small, medium or large companies linked to solar electricity. Chapter 11 provides a list of the regulatory framework worldwide, mostly a presentation of standards and regulations, a comparison between countries and a comment about its evolution. The book ends with Chap. 10 in which solar electricity will be put into the context of globalization, where on the one hand still a large amount of population lacks access to electricity while on the other hand solar electricity is now subject of speculation by investment funds and big multinationals. Climate change mitigation and the related international agreements are the closing subjects of the book.

Chapter 10
Socioeconomic Impacts of Solar Electricity

10.1 Cost of Ownership of Photovoltaic Systems

Total installed cost of a utility-scale PV system has experienced a constant reduction in the past 10 years; the global weighted average in 2010 was 4.71 USD/W_p, and in 2020 it dropped below the 1USD mark, reaching 0.88 USD/W_p, with even lower price cases (within the 5% percentile) reaching 0.57 USD/W_p [8]. It is important to emphasize that these data are calculated using real installed costs in different countries, and presented as an aggregate of a large variety of systems. The main driver for this reduction has been the technological improvement of all PV technologies, which have reached prices below 0.5 USD (2020 prices) per peak power (W_p). In Fig. 10.1, the evolution for crystalline silicon PV module price is shown, with two clear phases in the past 30 years. One initial phase is when cumulative production slowly advanced until it reached 1 GW_p of installed power in 2000; in this initial stage of lower production capacity, the learning rate was around 18% during the whole period. Then, in the years ranging from 2000 to 2006 approximately, a sudden increase in module demand, driven by subsidized policies in several countries, led to a shortage in poly-silicon production and the trend of price reduction stalled for a few years (even with a slight momentary increase of module price in some countries) until new silicon production capacity was installed in China to supply the new module factories that were rapidly built in this country allowing world production capacity to reach a cumulative production of 10 GW_p by 2007 and close the gap between demand and supply by 2010 [2]. When the new production capacity became fully operative, a new trend in price reduction with steeper slope started and has been kept during the past 10 years, with a learning rate double than before, reaching 41% and not showing signs of wearing when total installed capacity is about to reach 1 TW_p [10]. The average learning rate for the whole 30-year period has been 21%.

All technologies have experienced a steady learning curve that appears displaced almost parallel one respect to the other in the horizontal axis (which represents cumulative production) and have a similar slope as can be seen in Fig. 10.2. The

A. Urbina, *Sustainable Solar Electricity*, Green Energy and Technology,
https://doi.org/10.1007/978-3-030-91771-5_10

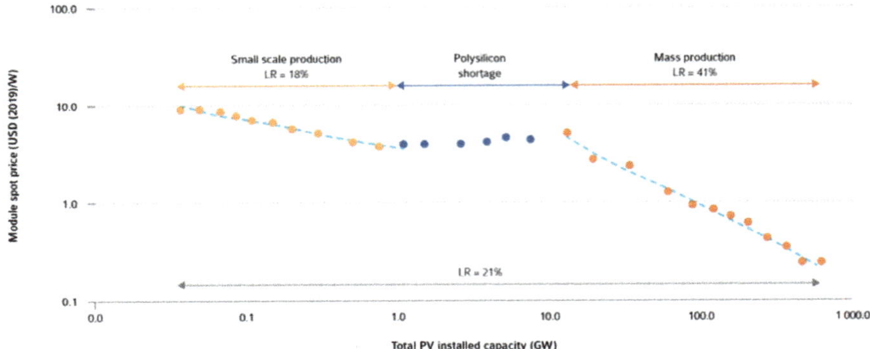

Fig. 10.1 PV modules spot price learning curve (1992–2020). *Source* IEA-PVPS Trends in PV Applications 2020 (Reproduced with permission from [10])

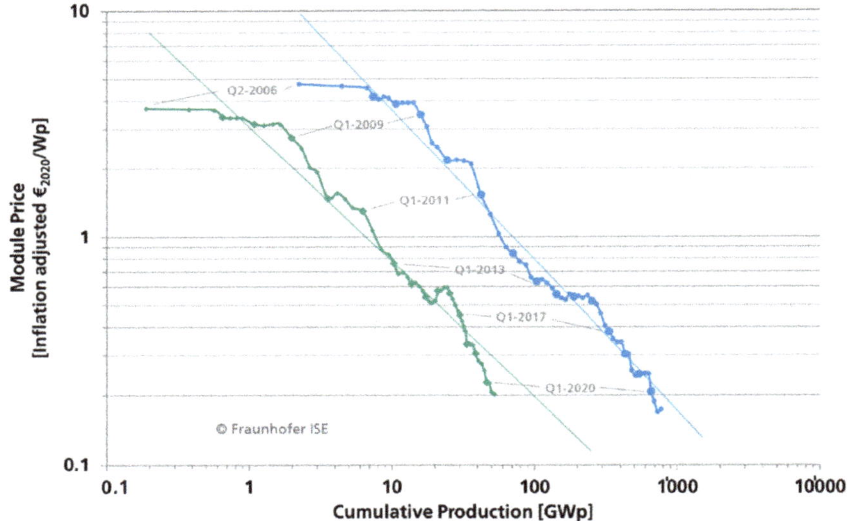

Fig. 10.2 PV modules price learning curve (2006–2020) of crystalline silicon and thin film technologies. *Source* Fraunhofer-ISE PV Report 2021 with estimated data from different sources (Navigant Consulting, EUPD, pvXchange for 2006–2010; and IHS Markit for 2011 onwards; reproduced with permission from: [4])

learning curve of crystalline silicon with a cumulative production at the end of 2020 of 773 GW_p and thin film technology with cumulative production of 52 GW_p (also at the end of 2020) are presented; the resulting learning rates are 32 and 30, respectively, delivering prices at the end of the curve presented in the graph lower than 2 euro cents per W_p for both technologies.

Table 10.1 Average solar PV module prices by technology and manufacturing country sold in Europe in 2020 (the country in parenthesis is where the module was manufactured)

PV technology (2020 USD/W_p)	Average	Cheapest
Crystalline Si (Germany)	0.523	0.271
Crystalline Si (Japan)	0.616	0.271
Crystalline Si (China)	0.535	0.271
Bifacial Si	0.407	0.381
Thin film a-Si/μ-Si	0.864	0.283
Thin film CdTe	0.522	0.281
Low cost (emerging)		0.192

Data source IRENA PV Power Costs Database 2021 [8]

A summary of the PV module cost for the commercial and emerging technologies is provided in Table 10.1 for 2020, according to the most recent IRENA average data. All technologies have reached prices below 0.5 USD/W_p, with reductions of about 70% in the past 15 years.

When the costs of other Balance of System (BoS) components of the PV system are added, with the inverter cost as the more important contribution, the total price of PV systems rises to a range from 2.5 to 0.3 USD/W_p (in 2020), with a clear difference between roof-top or BIPV residential systems and utility-scale large PV plants. The range goes from 2.5 to 0.5 USD/W_p for residential scale and from 1.5 to 0.3 USD/W_p for utility scale, with significant differences among countries as can be observed in Table 10.2 and Fig. 10.3. The breakdown of costs also differ among countries and has varied along time; in 2020, the cost of PV modules approximately represent 50% of hardware cost on average, with the other half for the balance of system costs, mainly inverter (20% at a cost of around 50 USD/kW) and the rest of BoS (30% for wiring, racking, transformer and grid connection, monitoring and control, safety switches and security). PV modules have reduced its price in recent years and therefore its share in the total cost of hardware components of the PV system has constantly decreased, while inverters and other BoS have also reduced its cost, but not so fast and therefore its share in total cost has increased. At an average price of 30 USD/kW, there is still room for cost reduction in inverters, which are experiencing a fast change in design and materials use in recent years. The total cost of systems decreased during the past decade between 69% and 88% depending on the countries of installation [8].

To these hardware costs, additional "soft" costs for system design, transport, installation, financing costs, overheads (margins), permits and taxes, cost of applying for incentives and customer acquisition provisions must be added; an economy of scale is achieved for the soft costs in large utility plants (which represent around 20% of total costs) in comparison to smaller systems (around 30%, for BIPV even larger due to extra design and installation costs); in both cases, variations among countries may be large due to different engineering costs, overheads, regulations and

Table 10.2 Detailed breakdown of utility-scale solar PV total installed costs in the top five countries with most cumulative installed capacity in 2020, the European (1) average (calculated for 22 countries including the Russian Federation and representing more than 99% of European cumulative capacity) and the World (2) average (calculated for 37 countries representing more than 99% of cumulative capacity). *Data source* IRENA PV Power Costs Database 2021 [8] and IRENA Renewable capacity statistics 2021 [9]

		China	USA	Japan	Germany	India	Europe (1)	World (2)	
Capacity	Cumulative (2020)	253.4	93.2	71.4	53.9	47.4	164.8	760.4	
(GW)	Installed (in 2020)	48.2	19.2	11.1	8.2	4.9	21.2	139.4	
Total cost	**2020 USD/kW**	**650.7**	**1,100.6**	**1,832.1**	**699.6**	**595.9**	**919.6**	**971.2**	
Category	Cost component								
Modules	Modules	258.0	356.1	366.9	234.0	223.7	304.8	307.5	307.5
BoS	Inverters	33.8	67.0	132.0	32.6	33.9	45.0	50.0	
	Racking and mounting	16.2	85.6	81.5	57.1	56.6	79.0	78.8	
	Grid connection	56.9	79.5	101.2	67.0	33.2	63.8	65.5	
	Wiring	17.0	64.6	58.0	36.8	38.0	35.5	44.3	
	Safety and security	9.5	27.3	22.3	8.9	34.0	19.6	23.3	
	Monitoring and control	2.5	14.4	18.4	4.8	2.3	6.3	8.9	270.7
Installation	Mechanical	61.6	171.4	485.6	60.0	29.2	91.4	99.1	
	Electrical	41.0	27.5	316.6	46.9	20.4	76.9	73.3	
	Inspection	9.5	7.7	45.2	11.6	5.6	18.1	17.9	190.4
Soft costs	Margin	67.3	125.2	96.0	91.2	23.6	108.3	111.3	
	Financing costs	41.4	14.6	18.2	3.9	52.4	15.7	25.6	
	System design	6.3	28.2	4.1	9.0	12.1	14.8	16.3	
	Permitting	9.2	6.5	39.1	19.5	13.5	29.1	29.8	
	Application	14.2	17.5	38.8	5.1	7.6	5.4	12.3	
	Provision	6.3	7.6	8.2	11.1	9.9	5.9	7.2	202.5

tax policies [10]. The O&M cost at utility scale differs strongly among countries, and IRENA uses for its calculations 17.8 USD/kW$_p$ for OECD countries and 9 USD/kW$_p$ for non-OECD countries for an "all-in" O&M costs that include insurance and asset management (concepts that are usually not reported in O&M surveys) [8].

A detailed breakdown of costs for PV systems in 37 countries representing more than 99% of cumulative installed capacity is available in the IRENA Renewable Power Cost Database [8], a summary for the top five countries with more cumulative capacity in 2020, and European and World averages are presented in Table 10.2 and Fig. 10.3.

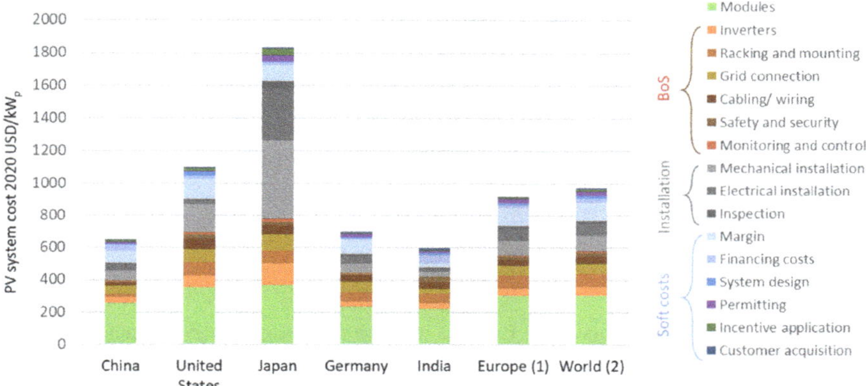

Fig. 10.3 Detailed breakdown of utility-scale solar PV total installed costs in the top five countries with the most cumulative installed capacity in 2020, the European (1) average (calculated for 22 countries including the Russian Federation and representing more than 99% of European cumulative capacity) and the World (2) average (calculated for 37 countries representing more than 99% of cumulative capacity). *Data source* IRENA PV Power Costs Database 2021 [8] and IRENA Renewable capacity statistics 2021 [9]

A graphical summary of the data included in Table 10.2 is shown in Fig. 10.3. A strong variation of costs can be observed, for example, leading to more than the triple total cost difference between China and India with respect to Japan, with Germany and the United States in an intermediate position; this variation does not depend on cumulative capacity (or even on installed capacity in the year for which data of costs were collected, 2020), on the contrary, it is the variations in soft costs and installation (also related to labour cost), the factor that has a larger impact on total system cost. When an average is calculated for 22 European countries and 37 World countries (in both cases representing more than 99% of cumulative capacity, respectively) with the data available at the IRENA Renewable Power Database (2021) [9]; the obtained breakdown is very similar, and the share for the broad categories (Modules, BoS, Installation and Soft Costs) is shown in the pie chart of Fig. 10.4.

If as expected PV module prices become even cheaper, and similarly inverter prices are further reduced, the breakdown of total costs for a PV system will be roughly less than one-third for PV modules, less than one-third for BoS components and more than one-third equally distributed in installation and other soft costs. The pie chart in Fig. 10.4 for the average world data is representative of this trend, despite variations among countries that can be large due mainly to very different installation and soft costs.

The values presented in Table 10.2 for the capital cost of photovoltaic systems in different countries can be compared with those estimated by the IEA for its World Energy Model 2020 [7]; for example, in the stated policies scenario, the starting point in 2019 is very similar to the values indicated in the table, in all cases slightly higher: for example, the USA has the higher cost per installed capacity at the system level,

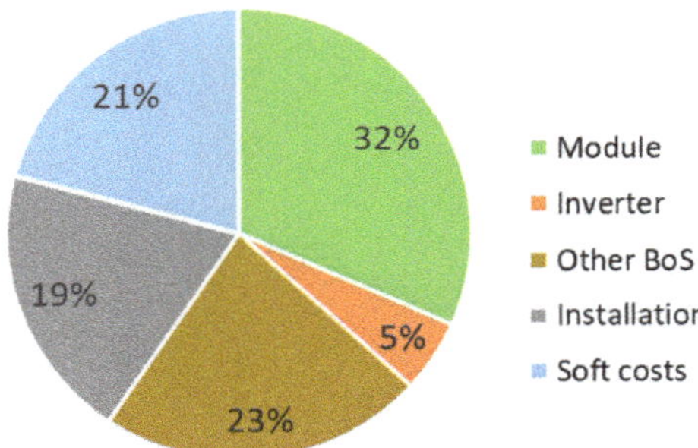

Fig. 10.4 Share of utility-scale solar PV total installed costs for aggregated categories of cost for the 2020 World average (calculated for 37 countries representing more than 99% of cumulative capacity). *Data source* IRENA PV Power Costs Database 2021 [8] and IRENA Renewable Capacity Statistics 2021 [9]

Table 10.3 Estimation of investment, area and staff requirement for a 10 GW PV factory

	Capital expenditure (million euro)	Total manufacturing area m^2	Staff 24/7 (5 shift) max. persons 24/7
Ingot/Wafer	570	210,000	2,100
Cell	970	140,000	2,700
Module	395	150,000	2,700
Total	1,935	500,000	7,500

Source Estimated data by scaling up from a factory size of 1 GW, [4]

1220 USD/kW, the European Union 840 USD/kW, China 790 USD/kW and India 610 USD/kW. The projected costs for 2040 are reduced to 680 USD/kW for the USA, 490 USD/kW for the Europeon Union, 450 USD/kW for China and 350 USD/kW for India, values that can be considered a cap, since the stated policies scenario is a baseline, and it is expected to be surpassed with regards to the penetration of renewable sources in the electricity mix of all countries worldwide.

PV technology has already proven that it is the cheapest source of energy in many countries. This competitive position can be reinforced if coordinated efforts to up-scale production facilities size is accomplished; this has been the case for China, and other efforts in this direction have been launched at the European level: the Fraunhofer Institute for Solar Energy Systems claims that it is possible to work out a coordinated implementation concept via the European Commission through cooperation among countries (such as France–Germany, Germany–Poland, Croatia–Germany…) and other regions, in order to build a factory with a 10 GW PV production capacity per year and it has calculated the costs, the required manufacturing area and staff (see Table 10.3), and it claims that considering today's overall political situation, the chance is high to find investors and build the factory [4].

10.2 The Cost of Solar Electricity: A Steady Learning Curve

The strong reduction in system cost and the improvement in cell efficiency and lifetime have contributed to a constant reduction in the price of electricity produced by a PV system throughout its lifetime.

According to the International Renewable Energy Agency (IRENA), the global weighted average levelized cost of electricity (LCOE) of utility-scale photovoltaic (PV) plants declined by 85% between 2010 and 2020, from 0.381 USD/kWh to 0.057 USD/kWh in 2020. The year- on-year reduction that year was 7%. In Fig. 10.5, the time evolution is presented, showing the average value and the 5 and 95% percentiles to illustrate variations, interestingly, the variations only have a weak correlation with the size of the PV plant, with small (≤ 1 MW$_p$) or large (≥ 300 MW$_p$) systems randomly distributed along the column representing each year, although larger plants tend to accumulate in the middle to lower part of the column.

The range of LCOE that can be obtained with a PV system can vary depending on irradiance and temperature ambient parameters of the location where the PV system is operating throughout its lifetime and also on the global performance ratio which depends on different contributions to losses; even if the cost of ownership of the PV system is the same, the LCOE may vary as shown in Fig. 10.6 for different countries and for three retail prices.

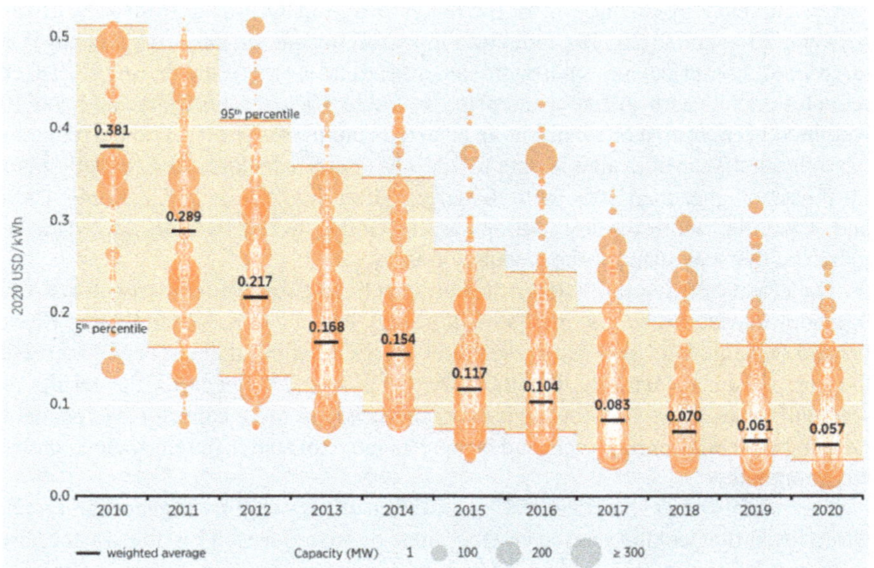

Fig. 10.5 Global utility-scale solar PV project levelized cost of electricity (LCOE) and range evolution from 2010 to 2020, showing 5 and 95% percentiles and representing PV plant size. Reproduced with permission from Ref. [8]

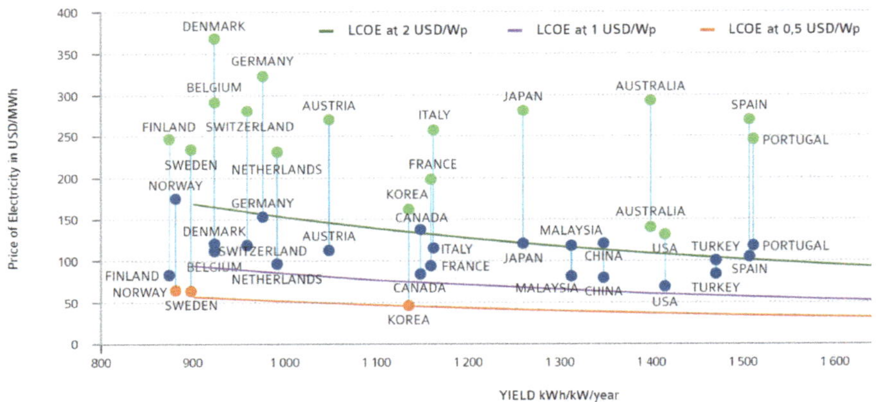

Fig. 10.6 LCOE of PV electricity as a function of solar irradiance and retail prices in key markets. The country yield (solar irradiance) here shown must be considered an average. Figure from IEA-PVPS Task 1 report "Trends in PV applications" (2020). Reproduced with permission from Ref. [10]

The main drivers for LCOE reduction in the past years have been the drop of PV module price (a 46% share of total reduction), followed by inverters (29% share), with also an important contribution of aggregated soft costs, which varies strongly among countries; this contribution to LCOE reduction is represented in Fig. 10.7.

On a country-by-country basis, the reductions in LCOE range from 77% to 88% between 2010 and 2020. The reduction in LCOE during the past 10 years has been large in all countries, although with some differences where the drivers for LCOE reduction acted with different intensity. In Table 10.4, the utility- scale solar PV weighted average cost of electricity in selected countries is shown, and the reduction from the first year of available data to 2020 has been calculated; the reduction is in most cases higher than 80%, with the largest drop in LCOE in India, Korea, China and Australia, while the lower drops are more due to the lack of older data and therefore the year span for the calculation is shorter.

The effect of using the Value Adjusted Levelized Cost of Electricity (VALCOE) indicator introduced by the International Energy Agency in its World Energy Model [7] and explained in Chap.3 is a correction of LCOE values that in general increases the cost of solar electricity. For the IEA Stated Policies Scenario, the results are summarized in Table 10.5 for the four countries where more capacity is expected to be installed in the coming years and is representative of four different socioeconomic environments.

The model shows that solar PV VALCOE is always slightly higher than LCOE, while fossil fuel technologies keep their value or are reduced. This adjustment plays against solar PV in this scenario which is a baseline based on stated policies. If other IEA scenarios more optimistic in terms of renewable capacity deployment are considered (for example, the Net Zero Emissions by 2050), the VALCOE calculation would deliver better results for renewable electricity costs. But it is convenient to

Table 10.4 Utility-scale solar PV weighted average cost of electricity in selected countries, 2010–2020

LCOE (2020 USD/kWh)	2010	2011	2012	2013	2014	2015	2016	2017	2018	2019	2020	Change[a] (%)
Australia	0.380	0.393	0.242	0.142	0.119	0.104	0.081	0.091	0.076	0.074	0.057	–85
China	0.305	0.248	0.185	0.148	0.113	0.086	0.076	0.068	0.053	0.051	0.044	–86
France	0.355	0.371	0.402	0.236	0.161	0.116	0.094	0.086	0.075	0.070	0.064	–82
Germany	0.338	0.292	0.240	0.194	0.146	0.118	0.105	0.099	0.096	0.079	0.066	–80
India	0.309	0.199	0.171	0.173	0.114	0.076	0.072	0.067	0.049	0.039	0.038	–88
Italy	0.380	0.366	0.186	0.167	0.140			0.072	0.064	0.060	0.058	–85
Japan		0.402	0.299	0.250	0.226	0.166	0.171	0.138	0.131	0.124	0.119	–71
Netherlands							0.125	0.145	0.113	0.111	0.107	–14
Republic of Korea	0.503	0.430	0.175	0.210	0.164	0.153	0.148	0.101	0.084	0.079	0.061	–88
Spain	0.292	0.206	0.168	0.130	0.126	0.075				0.049	0.046	–84
Turkey							0.103	0.097	0.076	0.068	0.052	–50
Ukraine		0.448		0.338				0.107	0.108	0.073	0.070	–84
United Kingdom	0.479	0.453	0.272	0.226	0.189	0.150	0.146	0.122	0.128	0.097	0.084	–82
United States	0.204	0.240	0.208	0.207	0.143	0.133	0.121	0.089	0.069	0.059	0.057	–72
Vietnam							0.166	0.243		0.072	0.065	–61

[a]Change from first year of data to 2020 (%)

Data source IRENA Renewable Cost Database (2021) [8]

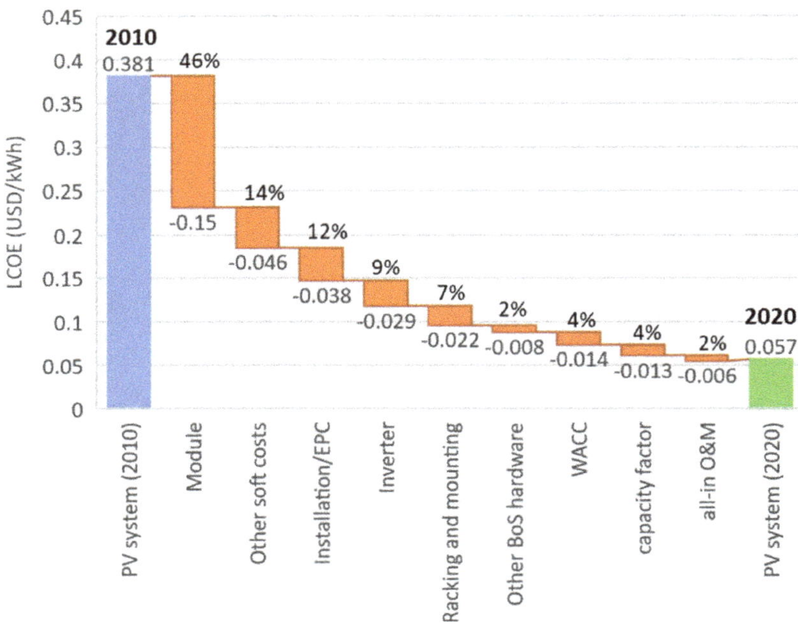

Fig. 10.7 Drivers for LCOE reduction in the past 10 years, showing the absolute reduction in USD/kWh and percentage of contribution of each driver to the total reduction (EPC: Engineering, Procurement and Construction costs ; WACC: Weighted Average Cost of Capital; O&M: Operation and Maintenance). *Data source* [8]

consider the baseline scenario for the discussion because despite the adjustment, solar electricity becomes the cheapest option in three of the considered countries that are representative of three regional socioeconomic environments; only in the case of the European Union, gas is cheaper (and this can be discussed in the view of new regulations for the energy transition in the European Union since it may seem a contradiction for a region with the advanced environmental protection legislation of the European Union and the strong commitments for greenhouse gas emission reductions).

10.3 The Cost of Electricity Storage in Batteries

Batteries will be a strategical part of global energy transition since storage will be required at very different scales. The most rapid development has been the use of batteries in electric vehicles that can store solar electricity for their own consumptions for moving the vehicle or as a buffer storage when the vehicle is parked and can exchange electricity with the grid. The cost of stationary storage is today higher than mobile storage, mainly due to the need of including extra components for battery installation (including packing and health and safety measurements) and more costly

Table 10.5 Levelized Cost of Electricity (LCOE) and Value Adjusted Levelized Cost of Electricity (VALCOE) calculated by the International Energy Agency for the Stated Policies Scenario of the World Energy Model 2020 for Solar photovoltaic, Wind onshore, Coal and Gas (Combined Cycle Gas Turbine (CCGT))

USD/MWh	LCOE 2019	VALCOE 2019	LCOE 2040	VALCOE 2040
China				
Solar PV	55	55	30	55
Wind onshore	65	65	55	60
Coal	50	50	70	60
Gas CCGT	75	70	100	95
USA				
Solar PV	75	80	40	55
Coal	75	75	75	75
Gas CCGT	50	45	65	65
European Union				
Solar PV	85	85	50	80
Wind onshore	75	80	65	80
Coal	120	110	150	125
Gas CCGT	65	50	110	75
India				
Solar PV	35	40	20	50
Wind onshore	55	60	50	55
Coal	55	55	55	50
Gas CCGT	60	60	85	65

Data source [7]

power management and battery protections to deal with more demanding charge and discharge cycles.

A calculation of the levelized cost of energy storing capacity of the battery (averaged throughout battery lifetime) provides an economical comparison of different technologies and poses clear requirements for any new technology in order to penetrate the market. In the United States of America, 869 MW of utility- scale battery capacity was installed at the end of 2018, with an electricity storage capacity of 1236 MWh. The average cost of this storage fell 71% during the past years, and reached 635 USD/kWh in 2020; the decline in price is driven by the development of lithium-ion batteries, which represent 90% of the market and have experienced a steep rate of decline in costs in the past 5 years. Similar trends have been observed in other countries, for example in Germany, where for small-scale, residential and commercial buildings; battery storage is slightly more expensive, with prices around 776 USD/kWh in 2020. Nevertheless, and despite variations among countries, the cost of electricity storage in batteries is declining at strong rates and reached competitive prices in 2020 (see Table 10.6).

Table 10.6 Average prices for electricity storage (net capacity) with lithium-ion battery systems (USD/kWh)

Country	USD/kWh	Year	Scale
USA	635	2018	Utility
Australia	670	2020	Residential
UK	752	2021	Residential
Germany	776	2020	Residential
France	955	2021	Residential
Italy	1069	2021	Residential

Data source IRENA, EUPD Research GmbH, Solar Choice [8]

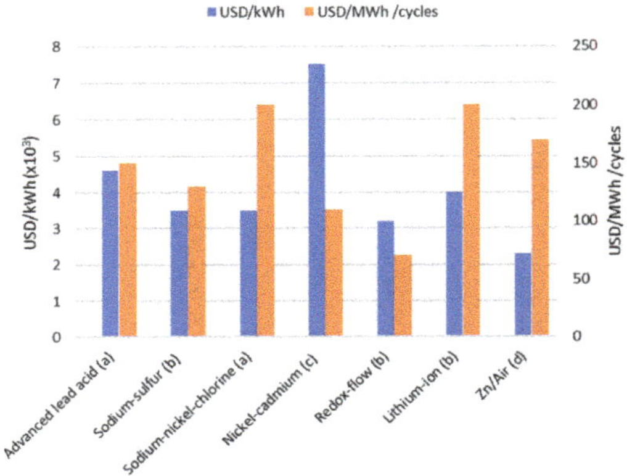

Fig. 10.8 Comparison of capital cost for power (blue, left axis) and energy storage (orange, right axis) for several battery technologies. Data from references as indicated in labels: **a** [1], **b** [3], **c** [11] and **d** [12]. Reproduced with permission from [12]

The cost of electricity storage in batteries will depend on the cost of ownership of the battery and the number and efficiency of cycles during its lifetime. All technologies deliver similar cost per stored electricity as indicated in Fig. 10.8.

10.4 Employment Opportunities Linked to the Solar Electricity Sector

At the end of 2019, the PV sector employed up to 3,5 million people globally, of which 2,2 million work in China clearly leading the PV job intensity in the world; India and the United States of America are following at great distance with about a quarter-million people employed in the PV sector, followed by Europe (170,000) and Japan (74,000), with much lower numbers in the rest of the world at the end of 2019. An estimated 1,3 million were employed in the upstream part (PV and BoS manufacture, more concentrated in a few countries), including materials and equipment, while 2,2 million were active in the downstream part (system construction, including O&M, more distributed where the systems are deployed, strongly varying in time depending on local policies and PV market development). The job intensity is very different for upstream or downstream activities; on average, upstream workload require around 10 full-time equivalent (FTE) jobs per MW_p of produced PV modules, while downstream workload demands 20 FTE per MW_p of installed PV capacity, with differences depending on the size of the system, with small-scale systems generating more jobs than utility-scale large plants. In the whole PV value chain, the job quality is good, with more focus on technical expertise in the upstream (and specially in the research and development stage) and more on manual skills in O&M operations [10].

The report about Sustainable Recovery of the International Energy Agency analysed the important impact that the COVID-19 crisis had produced on energy access [6], where lockdown measures have cancelled or reduced severely new development projects putting off-grid development at risk in one of the PV sectors with more decentralized creation of jobs. More than one thousand small companies employ around half a million people, and this job is now at risk if investment oriented to recovery electrification projects is not delivered. The policies that are announced in developed countries to dynamize the economy (Next Generation funds in Europe, the Infrastructure Bill announced in the USA) should also take into account the importance of maintaining or even enhancing international cooperation projects, and especially those oriented to rural electrification, since it will have a strong impact on job maintenance of a growing sector that is now at risk.

If universal access to electricity is kept as a priority in the world recovery, the decentralized photovoltaic systems could create around 900,000 jobs a year within the next 3 years, as it is estimated in the Sustainable Recovery report cited above, where every 100 solar home systems could generate the equivalent of 20 full-time induced jobs (although mostly informal) with half of them for women, improve health and education services and increase agricultural productivity and hence food security [5].

References

1. Alotto P, Guarnieri M, Moro F (2014) Redox flow batteries for the storage of renewable energy: a review. Renew Sustain Energy Rev 29:325–335. https://doi.org/10.1016/j.rser.2013.08.001, http://www.sciencedirect.com/science/article/pii/S1364032113005418
2. Bye G, Ceccaroli B (2014) Solar grade silicon: Technology status and industrial trends. Solar Energy Mater Solar Cells 130:634–646. https://doi.org/10.1016/j.solmat.2014.06.019, https://www.sciencedirect.com/science/article/pii/S0927024814003286
3. Dehghani-Sanij AR, Tharumalingam E, Dusseault MB, Fraser R (2019) Study of energy storage systems and environmental challenges of batteries. Renew Sustain Energy Rev 104:192–208. https://doi.org/10.1016/j.rser.2019.01.023, http://www.sciencedirect.com/science/article/pii/S1364032119300334
4. Fraunhofer-ISE (2021) Photovoltaics Report 2021. Tech. rep., Fraunhofer Institute for Solar Energy Systems, ISE, Germany. https://www.ise.fraunhofer.de/content/dam/ise/de/documents/publications/studies/Photovoltaics-Report.pdf

5. GOGLA (2020) Global Off-Grid Solar Market Trends Report 2020. Tech. rep., GOGLA - Lighting Global - World Bank. https://www.lightingglobal.org/wp-content/uploads/2020/03/VIVID%20OCA_2020_Off_Grid_Solar_Market_Trends_Report_Full_High.pdf
6. IEA (2020a) Sustainable Recovery. Tech. rep., International Energy Agency - World Energy Outlook Special Report, world Energy Outlook Special Report in collaboration with the International Monetary Fund
7. IEA (2020b) World Energy Model. Documentation. Tech. rep., International Energy Agency. https://www.iea.org/reports/world-energy-model
8. IRENA (2021a) Renewable Power Generation Costs in 2020. IEA-IRENA. https://www.irena.org/publications/2021/Jun/Renewable-Power-Costs-in-2020
9. IRENA IREA (2021b) Renewable Capacity Statistics 2021. IEA-IRENA. https://www.irena.org/publications/2021/March/Renewable-Capacity-Statistics-2021
10. Masson G, Kaizuka I (2020) Trends in Photovoltaic Applications 2020. Tech. Rep. Report IEA-PVPS T1-38:2020, International Energy Agency - Photovoltaic Power Systems Programme - Technology Collaboration Programme, iSBN 978-3-907281-01-7
11. Posada JOG, Rennie AJR, Villar SP, Martins VL, Marinaccio J, Barnes A, Glover CF, Worsley DA, Hall PJ (2017) Aqueous batteries as grid scale energy storage solutions. Renew Sustain Energy Rev 68:1174–1182. https://doi.org/10.1016/j.rser.2016.02.024, http://www.sciencedirect.com/science/article/pii/S136403211600232X
12. Santos F, Urbina A, Abad J, López R, Toledo C, Fernandez Romero AJ (2020) Environmental and economical assessment for a sustainable Zn/air battery. Chemosphere 250:126273. https://doi.org/10.1016/j.chemosphere.2020.126273, GotoISI://MEDLINE:32120147, type: Journal Article

Chapter 11
Standardization and Regulations for PV Technologies

At least three regulatory levels for the production, installation, operation and end of life of photovoltaic systems can be considered. Additionally, the Life Cycle Assessment methodology is also regulated by standards. In this chapter, the three levels are presented. First, a technical approach where the international technical standards of different standardization organizations are cited and briefly commented; the technical documents are accessible upon payment of a fee, unless some country organization provides free access to national standards that are identical or very similar to the international ones. The second level refers to regulatory frameworks (acts, directives and orders) that are mandatory in several regions or countries, and the third one is devoted to concepts that are still under development but that could be part of future broader regulatory frameworks in different regions (especially in Europe). Each section of this chapter focuses, respectively, on these three levels.

11.1 International Technical Standards for Photovoltaic Technology and Life Cycle Assessment

In this section, the main international technical standards regulating photovoltaic technology and life cycle assessment are briefly commented. The regional or national standards are adapted to international standards and sometimes the original document (or a very similar one) is freely available. Otherwise, its full access requires a fee payment per document or subscription to groups of standards.

A. Urbina, *Sustainable Solar Electricity*, Green Energy and Technology,
https://doi.org/10.1007/978-3-030-91771-5_11

11.1.1 International Organization for Standardization

The **International Organization for Standardization (ISO)** provides the wider overarching collection of standards (23,912 in 2021), including the ISO 14000 family for Environmental Management (with the subfamily of ISO14040 for Life Cycle Assessment), the ISO 45000 family for occupational health and safety for better and safer working conditions (developed in collaboration with USA-OSHA and the International Labour Organization, ILO) or ISO 50001 for Energy Management with a strong focus on energy efficiency and savings.

The Life Cycle Assessment methodology is regulated by two ISO standards (and complemented with sectorial recommendations to analyse the impacts on some categories, especially water). They are the following:

ISO 14040:2006. Environmental management—Life cycle assessment —Principles and framework. It describes the principles and framework for life cycle assessment (LCA) including definition of the goal and scope of the LCA, the life cycle inventory analysis (LCI) phase, the life cycle impact assessment (LCIA) phase, the life cycle interpretation phase, reporting and critical review of the LCA, limitations of the LCA, the relationship between the LCA phases, and conditions for use of value choices and optional elements. It covers life cycle assessment (LCA) studies and life cycle inventory (LCI) studies. Approved in 2006, reviewed and confirmed in 2016.

ISO 14044:2006. Environmental management—Life cycle assessment —Requirements and guidelines. It specifies requirements and provides guidelines for life cycle assessment (LCA) (described in the ISO 14040 standard). It extends the definitions list and clarifies issues related to recycling and how to avoid double accountability of processes and outputs related to recycling activities. It also provides case study examples with models for spreadsheets. Approved in 2006, reviewed and confirmed in 2016.

ISO 14046:2014. Environmental management—Water footprint—Principles, requirements and guidelines. It specifies principles, requirements and guidelines related to water footprint assessment of products, processes and organizations based on life cycle assessment (LCA). It provides principles, requirements and guidelines for conducting and reporting a water footprint assessment as a stand-alone assessment, or as part of a more comprehensive environmental assessment. Approved in 2014, reviewed and confirmed in 2020.

On the other hand, ISO standards for photovoltaics are scarce (in comparison to what is regulated by other organizations), they are developed by the technical committee TC-180 and the focus is on setting references for solar spectral irradiance (ISO 9845-1:1992 with a new standard under development) and on irradiance measurement instruments and methods (ISO 9060:2018 and ISO 21348:2007 which were reviewed and confirmed in 2021). The standard ISO 15387:2005 (reviewed and confirmed in 2021) was devoted to calibration and measurements of single-junction solar cells for space applications (under AM0 spectral irradiance). They also published one standard related to specification for glass to be used in building integrated photovoltaic (BIPV) applications (ISO/TS 18178:2018 and an extension with focus on module recycling for BIPV which is under development ISO/TS 21480).

11.1.2 International Electrotechnical Commission

The **International Electrotechnical Commission (IEC)** has developed a broad range of standards for electric and electronic products (more than 10,000 in 2021). The IEC Technical Committee TC-82 for "Solar photovoltaic energy systems" is responsible for writing all IEC standards related to photovoltaic technology since the early 1980s. The standards are constantly updated, and new ones are prepared by working groups to include new technical developments either in the manufacture of new types of PV modules or in the instruments used for their characterization. Also, balance of system components, including batteries, are regulated by standards. There are currently 169 published IEC standards by TC-82 related to photovoltaic technology, and work is in progress for 69 more (new ones or revisions). This set of standards is the most broadly used by the scientific community and technicians in research centres and companies.

The full set of IEC standards related to photovoltaic technology can be found on its website which has an efficient search engine; additionally, a compilation with comments can be found in the European Commission Joint Research Centre report on standards for the assessment of the environmental performance of photovoltaic modules, power conversion equipment and photovoltaic systems [5]. The most important series of IEC standards for PV is the **IEC 60904**, with 11 active parts devoted to photovoltaic devices: Measurement of photovoltaic current–voltage characteristics in natural or simulated sunlight, applicable for a solar cell, a subassembly of cells or a PV module (1); details for multijunction photovoltaic device characterization under concentrated or non-concentrated light (1.1); details for bifacial devices (1.2); requirements for reference devices (2); principles designed to relate the performance rating of PV devices to a common reference terrestrial solar spectral irradiance distribution, which is provided in the standard in order to classify solar simulators according to the spectral performance requirements (3); procedures for calibration traceability to SI units for reference cells (4); description of the open-circuit voltage method for the determination of the equivalent cell temperature (ECT) of PV devices with the purposes of comparing their thermal characteristics, determining the nominal operating cell temperature (NOCT) and translating measured I-V characteristics to other temperatures (5); computation for spectral mismatch corrections (7); requirements for the measurement of the spectral responsivity of both linear and non-linear photovoltaic devices for the determination of recombination processes: it is of special interest for research work (8 and 8.1 with focus on multijunction cells); methodologies for classification of solar simulators, which are labelled and classified as A+, A, B or C based on criteria of spectral distribution match, irradiance non-uniformity in the test plane and temporal instability of irradiance (9); methods to analyse the dependence of any electrical parameter of a PV cell with respect to a test parameter and to determine the degree at which this dependence is close to an ideal linear function and how to deal with deviations from linearity (10); methods to capture electroluminescence images of PV modules and how to interpret it (13); guidelines for measurements of the maximum power output of single-junction PV modules and for reporting at standard test conditions (STC) in industrial production line settings, with the aim to have consistent practices across the industry (14).

This IEC 60904 covers many issues related to photovoltaics summarized above, and it is the most broadly used set of recommendations. It is complemented with other standards, and a brief selection out of the 169 active ones is included here: IEC 60981 with procedures for temperature and irradiance corrections to measured I-V cell and module characteristics; IEC 61215 for the design and qualification of PV modules for terrestrial applications and long-term operation in open-air climates, including test requirements for different technologies: c-Si (1.1), CdTe (1.2), a-Si (1.3), CIGS (1.4) and a more recent publication for organic, dye-sensitized and perovskite cells (IEC TR 63228:2019); IEC 61683 with guidelines for measurement of efficiency of power conditioners; IEC 61724 for monitoring PV system performance (including a capacity and energy evaluation methods); IEC 61730 for safety qualification, with a description of the fundamental construction requirements for PV modules in order to provide safe electrical and mechanical operation and the methods and requirements to carry out the safety tests; IEC 61853 for module performance testing and energy rating, including irradiance and temperature performance measurements and power rating (1), spectral responsivity, incidence angle and module operating temperature measurements (2), energy rating of PV modules (3) and standard reference climatic profiles with climatic data sets (4) with the purpose to define a methodology to determine the PV module energy output and the climatic-specific energy rating for a complete year at maximum power operation for the reference climatic profile(s); IEC 62093 for balance of system components design qualification; IEC 62108 for concentrator PV modules and assemblies; IEC 62109 for safety issues related to power converters; IEC 62446 for inspection, tests, commissioning and maintenance of grid-connected systems; IEC 62458 with recommendations for PV array design and construction including wiring, electrical protection devices, switching and earthing provisions; IEC 62670, IEC 62688, IEC 62789 and IEC 62989 for concentrator PV systems; the broad family of IEC 62788 provides several recommendations for characterization of all kinds of materials involved in the PV system, either cells, encapsulation, glass, backsheet, edge seals (including degradation tests of which potential-induced degradation is treated in more detail with the IEC 62804, glass properties and degradation in the IEC 62805 and crystalline silicon cell light-induced degradation in IEC 63202); solar trackers are covered by IEC 62817; connectors in IEC 62852; maximum power point trackers in IEC 62891; inverters in IEC 62894. Also, a set of different tests for PV system construction quality control (in factory and on-site) and commissioning as well as maintenance procedures are described broadly in several standards.

Regarding stand-alone PV systems which include a battery for electricity storage, several standards have been approved. First, to regulate system design and battery function: IEC 62124 for stand-alone PV system design recommendations and PV performance evaluation (including battery testing and recovery after periods of low state-of-charge) in a variety of climatic conditions, and IEC 62509 for battery charge controllers. But secondly, a whole set of standards has been developed for rural electrification with PV systems: the IEC 62257 family.

Another important family of IEC standards is **IEC 62257**, which introduces a methodology for implementing rural electrification using autonomous hybrid renewable energy systems for setting up decentralized rural electrification in developing countries or in developed countries. There are 11 active parts for this standard with an ambitious objective of covering all stages of this kind of PV projects; after an introduction (1), it includes recommendations for the analysis of the socioeconomic conditions of the rural area where the decentralized electrification project is going to be implemented (2); project development and management, covering from contractual issues with stakeholders to technical test on the PV system and even requirement for recycling and environmental protection (3); recommendations for system selection and design for isolated systems in order to meet end-user identified needs (4); protection against electrical hazards (5); proposes a methodology to achieve the best technical and economic conditions for acceptance, operation, maintenance and replacement of equipment and complete system life cycle (6); a broad approach to other renewable systems or hybrid generators (7); the selection of batteries and the possible use of car flooded lead–acid batteries for a second life in PV systems are also addressed with regulations for tests to check if new or used car batteries can be used in PV stand-alone systems assuring a proper operation of the system, therefore creating a battery management system which covers the whole cycle including end of life (8); the other approach for rural electrification with micropower systems and microgrids are also addressed (9); and finally, recommendations of visual inspection of silicon modules for battery maintenance (10) and recommendations for laboratory tests for lamps and other appliances to be used in off-grid rural electrification (12).

A new standard has been recently approved, the **IEC 62994** (2019) devoted to environmental health and safety risk assessment of PV systems throughout its lifetime; it proposes a method to characterize and evaluate potential adverse impacts to human health or environment and make it possible to take measures to reduce them in the first attempt by a standardization organizations to tackle the sustainability evaluation at a technical level in the manufacture, use and end-of-life phases of PV systems by combining life cycle assessment and risk assessment methods (for routine and non-routine operation), and additional guidance for risks in PV systems in buildings addressed fire hazards in the more recent IEC TR 63226 document (2021).

11.1.3 Other International and National Standardization Organizations

The **Institute of Electrical and Electronics Engineers (IEEE)** has a Standards Coordinating Committee SCC-21 on Fuel Cells, Photovoltaics, Dispersed Generation, and Energy Storage whose standardization work focused on grid connection and minigrid quality of supply with distributed energy sources (IEEE Std. 1547 series 1–7, updated in 2020 and IEEE Std. 2030.9-2019 for microgrids), for smart grid interoperability (IEEE Std. 2030–2011), for discrete and hybrid energy storage systems that are integrated with the electric power infrastructure, including end-use applications and loads (IEEE Std. 2030.2-2015) and for design, operation, maintenance, integration and interoperability, including distributed resources interconnection, and of stationary or mobile battery energy storage systems (IEEE Std. 2030.2.1-2019).

The **American Society for Testing and Materials (ASTM International)** is another organization that provides more than 13,000 standards, of which around 300 are devoted to photovoltaic technology with recommendations which range from materials quality, monitorization, maintenance or safety requirements which provide a good complement to other regulations (work of Subcommittee E44.09 on Photovoltaic Electric Power Conversion, which is part of Committee E44 on Solar, Geothermal and Other Alternative Energy Sources). For example, their standards for the mechanical integrity of modules (ASTM E1830—15(2019)), standard tests for hot spot protection (ASTM E2481—12(2018)), fire prevention (ASTM E2908—12(2018)), insulation integrity and ground path integrity (ASTM E1462—12(2018)) or hail impacts (ASTM E1038—10(2019)) are very useful and cannot be found in other standards at this detailed level.

Several countries have their own institutes for standards which grant access to international ISO or IEC standards, develop their own standards and provide consultancy services to companies and institutions worldwide. The most important are the British Standards Institution (BSI), the Deutsches Institut für Normung (DIN) and the Japanese Standards Association (JSA). Although the main reference for standards regarding solar electricity is the IEC and the most important standards are those cited above, the national bureaus provide complementary details to many of the international standards, with detailed worked examples and usually, the compliance with national standards in the manufacturing of PV modules or design and construction of PV systems is compulsory.

11.2 Regulatory Frameworks for Production, Recycling and End of Life of PV Modules

Production of photovoltaic modules are not specifically regulated in any legislative framework, with the exception of recent Chinese policies. For the rest of the world, the environmental impact regulations that in general apply for any industrial activity are the ones that should be applied to the PV industry; even in some cases, the PV industry is exempted to comply with them. Very different is the framework for recycling and end of life of PV modules, which are treated as electrical and electronic equipment and it is strongly regulated in several countries; but there are comparatively very few standards with recommendations for PV recovery, reuse and recycling. PV modules and other balance of system components at their end of life are considered E-Waste.

In 2017, the Global E-Waste Report (United Nations University) warned that the waste from electrical and electronic equipment have been 44.7 million tons in the year 2016 (around 6 kg per person on a world average basis) [2]. And for 2021, it is estimated to surpass 50 million tons. Photovoltaic panels (and inverters) were only partially included in the calculation, since data gathering was difficult, with only a few countries recording the amount of decommissioned panels, and only partially.

Despite this growing problem, none of the IEC standards for PV technology does cover recycling issues, with a few exceptions: IEC TR 62635(2012) with guidelines for end-of-life information provided by manufacturers and recyclers and for recyclability rate calculation of electrical and electronic equipment, and more specifically for PV technology, the IEC 62257-3 (2015) which includes recommendations for recycling of components and environmental protection in rural electrification projects with PV systems and the one devoted to health and safety issues IEC TS 62994 (2019), both commented in the previous section.

A few ISO standards can be found for recycling of materials, and those are strongly focused on rare-earth elements. The ISO recognizes the need to extend regulations to recycling and recovery of materials and has started a programme to analyse the current situation and provide recommendations. The first task was devoted to plastics in 2008 and reviewed in 2018 (the results of the study were published as a standard ISO 15270); a more recent document (ISO/TR 23891:2020) gives a brief overview of the current situation in plastic recycling systems, compiles relevant existing standards and provides a short description of different recycling techniques and methods for recovery of plastic waste arising from pre-consumer and post-consumer sources; it aims to identify the necessity of standards in the plastic recycling system and give direction for the adoption of regional standards and/or the development of new and existing standards. Curiously, there is one single standard for recycling of glass to be used in BIPV applications (document ISO/TS 21480 under discussion, to be approved in 2021). More recently, ISO has developed a set of standards devoted to recycling of rare-earth elements (REE) (ISO 22450:2020, ISO/TS 22451:2021, ISO 22453:2021) and focused, respectively, on the requirements for providing information on industrial waste and end-of-life products (including a classification scheme), on measurement methods for quantifying REEs in industrial wastes and end-of-life products in solid, solid–liquid mixture or liquid forms, and finally on methods of information exchange between waste handlers and recyclers for REEs contained in industrial waste and end-of-life products. The aim is to increase the amount of REEs recovered from recycling and thus reducing the dependency on mining production and the associated geopolitical risks. These standards have limited impact on the PV industry, where the use of plastic is low and the content of REEs in PV modules is almost non-existent (although the dependence is higher in electronic equipment of

BoS). Another ISO standard is more relevant for the end-of-life stage of PV modules and may have an impact on "design for recycling" of components; it is the technical document ISO 8887-1:2017 (Design for manufacturing, assembling, disassembling and end-of-life processing) which was written with the aim to provide multiple life cycles to manufactured products; interestingly, it extends beyond specifications for manufacturing and assembling of products and incorporates guidance on the ultimate reusing, recovering, recycling and disposing of the components and materials used in the manufacturing process and during the operational lifetime (including maintenance and spare parts); a second part for this standard is under development.

In the following subsections, a brief summary of regulatory frameworks is presented for the main countries where either PV manufacture or PV installation is being carried out.

11.2.1 China

China has become the main PV manufacturer worldwide as was presented in Chap. 1. It is also the main PV installer in the world. Comparative LCA studies have shown that cumulative energy demand for PV module production was higher due to lower grid efficiency (on top of the electricity mix which includes an important contribution from coal plants) [10, 21]. Additionally, environmental impact regulations have been weak for Chinese industrial production. An effort was initiated by the Ministry of Industry and Information Technology since 2013, and reinforced in the more recent 14th Five Year Plan, with the aim to set standard conditions for the photovoltaic industry and promote a "healthy development" of the industry [12, 13]. Enforcement of central policies by local authorities are often weak, and the impact of these regulations is slow to penetrate all the supply chain for the PV module industry, starting with the more than 20 poly-silicon Chinese manufacturers. On the other end of the industry, the installation, operation and end of life of PV systems installed in China are requested to comply with environmental impact assessment reports required by the Environmental Protection Law originally adopted in 1989 and amended in 2014; it is a strongly regulated system which requires permits by the National Development and Reform Commission or the provincial equivalent and subsequent authorization of the Ministry of Land and Resources (MLR), the Ministry of Environmental Protection (MEP) and the Ministry of Water Resources. For end-of- life and recycling treatment of installed PV modules, the lack of strong enforcement of the electronic waste regulations already approved has created important environmental and health problems in China, which on top of its own electronic waste (including PV modules) is the first world importer of this kind of waste; the problem is to tackle informal recycling activities and implement effectively an already strong legislation of the sector.

11.2.2 European Union

In the European Union, globally the second region in manufacture and PV cumulative installation, the use of hazardous substances in industrial manufacturing is regulated by the Restriction of Hazardous Substances in Electrical and Electronic Equipment (RoHS) Directive 2011/65/EU of the European Parliament and of the Council (new consolidated text in April 2021 [8]); but for now, PV module manufacture is exempted to comply with this directive (Article 2, 4.i, which indicates that *"The directive does not apply to photovoltaic panels intended to be used in a system that is designed, assembled and installed by professionals for permanent use at a defined location to produce energy from solar light for public, commercial, industrial and residential applications"*). Indirectly, the directive may affect PV systems, since electronic power equipment of BoS components are not included in Article 2, 4.i exclusion, referred only to PV "panels". Lead and cadmium are strongly restricted (and broadly used in electronic components); if in the future this directive is to be applied also to PV module manufacture, it may pose a restriction to the manufacture of CdTe technology since the maximum allowed concentration of cadmium is 0.01% (weight), while for lead-based perovskite technology the restriction would be compatible with PV manufacture since maximum allowed lead concentration is 0.1% weight, compatible with current technology.

On the other hand, when considering end-of-life stages, the most important regulation affecting recycling of electrical and electronic equipment, which include PV system components, is the Directive 2012/19/EU of the European Parliament and of the Council on Waste Electrical and Electronic Equipment (WEEE) [9]. Its declared purpose is (Article 1)

> This Directive lays down measures to protect the environment and human health by preventing or reducing the adverse impacts of the generation and management of waste from electrical and electronic equipment (WEEE) and by reducing overall impacts of resource use and improving the efficiency of such use in accordance with Articles 1 and 4 of Directive 2008/98/EC,[1] thereby contributing to sustainable development.

Photovoltaic panels are mentioned explicitly in Articles 5 and 7 and included in the list of Annex I (more detailed in further annexes) clearly stating that the WEEE directive applies to the treatment of photovoltaic modules until their end-of-waste status is met or fractions of the photovoltaic modules are sent for recycling, recovery or disposal. The directive is mandatory for all European Union Member States, and it has ruled important issues regarding electrical and electronic waste management:

- The design and production of electrical and electronic equipment to facilitate reuse, dismantling and recovery of WEEE, its components and materials, is pro-

[1] Directive 2008/98/EC of the European Parliament and of the Council of 19 November 2008 on waste and repealing certain Directives; article 1 sets the subject and scope as "protect the environment and human health by preventing or reducing the adverse impacts of the generation and management of waste and by reducing overall impacts of resource use and improving the efficiency of such use"; article 4 establishes a waste hierarchy (a) prevention, (b)preparing for re-use, (c) recycling, (d) other recovery, e.g. energy recovery, and (e) disposal.

moted. It points to ecodesign requirements that are currently under discussion in the European Commission.

- Minimum WEEE collection rates are established. They must rise from 45% in 2016 to 65% in 2019, calculated as the average weight of electrical and electronic equipment placed on the market in the three preceding years, or alternatively 85% of the weight of electrical and electronic waste generated on the territory.
- Collection and transport of separately collected WEEE should be carried out in a way which allows optimal conditions for preparing for reuse, recycling and the confinement of hazardous substances. Reuse is given high priority and all treatment must be "proper", a definition which is left for future regulation of Member States that shall introduce certified environmental management systems.
- The European Commission request the European Standardisation Organisations to develop European standards for the treatment of WEEE (including recovery, recycling and preparing for reuse). This formal request aims at developing a "harmonized standard" which according to European Regulation (EU) 1025/2012 (article 2) is a standard recognized by any of the European Standardisation Organisations (CEN, CENELEC or ETSI.[2]) This has prompted important efforts to extend these standardization procedures at the European level as explained below. An already achieved practical consequence is that for the end-of-life phase of a PV system, a decommissioning plan is becoming a requirement for large systems, and facilities and processes are now being developed to handle modules and ensure proper treatment according to this directive.
- Appropriate inspections and monitoring to verify the proper implementation of this directive are left to Member States future regulations. The inspections should include shipments which export WEEE, an increasing activity which is producing the problem of uncontrolled dumps of toxic waste in developing countries.

This European regulation has put the European Union in a leading position with regard to recycling of electrical and electronic equipment in general and photovoltaic technology (modules and BoS) in particular. The global WEEE collection rate was 42% in 2018 in the European Union (the most recently available statistic for 28 member states, if the UK is excluded, for EU 27 member states the rate goes down to 38.9% according to Eurostat estimates). For category 4 WEEE (consumer equipment an photovoltaic panels), it is even lower with a European average of 14% (27 member states in 2020), but although waste collection rates are lagging behind objectives in many Member States, the policy has created a good framework for investment in recycling pilot plants (those presented in Chap. 9) and promoting research projects to develop and improve recycling methods for the different PV technologies. Nevertheless, the WEEE Forum[3] calls on legislators and policymakers to stop applying

[2] European Committee for Standardization (CEN), European Committee for Electrotechnical Standardization (CENELEC) and European Telecommunications Standards Institute (ETSI); only standards developed by these three European Standards Organizations (ESOs) are recognized as European Standards.

[3] WEEE Forum, a not-for-profit association of 43 WEEE producer responsibility organizations across the World, https://weee-forum.org/.

collection targets based on the "put on the market" methodology to PV panels and lay down realistic separate collection targets specifically aimed at PV panels that should be considered as a separate category (to be introduced in the European WEEE directive).

11.2.3 United States of America

The production of photovoltaic modules in the United States is regulated by the federal Clean Air (1970) and Clean Water (1972) Acts that are applied to any industrial production. The Federal Government approved these pieces of legislation due to the inactivity of state legislatures which did not regulated the emissions to air and water at the state level, a regulation vacuum which had led to serious pollution problems. These acts, together with the Resource Conservation and Recovery Act (RCRA) (1976) and the Toxic Substances Control Act (1976), regulate the emissions to air and water of any substance that may create a potential harm to the environment and human health. The study and classification of substance potential risks is carried out by the EPA and OSHA, federal agencies mentioned in Chap. 5. The photovoltaic industry in the USA is required to comply with the regulations if they use any of the substances listed by EPA, but there is a lack of enforcement mechanism that leaves to company's decisions the real application of these regulations.

By contrast with the European Union, the United States of America has no federal law requiring electrical and electronic waste collection, reuse or recycling (only acts for general waste management and potentially hazardous waste should be applied); instead, 25 states have passed their own WEEE legislation, with a broad range of scope and objectives, ranging from more strict rules in California to other very light-touch non-compulsory regulation aimed at voluntary compliance from end-users and companies. The result is that around only 22% of WEEE is accounted for [20]. The recycling procedures are regulated by the Resource Conservation and Recovery Act (RCRA) mentioned above, with specific Regulations, for example, for silicon tetrachloride, a by-product of poly-silicon production which is now recovered and recycled, but most substances are not regulated. Since 2012, a Toxicity Characteristics Leach Procedure (defined by EPA) has to be applied to the manufacturing and recycling processes of any new photovoltaic technology in order to evaluate its potential toxicity level.

Many countries develop their own national standards and in many cases go beyond international regulations with some requirements for quality or sustainability of products. An example is the American National Standards Institute (ANSI) in collaboration with NSF International has developed the standard NSF/ANSI 457-2019 focused on "Sustainability Leadership Standard For Photovoltaic Modules And Photovoltaic Inverters" [14]. The USA also launched the initiative called "Energy Star: Guidelines for Energy Management", focused on energy savings in the built environment, both for green buildings (residential or commercial) and for industrial energy management.

11.2.4 Other Countries

In the United Kingdom, regulations are implemented by national legislation, in particular the Environmental Permitting Regulations 2010 No. 675, which applies in England and Wales and obliges any industrial operator to request a permit for certain specified activities with the aim to minimize impacts on the environment and human health. It is a very general regulation but curiously contains specific treatment for electrical and electronic equipment (Schedule 12), titanium dioxide (Schedule 17, which may affect emerging PV technologies) and batteries and accumulators (Schedule 19). Similar legislation applies in Scotland and Northern Ireland. The European Union WEEE directive was transposed to the UK legislation and will effectively be applied even after Brexit.

In India, the environmental and human health impacts of industrial production are (weakly) regulated at the state level under the federal framework of the Environmental Protection Act (1986). Only large PV plants, in excess of 50 hectares of land occupation, are required to comply with environmental impact assessments. The electronic waste management in India is becoming an important problem since it is a country importing increasing amounts of electronic waste; attempts to regulate the sector started with the E-Waste Management and Handling Rules that were introduced in 2011 and were updated with the Solid Waste Management Rules and the Hazardous and Other Waste Rules in 2016. Any company recycling electronic waste must get a permit after inspection and approval by a board (at the state level). Nevertheless, the growing informal sector for electronic equipment recycling escapes regulation and the efficiency of the licence scheme is reduced. The regulations have to be applied on the basis of toxicity of PV panel components, and the problem with this approach is that the toxicity of PV modules is usually not declared and therefore the PV modules are not correctly treated at end of life [1, 15].

Despite the fact that in Japan PV module waste is expected to reach 1 million tons by 2030 and 7 million tons by 2050, there is no specific regulation for end-of-life management of PV modules [19]. Japan has a strict Law for the Recycling of Specified Kinds of Home Appliances (LRHA) since 2009 which puts responsibility and costs of recycling on consumers; recycling of electronic appliances has reached around 50%, but still a large part of electronic waste generated in Japan is exported to other countries. The Japan Photovoltaic Energy Association has proposed a roadmap for end-of-life management of PV modules that will be implemented by companies on a voluntary basis and has triggered some research and development effort for PV recycling technologies (for example, Mitsubishi Materials Corporation, Toho Kasei Co., Ltd., Hamada Corporation and Shinryo Corporation). The PV CYCLE organization has launched a Japanese chapter in collaboration with the Akita Prefectural Resources Technology Development Organization in 2021; this private foundation, specialized in mineral resources, aims to exploit the waste management market to recover materials from PV panels in the context of an increasing circular economy in Japan.

Like Japan, South Korea lacks a specific regulation for PV module end of life, and it is a big exporter of electronic waste to other countries. The regulations applied to photovoltaic waste are the Enforcement Rule of Wastes Control Act (Act 14783, 2017) which aims to reduce the amount of generated waste and sets recovery and recycling targets. The promising market for PV recycling in Korea is moving some companies to develop new technology and invest in pilot plants; for example, the Korea Institute of Energy Research (KIER) has announced in August 2021 a "non-destructive" technology that is claimed to recover 100% of glass and 20% of silicon to be reused in the production of new solar cells. The South Korean government (Ministry of Trade, Industry, and Energy) has built a PV recycling centre in North Chungcheong [11].

An increasing problem that affects countries already mentioned (China, India) but which is extended to many other countries with very weak environmental protection frameworks in Africa, Latin America and South East Asia is the legal or illegal export of electronic waste to be treated in countries where the cost is low, labour is exposed to health risks and substances are released to the environment because it is an activity which is not regulated or, if regulated, no enforcement is effective. This is often called a "race to the bottom" in the search of maximizing benefits, or simply, to avoid the cost of externalities that in this case are externalized also physically out of the country which generates the waste. Since the PV industry is set to grow in the next 2 years up to the Tera Watt scale of cumulative installed capacity, it is urgent to create a global framework to avoid this "race to the bottom" with respect to PV module recycling. Ideally, the modules should be recycled in the country where the PV system has been installed (with the additional benefit of avoiding emissions related to long transport routes), but there is no legislation to apply this recommendation in any country. Only a true global approach to solar electricity worldwide will avoid the worst effects of the externalization of electronic waste processing, including PV modules and other balance of system components.

11.3 Ecodesign, Ecolabelling and Green Public Procurement

A more broader scope is also being developed encompassing circular economy concepts, such as ecodesign, ecolabelling and green public procurement. This kind of market-related tools are considered an efficient substitute to transnational regulations which are difficult to negotiate and even more difficult to implement. These tools create a differentiation between products that are not obtained by normative regulations but by stakeholder- and end-user-informed choices; ecolabelling is more related to individual choices, ecodesign to industry-led voluntary actions guided by corporate social responsibility and green public procurement (or similar public purchasing tools) to institutional policies.

The European Commission has published in 2021 two reports, the first one is a discussion paper on potential ecodesign requirements and energy labelling schemes [7], and the second one is a preliminary report of the Joint Research Centre devoted to exploring the options and feasibility of EU Ecolabel and Green Public Procurement (GPP) criteria [4], in both cases for solar photovoltaic modules, inverters and whole systems.

11.3.1 Ecodesign

Regarding ecodesign, despite standards ISO 8887-1:2017, more general, and IEC TR 62635:2012 which indirectly focus on ecodesign recommendations for electrical and electronic equipment, there is still no ecodesign approach in the manufacturing lines of PV modules or BoS components. The reuse or recycling approach is not applied to the manufacture of commercial products, although the first steps for an ecodesign approach are being applied at the research level for emerging technologies.

Life Cycle Assessment of PV technologies provide useful results for ecodesign considerations, because it identifies the materials and processes that are more harmful to the environment and human health or that have more embedded energy and therefore guide future research and development to tackle the more demanding issues. The design of next generation devices and processing routes benefits from this information. For example, many studies point to the electricity consumption during metal deposition as the main contributor to energy embedded in thin film PV module manufacture, and future ecodesign developers should look for alternatives to these methods [3]. This recommendation is coincident with what is observed for emerging organic and hybrid technologies, and alternatives have been proposed and analysed from an LCA perspective [6]. Another example is the design of encapsulant materials that can be either more easy to separate from the cells at the end of life and then facilitate crystalline cell recovery and recycling of materials, or in a different approach, the encapsulation is used as an active material that at the end of life of the cell captures the toxic (or scarce) materials whose release to the environment should be avoided, thus facilitating a recovery process.

These ecodesign considerations must be extended to the manufacture of the balance of system components of PV systems, going beyond what is recommended in the IEC TR 62635:2012 technical report and focusing on life extension of the components (for which systematic assessment of failures is a useful tool) and reduction of embedded energy and materials in the future design of inverters, support structures and batteries for stand-alone systems.

The investment in ecodesign can be boosted if Extended Producer Responsibility (EPR) for photovoltaic-related products (PV modules and balance of system) is included in national and regional legislation. The EPR schemes make manufacturers responsible for end-of-life treatment of the product they have manufactured (responsibility shared with end-users); if this responsibility is translated into liability and possible fines if objectives for recycling and low-impact land filling are not achieved, then companies will improve their manufacturing lines and include the ecodesign approach for longer lifetimes and better recyclability at end of life of the products that they manufacture. The PV industry can be a good example where these kinds of policies are implemented and evaluated.

11.3.2 Ecolabelling

Ecolabelling is nowadays very extended and applies to a broad range of products. There is potential for a new EU ecolabel product group devoted to PV systems, but it is not clear how the scheme could be implemented because much of the retail sales of modules and inverters are business-to-business products and when the end-user receives the product, the market decisions have already been taken by the companies which design and build the PV system. Therefore, it seems that the ecolabelling will be more effective for "do-it-yourself" kits where final end-users make the decision of buying the PV system in the retail shop; this will limit the ecolabelling scheme to small roof-top systems (around 5 kW_p). The European Union should take into consideration the recommendations of the NSF/ANSI 457-2019 standard and coordinate actions for an ecolabelling scheme on a global scale for PV systems.

Other schemes similar to ecolabelling are being implemented by private initiatives like the Solar Energy Industry Association (SEIA) in the USA, which has established a Solar Industry Environmental and Social Commitment programme; the companies can apply to be part of the programme if they comply with several key performance indicators that are related to environmental impact reduction and human health protection; the SEIA publishes a report about the "sustainability" achievements of the companies added to the commitment programme. All the schemes work on a voluntary basis, but the companies must be proactive and transparent in publishing their own evaluation reports. A similar scheme is used by the Silicon Valley Toxics Coalition (SVTC) which applies also to PV manufacturers. Solar Power Europe (formerly the European Photovoltaic Industry Association, EPIA) publishes a report with information about specific achievements with several technical parameters which inform about the progress towards a more sustainable production for all PV technologies.

It is now common practice to relate these parameters or indicators to the Sustainable Development Goals proposed by the United Nations Organization and voluntarily adopted by 195 countries in 2015 with the aim to end extreme poverty, reduce inequality and protect the planet by 2030. Within a few years, an evaluation of this kind of voluntary schemes and targets linked to corporate social responsibility should show if they have been useful to increase the sustainability of the PV industry or are just a new stage of greenwashing marketing.

11.3.3 Green Public Procurement

In some countries, local regulations for energy consumption of public bodies have included percentages for a minimum consumption of energy from renewable sources. But up-to-date, photovoltaic-related products have not been specifically considered for Green Public Procurement (GPP) policies; nevertheless, photovoltaic technology comprises a group of products that could be easily incorporated because they have a clear link to objectives related to GPP schemes: they provide greenhouse gas emissions that are "avoided" in a quantitative way that can be easily calculated, contribute to energy efficiency and reduce air pollution in cities where ambitious solar roofs programmes are implemented. Therefore, solar electricity is a good candidate for GPP policy. Acting as a positive feedback, inclusion of PV in GPP policies will contribute to supporting greater deployment and yield optimization of solar photovoltaic power, will promote the reduction of environmental impacts along the life cycle of solar photovoltaic systems and components (to make them more eligible in a competitive GPP selection within other renewable sources for electricity) and can contribute towards the achievement of grid parity for the cost of solar electricity by promoting the best practices in design optimization and component selection. GPP will supplement market options in the currently ongoing process that is stimulating innovation in module and inverter manufacture, in PV system design with better performance ratio and in more user-friendly monitorization schemes that improve operation and maintenance strategies.

In the European Union (EU), the revision of the Renewable Energy Directive (2009/28(EC)) established a target of 20% for the percentage of renewable energy consumption in the EU by 2020. Globally, the target has been accomplished on average, because some countries clearly surpassed this level, but many others are lagging behind. One tool to comply with the target for 2030 (32%, there are proposals to be increased to 40%) is the practical implementation of Green Public Procurement policies. It will require close collaboration between agencies of the European Commission, working on evaluation reports of the contribution to sustainability of a given product, such as the Eco-Management and Audit Scheme (EMAS) which is a management tool developed to help companies and other organizations to evaluate, report and improve their environmental performance, or the EU Product Environmental Footprint Category Rules (PEFCR) which complements conventional Life Cycle Assessment of the PV systems [16, 18]. The United Kingdom developed an economic tool, the Renewable Obligation Certificates (ROCs) that are granted to electricity suppliers which are compelled to reach a minimum target of renewable generation; these ROCs can be sold to other suppliers creating a secondary market trading system which acted as an incentive to the production of solar electricity. Suppliers either produce enough renewable electricity or purchase ROCs to cover the deficit; failing to do so, they have to pay a penalty.

In the United States of America, the Energy Policy Act (2005) has regulated at the federal level a minimum of 5% of energy consumption from renewable sources in all federal departments (to be implemented between years 2010 and 2012) and rising to

7.5% thereafter; this act led to a practical Clean Energy Contracting requisites for all public Power Purchase Contracts by federal agencies (with some exceptions, like the Department of Defence) but limited to the maximum time extension that PPCs may have (10 years) which is detrimental for renewable energies whose economic return maximization often requires longer payback times [17]. Photovoltaic technologies have EPBTs well below this limit, but in order to compete in this kind of contracts, margins are only sufficiently high with longer periods. Many states in the USA have set targets for Renewable Electricity Standards which compel electrical utility companies to produce at least a certain percentage of electricity from renewables, in some cases also including specific targets for solar electricity (California is the most ambitious with 33% in 2020 and Hawaii has proposed 40% by 2030).

In India, the Electricity Act (2003) created two instruments that are pushing for a renewable electricity penetration in the market via green procurement: one oriented to energy suppliers, which are obliged to purchase and distribute a specific amount of energy from renewable sources, the Renewable Purchase Obligation (RPO) and Renewable Energy Certificate, and the other one to state authorities that are compelled to adopt an RPO to their own energy purchases; these policies are being adopted by the Indian states at a different pace, with Rajasthan, Tamil Nadu, Karnataka and Punjab leading the way [17].

References

1. Bajagain R, Panthi G, An YJ, Jeong SW (2020) Current practices on solar photovoltaic waste management: an overview of the potential risk and regulatory approaches of the photovoltaic waste. J Korean Soc Environ Eng 42(12):690–708. https://doi.org/10.4491/KSEE.2020.42.12.690. Publisher: Korean Society of Environmental Engineers
2. Baldé CP, Forti V, Gray V, Kuehr R, Stegman P (2017) The Global E-waste Monitor—2017. Tech. rep., United Nations University (UNU), International Telecommunication Union (ITU) & International Solid Waste Association (ISWA), Bonn/Geneva/Vienna, iSBN Printed Version (UNU): 978-92-808-9053-2 / (ITU) 978-92-61-26311-9 ISBN Electronic Version (UNU): 978-92-808-9054-9 / (ITU) 978-92-61-26321-8
3. Chatzisideris MD, Espinosa N, Laurent A, Krebs FC (2016) Ecodesign perspectives of thin-film photovoltaic technologies: a review of life cycle assessment studies. Life Cycle, Environ, Ecol Impact Anal Sol Technol 156:2–10. https://doi.org/10.1016/j.solmat.2016.05.048, https://www.sciencedirect.com/science/article/pii/S0927024816301556
4. Dodd N, Espinosa N (2021) Solar photovoltaic modules, inverters and systems: options and feasibility of EU Ecolabel and Green Public Procurement criteria. Tech. Rep. KJ-NA-30474-EN-N, EU-JRC, European Commission-Joint Research Centre. https://www.doi.org/10.2760/29743, iSBN 978-92-76-26819-2 ISSN 1831-9424
5. Dunlop E, Gracia Amillo AM, Salis E, Sample T, Taylor N (2018) Standards for the assessment of the environmental performance of photovoltaic modules, power conversion equipment and photovoltaic systems. Tech. Rep. JRC111455, Joint Research Centre (European Commision). https://publications.jrc.ec.europa.eu/repository/handle/JRC111455, https://doi.org/10.2760/89830ISBN 978-92-79-86608-1
6. Emmott CJM, Urbina A, Nelson J (2012) Environmental and economic assessment of ITO-free electrodes for organic solar cells. Solar Energy Mater Solar Cells 97:14–21. https://doi.org/10.1016/j.solmat.2011.09.024, GotoISI://WOS:000300653800003, type: Journal Article

7. European Commission-Joint Research Centre (2021) Discussion paper on potential Ecodesign requirements and Energy Labelling scheme(s) for photovoltaic modules, inverters and systems. Tech. rep., European EU-JRC, European Commission-Joint Research Centre. https://susproc.jrc.ec.europa.eu/product-bureau/sites/default/files/2021-04/Discussion%20paper%20Ecodesign%20Photovoltaic%20Products.pdf
8. European Parlament and European Council (2011) Directive 2011/65/EU of the European Parliament and of the Council of 8 June 2011 on the restriction of the use of certain hazardous substances in electrical and electronic equipment. http://data.europa.eu/eli/dir/2011/65/2021-04-01, document 02011L0065-20210401
9. European Parlament and European Council (2012) Directive 2012/19/EU on waste electrical and electronic equipment (WEEE). https://eur-lex.europa.eu/legal-content/EN/TXT/?uri=CELEX:02012L0019-20180704
10. Fu Y, Liu X, Yuan Z (2015) Life-cycle assessment of multi-crystalline photovoltaic (PV) systems in China. J Clean Prod 86:180–190. https://doi.org/10.1016/j.jclepro.2014.07.057, http://www.sciencedirect.com/science/article/pii/S0959652614007859
11. Kim H, Park H (2018) PV waste management at the crossroads of circular economy and energy transition: the case of South Korea. Sustainability 10(10). https://doi.org/10.3390/su10103565
12. MIITa (2013) Several Opinions of the State Council on Promoting the Healthy Development of the Photovoltaic Industry. Tech. rep., Ministry of Industry and Information Technology. China. https://policy.asiapacificenergy.org/node/127/portal
13. MIITb (2013) Standard Conditions for Photovoltaic Manufacturing Industry. Tech. rep., Ministry of Industry and Information Technology. China. https://policy.asiapacificenergy.org/node/128/portal
14. NSF (2017) NSF 457 Sustainability Leadership Standard for Photovoltaic Modules. Tech. Rep. 457i1r1, NSF International. https://standards.nsf.org/apps/group_public/download.php/36153/JC%20Memo%20and%20Ballot%20457i1r1.pdf
15. Shinkuma T, Managi S (2010) On the effectiveness of a license scheme for E-waste recycling: the challenge of China and India. Environ Impact Assess Rev 30(4):262–267. https://doi.org/10.1016/j.eiar.2009.09.002, https://www.sciencedirect.com/science/article/pii/S0195925509001231
16. Stolz P, Frischknecht R, Wyss F, de Wild-Scholten MJ (2016) PEF screening report of electricity from photovoltaic panels in the context of the EU Product Environmental Footprint Category Rules (PEFCR) Pilots. Tech. rep., Treeze Ltd., Switzerland and SmartGreenScans, Netherlands. http://pvthin.org/wp-content/uploads/2020/05/174_PEFCR_PV_LCA-screening-report_v2.0.pdf
17. Sundaram S, Benson D, Mallick TK (2016) Solar photovoltaic technology production. Elsevier, Academic, Potential Environmental Impacts and Implications for Governance
18. Wade A, Stolz P, Frischknecht R, Heath G, Sinha P (2018) The Product Environmental Footprint (PEF) of photovoltaic modules-Lessons learned from the environmental footprint pilot phase on the way to a single market for green products in the European Union. Progress Photovolt: Res Appl 26(8):553–564. https://doi.org/10.1002/pip.2956. Publisher: John Wiley & Sons Ltd
19. Weckend S, Wade A, Heath G (2016) End-of-life management: Solar Photovoltaic Panels. Tech. Rep. Report Number: T12-06:2016, IRENA in collaboration with IEA-PVPS Task 12. https://www.irena.org/publications/2016/Jun/End-of-life-management-Solar-Photovoltaic-Panels, iSBN: 978-92-95111-99-8
20. WEEEForum (2021) Issues associated to photovoltaic panels and compliance with EPR legislation. Tech. rep., WEEE Forum. https://weee-forum.org/wp-content/uploads/2021/06/WEEE-Forum-PV-Panels-Issue-Paper-2021-Final.pdf
21. Yang D, Liu J, Yang J, Ding N (2015) Life-cycle assessment of China's multi-crystalline silicon photovoltaic modules considering international trade. J Clean Prod 94:35–45. https://doi.org/10.1016/j.jclepro.2015.02.003, http://www.sciencedirect.com/science/article/pii/S0959652615001079

Chapter 12
Solar Electricity and Globalization

Solar electricity has always been linked to development projects aimed to bring energy access to people in regions where there is no electricity grid and the only access to energy consists in burning biomass or using diesel generators. Electrification of rural areas in developing countries has always been an important application of photovoltaic technology; also, in countries with 100% electrification there are applications which required electricity in locations where the grid is not available: signalling, communications or remote buildings in isolated areas. Finally, satellites, the International Space Station (ISS) or other spatial exploration vehicles consume electricity generated by highly efficient photovoltaic modules. This was the original portfolio of solar electricity applications: locations where there is no electricity grid and the market competition are against a very expensive extension of grids to remote locations.

This sector of applications was dominant until the early 90s and it is still important, but has been relatively dwarfed by the huge deployment of photovoltaic systems in locations within the reach of the electricity grid. Initially, this new sector of PV development in competition with grid access required specific policies and economic incentives due to the high prices of PV systems and cost of solar electricity since direct competition in an unregulated market would have not been able to overcome the barriers faced by photovoltaics in developed countries in direct competition with commercial cheap grid electricity whose generation was locked to the burning of fossil fuels; the so-called "carbon lock-in". Two main drivers motivated the adoption of specific policies to promote solar electricity: first, the oil crisis in the 70s which demonstrated the vulnerability of the energy supply required to cover the always increasing demand for cheap energy by developed countries: the supply was cut for a few months and prices peaked to figures never seen before. The world suddenly faced the need to diversify the energy sources and slowly start to unlock the carbon lock-in. The second driver has been climate change: in 2021, it is considered as one of the main challenges faced by humankind; this second driver slowly started to be

A. Urbina, *Sustainable Solar Electricity*, Green Energy and Technology,
https://doi.org/10.1007/978-3-030-91771-5_12

evident at the same time that the oil crisis developed, but it was not common worrying beyond specialized circles until in the early 90s scientific evidence about the link between greenhouse gas (GHG) emissions and global warming was overwhelming and the first supranational policies to curtail GHG emissions were adopted, first the Kyoto Protocol (1997), later the Paris Agreement (2015).

These socioeconomic considerations go beyond the conventional approach of Life Cycle Assessment, and there is no methodological procedure to quantify its impacts on different categories. But the policies that are implemented to face the two big challenges mentioned above (lack of access to electricity and climate change) have strong impacts in all LCA categories. Photovoltaic technology provides a tool to tackle both problems simultaneously: providing access to cheap solar electricity (for people living with or without access to the grid) is a powerful tool for climate change mitigation.

12.1 World Electricity Consumption *Per Cápita*

The map of electricity consumed *per cápita* in each country presented in Fig. 12.1 shows large inequalities among countries; furthermore, the average figure representing each country often hides the large inequalities within the country, which are more difficult to measure, although generally, the lower the consumption, the more unequal within the country (many data commented in this section are obtained from the University of Oxford "Our World in Data" website https://ourworldindata.org/ which uses different data sources, in particular for energy data: BP Statistical Review of World Energy [4] and the World Development Indicators of the World Bank, https://databank.worldbank.org/). In 2020, there are more than 20 countries with an average electricity consumption *per cápita* lower than 100 kWh/year, all of them in Africa with the exception of Palestine, Afghanistan and Haiti, while there are 20 countries which consume more than 10,000 kWh/year: a factor 100 difference! In the middle, a majority of countries consume between 3,000 kWh/year and 7,000 kWh year *per cápita*, delivering a World average of 3,316 kWh in 2020; but a whole continent, Africa, is well below this average, with only 665 kWh/year.

The two most populated countries in the world, China (5,297 kWh/year in 2020, a huge increment from 382 kWh/year in 1985) and India (972 kWh in 2020, a more modest increment from 238 kWh in 1985), have diverged strongly in its statistics thus showing how a committed policy can make an impact. On the other hand, the countries showing lower electricity consumption in 2020 (<100 kWh/year *per cápita*) are almost the same as those in 1985 with only a few exceptions, one which is worth mentioning because part of its increment in electricity consumption has been driven by photovoltaic technology: Vietnam moved from 83 kWh/year in 1985 to 2,745 kWh/year in 2020, the largest relative change in the world, with an increment of +3,198% (followed by China with +1,287%). Although not so impressive in absolute terms, Cambodia was the third country in relative change, with +1,097% (moving from 38 kWh/year in 1985 to 460 kWh/year, still in the lower group).

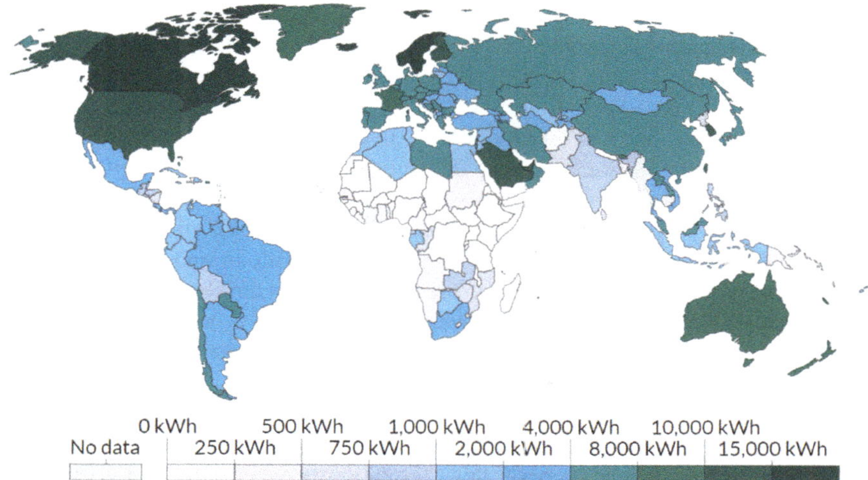

Fig. 12.1 Average annual electricity consumption *per cápita* in 2020, measured in kWh/year. *Source* online data server https://ourworldindata.org/ (University of Oxford, UK) with data from BP Statistical Review of World Energy (2021) [4]

Globally, there is a steady progression towards a higher electricity consumption *per cápita* in a majority of countries; specially for countries with lower levels which at different paces are moving to higher levels of consumption. Interestingly, there are also a group of countries which are reducing their electricity consumption: they can be divided into two very different subgroups, on the one hand, countries with low consumption levels which are further reduced due to war, conflict or severe economic crisis (for example, Syria, Yemen, North Korea, Ukraine, Namibia, Mozambique, Paraguay and Jamaica), and on the other hand, there are countries which are reducing their electricity consumption from already high levels in 1985 to still high but smaller levels in 2020, thus providing a demonstration that electricity consumption (and more generally, primary energy consumption) can be reduced without compromising economic development; examples of this subgroup are the United Kingdom, Canada, Denmark and Switzerland.

12.2 Access to Energy and Development

Access to energy has been linked to development throughout human history. The Human Development Index (HDI) was proposed in 1990 by the United Nations Development Program as an aggregated indicator to measure development. The HDI combines three dimensions: a long and healthy life measured by life expectancy at birth, an education index represented by a combination of mean and expected years of schooling and an econometric quantification of a "decent standard of living" measured by gross national income (GNI) *per cápita* [31, 43]. The values of HDI run from 0 to (ideally) 1, with the most developed country, Norway, reaching 0.975. The minimum access to energy for the "decent standard of living" varies according to different studies, but it is set around 40 GJ/year per person [18]. The curve representing HDI versus primary energy demand *per cápita* has been used for a long time to provide a clear correlation between both parameters, showing that an initial access to a few GJ per year and *per cápita* has a very strong effect on HDI, generating a steep increment up to HDI = 0.8 (achieved with around 100 GJ/year *per cápita*), while access over 100 GJ/year *per cápita* reduces the slope of the increment although it still contributes to an improvement in HDI; that is, to increase HDI from 0.8 to 0.9 requires (on average) another 100 GJ/year *per cápita* and any extra energy consumption has a low impact on HDI. Furthermore, the HDI of some countries has increased in recent years at the same time that the energy consumption *per cápita* has been reduced.

More recent approaches redraw this correlation between HDI and GJ/year *per cápita* by using the **energy footprint** defined as the energy consumed worldwide to produce the goods and services demanded by that country; the result is the graph presented in Fig. 12.2 where the actual energy consumption in a country is redistributed and assigned to the countries where the produced goods are consumed [2]. This approach has a strong effect in some countries, for example, China, which is a high

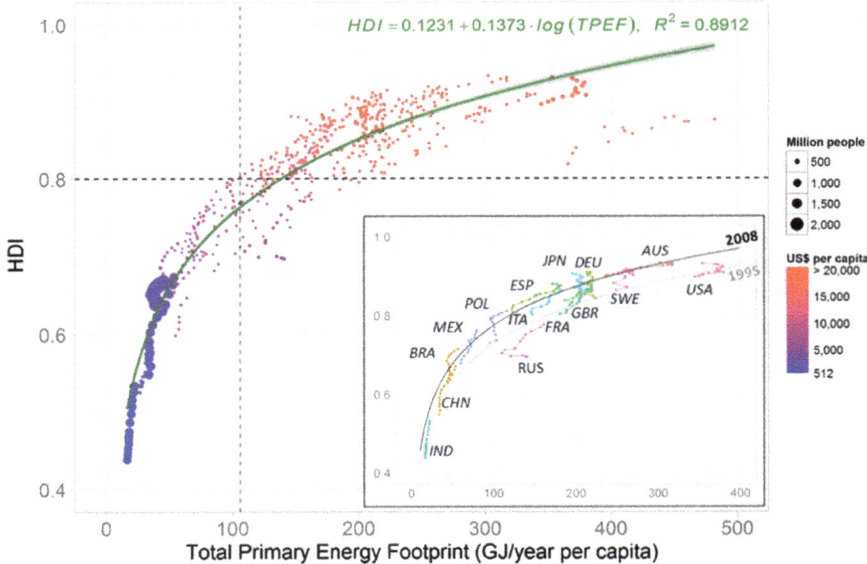

Fig. 12.2 Human development index versus Total Primary Energy Footprint(in GJ/year *per cápita*) and its evolution during 1998–2008. Reproduced with permission from reference [2]

consumer of energy but which exports many of the manufactured goods, while countries like United States of America or the European Union members are high energy consumers not only because they demand a high supply of energy but also because they consume a large quantity of imported goods that have been produced in other countries. Although methodologically the calculation is more complex, the footprint approach shows that the conventional total primary energy demand assigned to any country (energy consumed in that country) underestimates the energy required to maintain a high level of development because a significant part of the energy used by emerging countries is being devoted to sustaining the welfare of developed countries with high HDI. International trade of goods encompasses a subtle international trade of energy consumption and therefore also of greenhouse gas emissions that are not represented by direct statistics of primary energy consumption.

12.3 Solar Electricity for Rural Electrification: When There is No Electricity Market

When the map of Fig. 12.3 is compared to the one shown in the previous section (Fig. 12.1), an approximate correlation becomes evident: the countries with lower electricity consumption *per cápita* are roughly those with lower access to electricity. Since 2015, the number of people without access to electricity fell for the first time below 1 billion, and since then this number has been slowly reduced because a big effort on rural electrification has been carried out in the past decades, delivering electricity at an average rate to around 300,000 people per day who got access for the first time in their lives. The increasing population demands that this effort must be kept or even accelerated to reduce the share of people without electricity, currently at a world average of 13% (in 1990, this percentage was 29%).

The situation is different depending on geographical regions. In sub-Saharan Africa, a subcontinent with average access to electricity of 42.79% of the population, there were around 462 million people without access to electricity in 2020. The access in 1990 was only 15.9% so the improvement has been large, but still more than half the population is excluded from electricity, and this population is steadily growing, thus indicating that despite the efforts in electrification, the challenge is growing. Ten countries of sub-Saharan Africa still have access to electricity lower than 20% and only the improvements in South Africa (84.2%) and Nigeria (59.3%), the two most populated countries, have improved the average. Fortunately, no country in Africa has reduced its access percentage between 1990 and 2020, indicating that at least the increment in population is being met with a commitment to improving electrification at the same pace.[1]

In India, 85% of the population has access to electricity, but it still leaves 200 million people without access despite efforts in rural electrification. The population

[1] Strictly speaking, only Djibouti reduced its access form 59.6% in 1990 to 51.7% in 2020, but with only 1 million inhabitants, the impact on the African average is minimal.

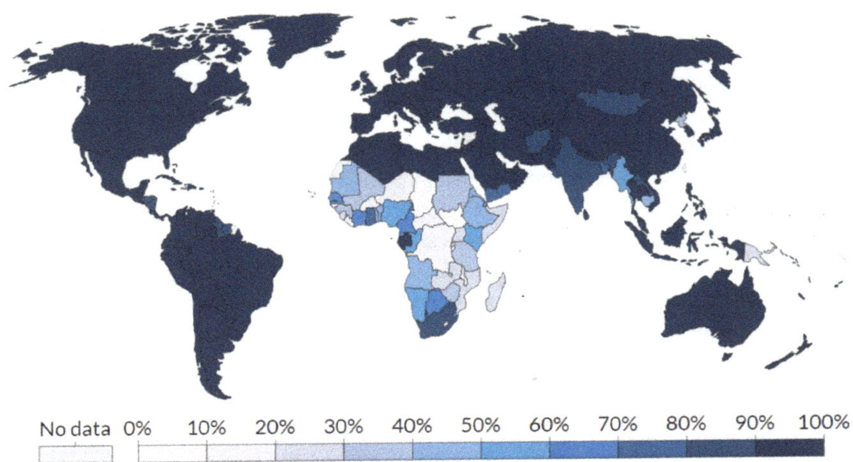

Fig. 12.3 Share of the population with access to electricity. *Source* online data server https://ourworldindata.org/ (University of Oxford, UK) with data from World Bank—World Development Indicators

growth rate is 1% annual (in 2019), posing a challenge to increment the share of people with access to electricity, which has strongly increased since 43.3% in 1990, but has recently declined since the all-time high of 88% was achieved in 2015.

By contrast, in China, full rural electrification was achieved in 2015, thanks in part to massive subsidies for small solar home systems, especially in Yunnan and other south-west regions, and to the extension of the electricity grid to cover almost 100% of the country. Since then, the Chinese government has promoted poverty alleviation policies in which the construction of small solar photovoltaic systems connected to the grid is a substantial part of the investment: the objective is to install two million systems (around 5 kW_p each) and sell the produced solar electricity to the grid, thus generating an extra income of about 3,000 Chinese Yuan (CNY) per household.

In Malaysia, rural electrification has progressed steadily, and it will reach 100% electrification in 5 years, with many locations where isolated systems will be installed, often in combination with other technologies in hybrid energy systems, such as PV-diesel generators or PV-micro-hydropower in small microgrids or solar home systems; as an example, the Sarawak Alternative Rural Electrification Scheme (SARES) that was launched in 2016 has a high rate of success with the electrification of almost 5,000 households in 192 villages so far [24].

In some high-income countries, the governments have launched policies aimed to provide cheap electricity or additional income in low-income households by subsidizing the installation of small or medium-size PV systems. As an example of such policies, Italy has launched "reddito energetico" (energy income) projects in some municipalities, where the PV system is installed on loan (at low cost) and used in a self-consumption scheme, which results in an important reduction in the electricity bill of the household; another example is Seoul, in Korea, where 30 kW_p PV systems have been built and are "shared" by 100 poor households, thanks to non-profit organization subsidies (around 300 W_p per household, although peak power consumption could be higher at some time intervals, thanks to smart meter control). Contrary to initial deployment of solar electricity as an expensive energy supply that required subsidies in electrified locations even in developed countries (the 1990s and early 2000s) and strong investment in developing countries for rural electrification where the electricity grid did not reach remote locations, it has become the cheapest option both in developed and developing countries, thanks to a combination of high grid electricity prices and a strong reduction in PV system prices. The break-even point for solar electricity has been definitively broken.

From a technical point of view, an important breakthrough was achieved in rural electrification with photovoltaic energy when a technical standard for Solar Home Systems (SHS) was proposed [12, 41]. This standard provides recommendations for the design and quality of components for small solar home systems in order to achieve a good performance ratio, low maintenance requirements and extend the operational lifetime of the systems which were usually much lower than the PV module lifetime. Importantly, this SHS standard was proposed for stand-alone systems which include a battery, the component that requires more maintenance and more replacements during the lifetime of the system; the recommendations of the standard are very useful for local installers and final users; furthermore, a good system design is often complemented with an analysis of local operational conditions, ageing of equipment or replacement of spare parts, which should also be effectively monitored in order to obtain information about best practices by using "dependability analysis methods" for case studies in the field: a full understanding of the complex interactions surrounding rural electrification is required to increase the success rate in this kind of development programs [10]. Millions of SHS of different sizes and configurations have been the backbone of rural electrification in remote livelihoods in developing countries during the past five decades, delivering a very large experience about hardware design and maintenance and the social organization that is often needed to keep the solar electricity running in these locations (user's networks, local markets, spare part replacements, second-hand equipment, etc.) [37]. A more recent family of standards was developed since 2015 by the International Electrotechnical Commission (IEC 62257) with recommendations for rural electrification using photovoltaic and hybrid systems.

Related to stand-alone systems for rural electrification, there is a large variety of small and medium photovoltaic systems designed for water pumping, either for human consumption or for irrigation purposes. Again, these kinds of systems were related to development programmes which were subsidized and involved local stakeholders to design, construct and maintain the systems that were called "the silent water" in contrast with the very noisy diesel generator which was (and still is) the principal equipment for water pumping. Several projects increased the practical knowledge required to improve the performance ratio of the photovoltaic/pump system and deliver the demanded water to the end-users [15]. Several field experiences exist and have been carefully evaluated to provide recommendations in order to improve future projects [8, 17, 29, 30].

The relationship between agriculture and photovoltaic technology goes beyond the solar electricity supply for pumping and irrigation. It is related to land occupation which is considered an environmental impact of large PV plants, especially in competition with the use of land for food production and has generated an increasing area of research called "agrivoltaics" [1, 11]. A redesign of PV systems is carried out to optimize the combined yield of the crop and the solar electricity production. The strategies go from developing new technologies with tuned band gaps for selective absorption that preserves the photosynthetically active regions of the solar spectrum required by the plants to simply increasing the height of the rack-mounting systems [9, 14, 36]. The PV systems can be installed on greenhouses, floating in irrigation dams or dispersed and sharing the land with the crop, in many cases also taking into account the impacts on landscape [23, 42]. Agrivoltaics is also related to the water-photovoltaic synergy and especially provide additional benefits in drylands that are prone to droughts and have high irradiance; the trade-offs between the agricultural activity and the solar electricity production may even affect the local microclimate [3]. A more detailed discussion of "agrivoltaics" has been presented in Sect. 7.5.

12.4 Mitigation of Climate Change: From Kyoto Protocol to Paris Agreement and Beyond

Climate change is the most pressing issue that the world is facing. It is beyond doubt that climate change is already occurring and that the main cause is anthropogenic. The Intergovernmental Panel on Climate Change (IPCC), which is the United Nations body for assessing the science related to climate change, has released on 9 August 2021 the results of Working Group I about "The Physical Science Basis" of the Sixth Assessment Report [22]. A selection of headline statements are as follows.

> A.1 It is unequivocal that human influence has warmed the atmosphere, ocean and land. Widespread and rapid changes in the atmosphere, ocean, cryosphere and biosphere have occurred.
> A.2 The scale of recent changes across the climate system as a whole and the present state of many aspects of the climate system are unprecedented over many centuries to many thousands of years.
> B.1 Global surface temperature will continue to increase until at least the mid-century under all emissions scenarios considered. Global warming of 1.5 °C and 2 °C will be exceeded during the 21st century unless deep reductions in carbon dioxide (CO_2) and other greenhouse gas emissions occur in the coming decades.
> D.1 From a physical science perspective, limiting human-induced global warming to a specific level requires limiting cumulative CO_2 emissions, reaching at least net zero CO_2 emissions, along with strong reductions in other greenhouse gas emissions. Strong, rapid and sustained reductions in CH_4 emissions would also limit the warming effect resulting from declining aerosol pollution and would improve air quality.

The need to reduce Greenhouse Gas (GHG) emissions was clear since many decades ago. The first international agreement aimed to reduce GHG emissions was achieved in 1997. After several years of negotiations in the United Nations

Framework Convention on Climate Change (UNFCCC), the **Kyoto Protocol** was adopted on 11 December 1997 [44]. It entered into force in 2005 after 55% countries representing at least 55% of emissions (1990 levels) ratified its signature; currently, it has been signed by 192 countries. 37 industrialized countries assumed individual targets for the reduction of greenhouse gas emissions; these targets add up to a reduction of 5% in the first five-year period of reductions (2008–2012) compared to 1990 levels. Although world emissions globally grew in the same period, the results may seem a success since the 37 countries which committed GHG reductions surpassed their aggregate commitments by an average 2.4 Gt of CO_{2eq}. But these apparently good results have mixed interpretations, since a more careful look at the data indicates that i) the former Soviet Union states achieved a 2.2 Gt CO_{2eq} reduction (–38%, a true collapse) due to the severe economic crisis which almost stalled its industrial output and the reductions were not due to a policy of energy efficiency derived from the Protocol, ii) the hypothetical participation of the USA (did not sign) and Canada (formally withdrew) would have reduced this over-achievement by a net 1 Gt CO_{2eq}, iii) nine countries achieved their targets by buying "carbon credits", which means that 0.3 Gt of CO_{2eq} per year of the claimed reductions were not true cuts because these countries did not domestically achieve their reduction commitments and iv) the Clean Development Mechanism (CDM) contributed with a further 0.3 Gt of CO_{2eq} annual emission reductions which are a result of accountability mechanisms not reflecting true net reductions (on top of that, the CDM mechanism has been contested by several beneficiary countries) [19, 39, 40].

Nevertheless, the Kyoto Protocol was a political success because it created international binding legislation that had to be obeyed by the signatory parties. And it was done so by all of them, including Canada, which signed the Protocol and preferred to withdraw (using the mechanisms and respecting the deadlines of the Protocol) instead of non-complying. Even if there is no overarching international authority with executive powers to enforce the Protocol, its mere existence has been enough for all Parties, which felt compelled to comply with the Protocol.

Even if the Kyoto Protocol only involved 37 countries, and the two biggest GHG emitters did not have reduction commitments (the USA did not sign, and China signed but it did not have to comply with reduction targets), it opened the road for more ambitious targets aiming to involve all countries in the World and which ultimately led to the Paris Agreement. Furthermore, the first period of the Kyoto Protocol was extended to a second commitment period which lasted from 2013 to 2020 and aimed to a reduction of 18% compared to 1990 levels (the Doha Amendment) and thus acted as a bridge to cover the period before the next agreement entered into force.

The **Paris Agreement** is a legally binding international treaty on climate change. It was adopted by 196 Parties at the UNFCCC 21st Conference of the Parties (COP21) in Paris on 12 December 2015 and entered into force on 4 November 2016. It aims to limit global warming as declared in Article 2.a [45].

> Holding the increase in the global average temperature to well below 2 °C above pre-industrial levels and pursuing efforts to limit the temperature increase to 1.5 °C above pre-industrial levels, recognizing that this would significantly reduce the risks and impacts of climate change.

In order to achieve this ambitious target, all countries should aim to peak their GHG emissions as soon as possible and then start reductions with the long-term target of "carbon neutrality". A few countries claim they had already peaked before 1990 since the Kyoto Protocol compliance already reduced its emissions below the level of 1990, but most countries were not required by the protocol to reduce their emissions. These short-term and long-term targets demand ambitious efforts to mitigate climate change, adapt to its effects and at the same time reduce poverty and accomplish the United Nations Sustainable Development Goals.

All countries must submit their plans for climate action in 2020 (the deadline is now 2021, since the COP26 to be held at Glasgow was postponed, and still not celebrated at the time of writing). These plans are known as **nationally determined contributions** (NDCs). The NDCs are specific plans, which should quantify the aimed reductions and the policies to be implemented in order to accomplish the objectives; it therefore requires technical analysis and quantification of emissions from all sectors, and in particular an analysis of future energy demand and supply, with well-defined targets for renewable energy capacity. This NDCs should be complemented with long-term strategies, sometimes defined only with very general statements. The NDCs declared so far and evaluated by the UNFCCC fall far short in order to achieve the proposed targets in the Paris Agreement despite a recent update of commitments for 2025 and 2030.

The quantitative data provided by the UNFCC 2021 report indicates that

> 10. Total GHG emission levels resulting from implementation of the targets communicated in the new or updated NDCs are projected to be around 14.04 Gt CO_{2eq} in 2025 and around 13.67 Gt CO_{2eq} in 2030, which is about 0.3% (38 Mt CO_{2eq}) lower for 2025 and about 2.8% (398 Mt CO_{2eq}) lower for 2030 than the total emission levels according to the Parties' previous NDCs.
> 11. The Parties' total GHG emissions are, on average, estimated to be:
> (a) By 2025, 2.0% higher than the 1990 level (13.77 Gt CO_{2eq}), 2.2% higher than the 2010 level (13.74 Gt CO_{2eq}) and 0.5% higher than the 2017 level (13.97 Gt CO_{2eq});
> (b) By 2030, 0.7% lower than in 1990, 0.5% lower than in 2010 and 2.1% lower than in 2017.

After the results presented in paragraphs 10 and 11 reproduced above, the UNFCC report provides a very strong statement indicating that [46]:

> According to the SR1.5,7[2] to be consistent with global emission pathways with no or limited overshoot of the 1.5 °C goal, global net anthropogenic CO_2 emissions need to decline by about 45% from the 2010 level by 2030, reaching net zero around 2050. For limiting global warming to below 2 °C, CO_2 emissions need to decrease by about 25% from the 2010 level by 2030 and reach net zero around 2070. Deep reductions are required for non-CO_2 emissions as well. Thus, the estimated reductions referred to in paragraphs 10–11 above fall far short of what is required, demonstrating the need for Parties to further strengthen their mitigation commitments under the Paris Agreement.

The world's primary energy consumption in 2020 generated 12,5 Gt of CO_{2eq} emissions, slightly below the amount of 2019 (–3.3%, a decrease that can be attributed

---

[2] This is one of the climate scenarios developed by the International Panel of Climate Change in its 2018 special report [21].

both to lower energy consumption in 2020 due to COVID-19 pandemic and to the largest share of renewable energy generation in the electricity sector). This value is expected to rise in the coming years, although the Paris Agreement aims to peak emissions "as soon as possible". The challenge will require a deep transformation of the energy sector, either on the demand side or on the supply side, in both cases definitively breaking the carbon lock-in of the past century and moving to a new energy paradigm based on a renewable lock-in. The oil crisis was the shock that initiated the move, but today it is the mitigation of climate change, let's call it the "global warming" shock, the more urgent challenge that humankind is facing. The discussion about how to avoid (or prepare) for the peak oil is no more the driver for change, the push for the energy transition is now coming from the discussion about how to achieve the peak of GHG emissions and adapt to the climate change that is already occurring. The problem is not running out of oil, the problem is how to keep this oil (and coal, and gas) underground. In a high-impact article, Christophe Mc Glade and Paul Ekins provided a clear conclusion: *"globally, a third of oil reserves, half of gas reserves and over 80% of current coal reserves should remain unused from 2010 to 2050 in order to meet the target of* 2 °*C"* [25]. The need for out-phasing fossil fuels has been emphasized by Fatih Birol, the International Energy Agency chief executive in the presentation of the recent Net Zero Emissions by 2050 report; he told The Guardian (18 May 2021)[3]:

> If governments are serious about the climate crisis, there can be no new investments in oil, gas and coal, from now, from this year.

The IEA has insisted that exploitation and development of new oil and gas fields must stop in 2021 and that no new coal-fired power stations can be built if the World is to stay within the safe limits of global warming and meet the goal of net-zero emissions by 2050 [20]. And again in 2021, P. Ekins et al. have updated the assessment of the amount of fossil fuels that would need to be left in the ground to allow for a 50 per cent probability of limiting warming to 1.5 °C; their results increase the previous calculation of 2015, and the new recommendation is that 58% for oil, 59% for fossil methane gas and 89% for coal must remain unextracted by 2050 (percentages refer to the 2018 reserve base). The new analysis assumes a stronger climate ambition and a more positive outlook about emerging technologies for renewable energies and zero-emission vehicles [34, 47].

To accomplish this challenge, renewable energies and in particular solar electricity from photovoltaic technology will provide an important contribution to emission reductions by replacing electricity produced with fossil fuels by solar electricity. The net balance of GHG emissions for photovoltaic technology is negative, and therefore they are always analysed as **avoided emissions**. It seems very difficult that all the countries move to the decisive actions reclaimed by Fatih Birol; for example, the UK has announced new investments in oil and gas exploration and China is building new coal-fired power plants. Nevertheless, the avoided emissions approach for GHG accountability is valid since energy consumption is set to grow in the coming years

---

[3] The Guardian, 18 May 2021. https://www.theguardian.com/environment/2021/may/18/no-new-investment-in-fossil-fuels-demands-top-energy-economist.

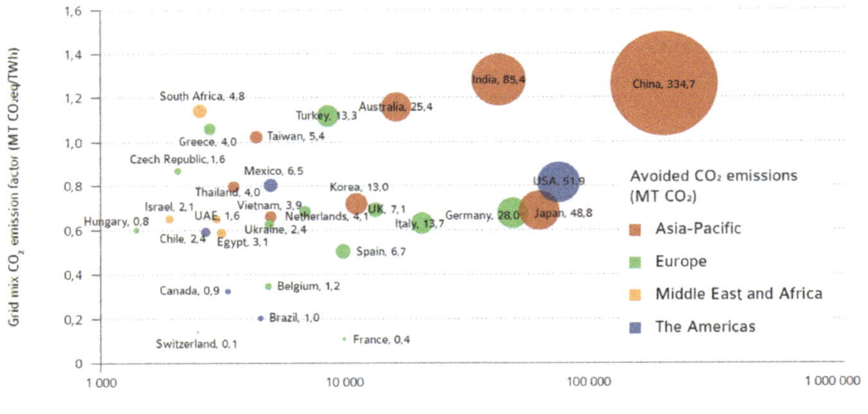

Fig. 12.4 Yearly CO_2 emissions avoided by PV in the first 30 countries in the ranking of cumulative installed PV and which represent in total around 97% of the global avoided emissions in 2019. *Source* IEA-PVPS Trends in PV Applications 2020, reproduced with permission from [24]

and facilitates the path for a fossil-fuel-free electricity generation as a first step to move to a completely fossil-fuel-free primary energy generation. In Fig. 12.4, a graphical presentation of avoided emissions in the first 30 countries in the ranking of cumulative installed photovoltaic systems is presented, with China clearly at the head by a double effect: it is the country with the most installed capacity and where the displaced electricity from the grid reduces more emissions since its grid mix has a strong contribution from fossil fuels and specially from coal.

In Chap. 6, the greenhouse gas emissions produced during the manufacturing of PV systems was discussed, and the amount of CO_{2eq} emissions per kWh of solar electricity production throughout the whole lifetime of the PV systems was presented and compared to other renewable and non-renewable energy production technologies. When climate change mitigation measurements are implemented, this accountability of produced emissions is the base for the calculation, but in the global balance, solar electricity also contributes by "avoiding" emissions that would have been emitted by consuming the demanded electricity that has been supplied by the PV system. This calculation is strongly dependent on the geographical location where the PV system is installed; on the one hand, the amount of produced electricity depends on irradiance and temperature; on the other hand, the grid electricity that is being substituted has a specific electricity mix which evolves in time. The calculation of the avoided emissions is therefore a challenging task that requires the assumption of scenarios where the electricity mix of a given country evolves depending on the amount of renewable energy share in the total installed capacity and on the grid efficiency.

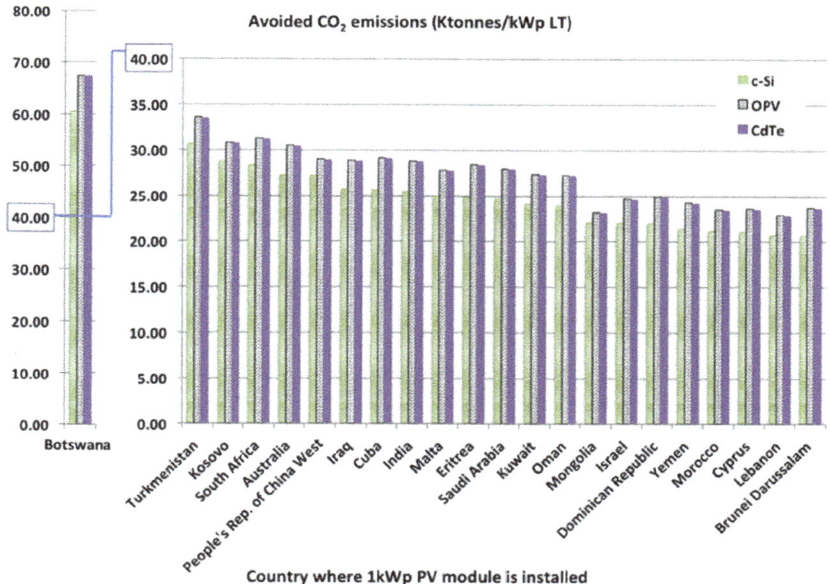

Fig. 12.5 Avoided emission (CO_{2eq}) dependence on the installation place for 1 kW_p of installed capacity throughout its lifetime for three different PV technologies: c-Si, CdTe and organic photovoltaics (OPV), assuming that their manufacture has taken place in China. Reproduced with permission from [38]

Several country combinations for the manufacture and installation of the PV systems can be considered, although the PV market today is dominated by Chinese manufacturers, and therefore Chinese electricity mix has a large impact on the embedded emissions of the PV modules; but also, since China is the country where most capacity is installed (especially in north western parts of the country), it also influences in terms of avoided emissions. In Fig. 12.5, the avoided emissions by PV modules of three different technologies assuming that they have been manufactured in China and installed in several countries are presented: the global geographical optimization for the decision where a PV system will be installed can avoid a much larger amount of GHG emissions with the same installed PV nominal capacity.

A graphical representation of the possibilities of the potential to avoid GHG emissions by a planned combination of manufacture and installation country selection is shown in Fig. 12.6; a few years ago, the PV manufacture was more distributed, with Europe, the USA, Japan and China having significant share, and this kind of proposal could have been (although ideally) considered. Today, production is dominated by China (more than 70% of market share since 10 years ago), and this fact strongly limits this kind of imagined policy [38]. Nevertheless, the picture is very dynamic, the electricity mix of countries are rapidly evolving and the market is facing a development of production capacity outside China. Dynamical models

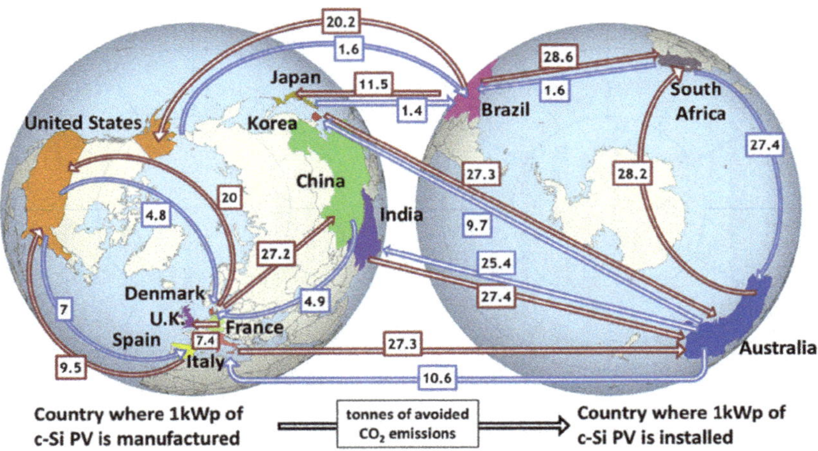

Fig. 12.6 Geographical dependence of the possibilities to avoid emissions by combining countries where modules are manufactured or installed. The numbers within the squares indicate the GHG avoided emissions (tonnes of CO_2eq/kW$_p$ of installed PV capacity throughout lifetime of the PV system) for a given combination of locations: the origin of the arrow is the manufacturing place and in the end is the installation location of the facility. Reproduced with permission from [38]

have been used to evaluate the benefit of technologies with a lower GHG footprint in achieving rapid GHG emission reductions for modules manufactured in Germany and India; in addition, it was demonstrated that short lifetime PV technologies, with a low GHG footprint, such as organic PV, can show greater emission reductions despite a higher levelized global warming potential (g CO_{2eq}/kWh of produced electricity throughout its lifetime) due to the need for more replacement of this short-lived technology compared to the 30 years considered for crystalline silicon technologies [13].

The real contribution of solar electricity to climate change mitigation will be determined by the photovoltaic capacity to be installed in each country in the coming years and the avoided emissions that the generated electricity will allow to displace from the electrical grids with its actual mix (which is constantly changing). In Chap. 1, the future scenarios of International energy Agency were presented, with the corresponding envisaged PV capacity to be installed in the coming years. These projections have taken into account some of the already known NDCs. It is possible that more ambitious targets will be announced in Glasgow, but nevertheless, the IEA NZE2050 scenario is sufficiently ambitious to be considered as a realistic cap for future PV deployment.

12.5 Geopolitics of Photovoltaics

In this final section, which could be expanded into a full book, only a few comments that fall well beyond a Life Cycle Assessment approach will be included.

One important advantage of renewable energies is that they are distributed around the world and that access to the resource should not be affected by geopolitical conflict. This is even more evident for photovoltaic technology, since the primary source of energy, the light coming from the Sun, can be collected and converted with good efficiency in any country. There is more or less yearly irradiation arriving from the Sun into any country, but even in countries at high latitudes with low irradiation levels can benefit from solar electricity.

Nevertheless, the energy conversion requires the manufacture of photovoltaic modules and balance of system equipment, whose material components are not equally distributed around the world. Although photovoltaic technologies are not strongly dependent on very scarce materials, there are risks associated with some specific technologies. In Chap. 5, Sect. 5.3, the materials embedded in solar cells or required for the manufacturing process were listed and the world annual production was commented. The geographical dependence of this production is analysed here in more detail for the scarcer materials included in Table 5.1.

In Table 12.1, the total production in 2019 (tonnes) and the share for the six main producers are shown (data from the British Geological Survey [5]). Those are the minerals identified as more critical for photovoltaic technologies.

For silicon-based technologies (crystalline or amorphous), the bottleneck for production due to mineral scarcity could arise from the use of silver in electrodes. Silver mining is highly distributed around the world, with México and Perú as the main producers, followed by China. Silver content could be reduced in tin–silver alloys or replaced by copper if scarcity becomes a serious risk in competition with other applications for silver.

The risk is higher for CdTe thin film technology; the production of both minerals is highly concentrated in China, which is the main supplier, with 37.4% for cadmium and 73.8% for tellurium. The market for CdTe could keep its modest share, but if global production grows, it may face a scarcity problem, exacerbated by the excessive dependence of the production in China, with tellurium being the main problem with production levels 40 times lower than cadmium (although data do not include undisclosed information from Australia, the USA, Germany and former Soviet Union countries that are believed to produce tellurium). Similarly, for other thin film technologies, like CIGS and III-V tandem cells, mineral production is highly concentrated in China, especially for gallium, a fact that represents a high risk since gallium is demanded by the microelectronic industry for the fabrication of high-mobility transistors; it is difficult that photovoltaic technology could compete in this market for modules other than for spatial applications; scarcity of germanium may also affect III-V technology because it is used as a substrate to the growth of the III-V tandems (as well as for amorphous silicon/germanium tandem technology); production of germanium is highly concentrated in China (90.5%). Finally, it is worth

Table 12.1 Share of production of critical minerals for PV technologies.

Mineral	Major producers	Share (%)	Total Production in 2019 (tonnes)
Silver	México	22.2	26,261
	Perú	14.7	
	China	13.1	
	Australia	5	
	Chile	4.9	
	Bolivia	4	
Cadmium	China	37.4	27,526
	Korea	17.8	
	Japan	6.9	
	Canada	6.5	
	Russia	6.5	
	Kazakhstan	5.1	
Tellurium	China	73.8	625
	Japan	8.8	
	Russia	8	
	Sweden	6.7	
	Canada	2.4	
	Bulgaria	0.5	
Indium	China	62.7	851
	Korea	11.7	
	Japan	8.2	
	Canada	7	
	France	4.7	
	Belgium	2.3	
Gallium	China	93.9	380
	Russia	3.4	
	Ukraine	1.1	
	Japan	0.8	
	Korea	0.8	
Selenium	China	31.8	4,264
	Germany	18.5	
	Japan	18.1	
	Belgium	4.7	
	Russia	4.3	
	USA	3.4	
Germanium	China	90.5	95
	Russia	6.3	
	USA	3.2	
Graphite	China	61.8	1.1 million
	Mozambique	10.1	
	Brazil	8.5	
	Madagascar	4.7	
	Korea (Dem. P. R.)	3.5	
	Turkey	1.5	

Data from [5]

mentioning graphite, a material that has been included in the strategic minerals list despite its high production rate (more than 1.1 million tonnes in 2019); graphite is a source for graphene that is being used for emerging organic and hybrid technologies; it is again an example of how dependence on Chinese production affects a majority of minerals of strategic interest.

The mining of minerals in several countries has been linked to bad working conditions and even to human right abuses, including child labour. The Organization for Economic Cooperation and Development (OECD) has published a report which analyses the situation around the world with a focus on high-risk areas with armed conflict or high level of violence; the report also provides recommendations with the aim to enhance global responsible supply chains of all minerals. The problem with these kinds of recommendations is that they are not followed and in many cases only contribute to an image washing for mining companies [32, 33]. The four minerals that are linked to the worst conflicts worldwide are tin, tantalum, tungsten and gold (the so-called T3G group of minerals) and have been the focus of attention for their role in fuelling ongoing conflict in recent years. Tin is the only one of the T3G group that affects the PV module manufacture supply chain (also at a much lower level, gold, which is used in emerging technologies for research cells), but many other minerals required for PV technology are linked to mining sites associated with violence in several African or Latin-American countries. Of special interest due to the importance of China in the supply of minerals for photovoltaic technology are the "Chinese Due Diligence Guidelines for Responsible Mineral Supply Chains" published by the China Chamber of Commerce of Metals, Minerals and Chemicals Importers and Exporters (CCCMC), which follows the "Guidelines for Social Responsibility in Outbound Mining Investments" and the "Chinese responsible mining guidelines" [7]. These guidelines were developed in a collaboration regulated by a memorandum of understanding signed by the CCCMC and the OECD in 2015; they are the basis of a commitment by Chinese mining companies to comply with OECD recommendations. The declared objective by the CCCMC with this guidelines is [7]

> ... providing guidance to all Chinese companies which are extracting and/or are using mineral resources and their related products and are engaged at any point in the supply chain of minerals to identify, prevent and mitigate their risks of contributing to conflict, serious human rights abuses and risks of serious misconduct, as well as to observe the United Nations Guiding Principles on Business and Human Rights during the entire life-cycle of the mining project.

The next step in possible geographical dependencies can be found in technological development PV module manufacture in the past decades. Although scientific knowledge about photovoltaic technology has been built in a broad international cooperation since the early 90s with leading universities and research centres in Europe, the United States, Japan and Australia, later joined by China and Korea (an in a minor degree other countries), the module production has been concentrated in a few regions not always coincident with the location where scientific development was carried out (for example, Australia has been a major country for the development of silicon technology, but never a big producer of PV modules); this knowledge is now being applied by Chinese manufacturers either by exploitation of patents or by

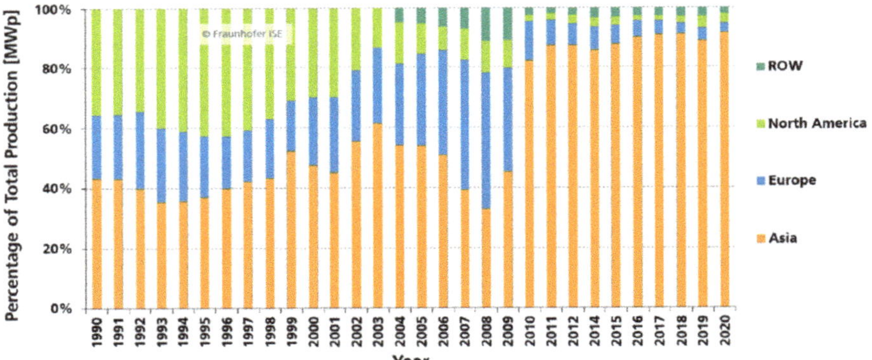

Fig. 12.7 Photovoltaic module production by Region 1990–2020 Percentage of Total MW_p Produced. *Data source* Up to 2004 Strategies Unlimited; 2005 to 2009: Navigant Consulting; since 2010: IHS Markit. Graph: PSE Projects GmbH 2021; reproduced with permission from Fraunhofer ISE Photovoltaic Report 2021 [16]

developing their own patents after applying the knowledge of many Chinese PhD and postdoctoral students who worked in world leading research centres and then moved to existing Chinese companies or founded new companies. Two clear phases can be identified in the time evolution of PV module manufacture shown in Fig. 12.7: from an earlier more distributed manufacturing capacity when the total volume was small (below 10 GW_p/year, roughly before 2006), the biggest share of production has concentrated in China since 2010; these two phases had an intermediate stage, roughly between 2006 and 2009 when poly-silicon supply was low (prices were higher), then the installed new factories multiplied by ten the production capacity, moving definitively to China the dominance of the poly-silicon market; the increasing output scale led to more concentration [6]. In 2013, the Chinese Ministry of Industry and Information Technology designed a production policy and approved a set of standard conditions and production objectives for photovoltaic module manufacture in which the supply chain for silicon PV technology was up-scaled and modernized with the result of a clear market dominance since then [26, 27]. Nevertheless, the picture is dynamic and correlates with shifts in the share of the market of different technologies: the USA was a major manufacturer of thin film CdTe modules that are now losing market share, while China is the larger manufacturer of silicon technology and since then this technology is still dominating the market thanks to its competitive advantages (performance and cost); the emergence of new technologies and the more distributed manufacturing facilities evidenced by the increment of the production share in the rest of the world may evolve again to a more diversified picture. Europe and Japan (and in a lesser degree the USA), big manufacturers up to 2008, will need a big investment effort to reinstall manufacturing capacity in their own land.

Clearly, the geopolitical issues regarding photovoltaic technology are focused on China. It will be a mistake to approach this reality in a confrontational way; either on the side of access to materials, or to manufactured modules, or to installed capacity, China is leading by the brute force of numbers. The initial response by other countries has been defensive, for example, the European Union, together with Japan and the USA, has filed a case in the World Trade Organization against China's quotas on rare earths and other minerals; in 2014, the WTO ruled out that the Chinese quotas were "designed to achieve industrial policy goals rather than conservation" and thus violated international law; these rulings may seem contradictory to the objective of diminishing impacts on mineral depletion categories. Later on, the European Union started negotiations with the objective of reducing import duties changing the strategy to a less confrontational one, and long-term supply contracts have also been signed. Despite these kinds of actions that may be effective in the short term, several authors consider that the long-term risks are still important due to an increment in the dependence on China that will hamper European competitiveness in global markets [28, 35].

An all-out commercial war affecting photovoltaic technology will disrupt supply chains and slow down future capacity deployment, strongly needed to increase solar electricity share in world energy consumption and thus reducing GHG emissions. On the contrary, a scientific, technological and commercial cooperation among countries will improve the technology and strongly contribute to optimizing manufacturing lines by reducing its impacts (the information to do so is coming from detailed LCA studies); the scientific cooperation has always created links between research communities, reduced barriers of mistrust and paved the way for more global cooperation among countries. The progress of photovoltaic technology, either at a fundamental level in emerging technologies or to improve the industrial output of already well-established technologies with the objective of reducing impacts and increase recyclability, has already been an international success and it shows the way forward.

References

1. Agostini A, Colauzzi M, Amaducci S (2021) Innovative agrivoltaic systems to produce sustainable energy: an economic and environmental assessment. Appl Energy 281:116102
2. In Arto, In Capellán-Pérez, Lago R, Bueno G, Bermejo R (2016) The energy requirements of a developed world. Energy Sustain Dev 33:1–13
3. Barron-Gafford GA, Pavao-Zuckerman MA, Minor RL, Sutter LF, Barnett-Moreno I, Blackett DT, Thompson M, Dimond K, Gerlak AK, Nabhan GP, Macknick JE (2019) Agrivoltaics provide mutual benefits across the food-energy-water nexus in drylands. Nat Sustain 2(9):848–855
4. BP (2021) Statistical Review of World Energy 2021 (70th edition). Tech. rep., Britihs Petroleum. https://www.bp.com/content/dam/bp/business-sites/en/global/corporate/pdfs/energy-economics/statistical-review/bp-stats-review-2020-full-report.pdf

5. British Geological Survey (2021) World Mineral Production 2015-2019. Tech. rep., British Geological Survey. https://www2.bgs.ac.uk/mineralsuk/statistics/wms.cfc?method=searchWMS, iSBN 978-0-85272-790-4
6. Bye G, Ceccaroli B (2014) Solar grade silicon: Technology status and industrial trends. Solar Energy Mater Solar Cells 130:634–646
7. CCCMC (2017) Tech. rep., China Chamber of Commerce of Metals, Minerals and Chemicals Importers and Exporters (CCCMC). http://www.cccmc.org.cn/docs/2016-05/20160503161408153738.pdf
8. Chandel S, Nagaraju Naik M, Chandel R (2015) Review of solar photovoltaic water pumping system technology for irrigation and community drinking water supplies. Renew Sustain Energy Rev 49:1084–1099
9. Cossu M, Cossu A, Deligios PA, Ledda L, Li Z, Fatnassi H, Poncet C, Yano A (2018) Assessment and comparison of the solar radiation distribution inside the main commercial photovoltaic greenhouse types in Europe. Renew Sustain Energy Rev 94:822–834
10. Diaz P, Egido MA, Nieuwenhout F (2007) Dependability analysis of stand-alone photovoltaic systems. Progress Photovolt Res Appl 15(3):245–264
11. Dinesh H, Pearce JM (2016) The potential of agrivoltaic systems. Renew Sustain Energy Rev 54:299–308
12. Egido MA, Lorenzo E, Narvarte L (1998) Universal technical standard for solar home systems. Progress Photovolt Res Appl 6(5):315–324
13. Emmott CJM, Ekins-Daukes NJ, Nelson J (2014) Dynamic carbon mitigation analysis: the role of thin-film photovoltaics. Energy Environ Sci 7(6):1810–1818
14. Emmott CJM, Roehr JA, Campoy-Quiles M, Kirchartz T, Urbina A, Ekins-Daukes NJ, Nelson J (2015) Organic photovoltaic greenhouses: a unique application for semi-transparent PV? Energy Environ Sci 8(4):1317–1328. https://doi.org/10.1039/c4ee03132f. Go to ISI://WOS:000352275500020, type: Journal Article
15. Fedrizzi MC, Ribeiro FS, Zilles R (2009) Lessons from field experiences with photovoltaic pumping systems in traditional communities. Energy Sustain Dev 13(1):64–70
16. Fraunhofer-ISE (2021) Photovoltaics Report 2021. Tech. rep., Fraunhofer Institute for Solar Energy Systems, ISE, Germany. https://www.ise.fraunhofer.de/content/dam/ise/de/documents/publications/studies/Photovoltaics-Report.pdf
17. Gao Z, Zhang Y, Gao L, Li R (2018) Progress on solar photovoltaic pumping irrigation technology. Irrig Drain 67(1):89–96
18. Goldembert J (2001) Energy and human well being. Tech. Rep. Report No.: HDOCPA-2001-02, Human Development Report Office (HDRO), United Nations Development Programme (UNDP). https://econpapers.repec.org/paper/hdrhdocpa/hdocpa-2001-02.htm
19. Grubb M (2016) Full legal compliance with the Kyoto Protocol's first commitment period—some lessons. Clim Policy 16(6):673–681
20. IEA (2021) Net Zero by 2050. A Roadmap for the Global Energy Sector. Tech. rep., International Energy Agency, net Zero by 2050 Interactive iea.li/nzeroadmap Net Zero by 2050 Data iea.li/nzedata
21. IPCC (2018) Global Warming of 1.5°c. Tech. rep., Intergovernmental Panel on Climate Change - United Nations Organization. https://www.ipcc.ch/sr15/, iPCC SR1.5
22. IPCC (2021) Climate Change 2021: The Physical Science Basis. Contribution of Working Group I to the Sixth Assessment Report of the Intergovernmental Panel on Climate Change. Tech. Rep. AR6-WG1, Intergovernmental Panel on Climate Change - United Nations Organization. https://www.ipcc.ch/report/ar6/wg1/#SPM. Cambridge University Press
23. Liu H, Krishna V, Lun Leung J, Reindl T, Zhao L (2018) Field experience and performance analysis of floating PV technologies in the tropics. Progress Photovolt Res Appl 26(12):957–967
24. Masson G, Kaizuka I (2020) Trends in photovoltaic applications 2020. Tech. Rep. Report IEA-PVPS T1-38:2020. International Energy Agency - Photovoltaic Power Systems Programme - Technology Collaboration Programme, iSBN 978-3-907281-01-7

25. McGlade C, Ekins P (2015) The geographical distribution of fossil fuels unused when limiting global warming to 2°C. Nature 517(7533):187–190
26. MIITa (2013) Several Opinions of the State Council on Promoting the Healthy Development of the Photovoltaic Industry. Tech. rep., Ministry of Industry and Information Technology. China. https://policy.asiapacificenergy.org/node/127/portal
27. MIITb (2013) Standard Conditions for Photovoltaic Manufacturing Industry. Tech. rep., Ministry of Industry and Information Technology. China. https://policy.asiapacificenergy.org/node/128/portal
28. Moss R, Tzimas E, Kara H, Willis P, Kooroshy J (2013) The potential risks from metals bottlenecks to the deployment of Strategic Energy Technologies. Spec Sect: Run TransitS Sustain Econ Struct Eur Union Beyond 55:556–564
29. Narvarte L, Lorenzo E (2010) Sustainability of PV water pumping programmes: 12-years of successful experience. Progress Photovolt Res Appl 18(4):291–298
30. Narvarte L, Lorenzo E, Aandam M (2005) Lessons from a PV pumping programme in south Morocco. Progress Photovolt Res Appl 13(3):261–270
31. Neumayer E (2012) Human development and sustainability. J Hum Dev Capabilities 13(4):561–579
32. OECD (2016) OECD due diligence guidance for responsible supply chains of minerals from conflict-affected and high-risk areas, 3rd edn. OECD Publishing, Paris. http://dx.doi.org/10.1787/9789264252479-en
33. OECD (2017) Practical actions for companies to identify and address the worst forms of child labour in mineral supply chains. Tech. rep., Organization for Economic Cooperation and Development (OECD), Paris. http://mneguidelines.oecd.org/Practical-actions-for-worst-forms-of-child-labour-mining-sector.pdf
34. Pye S, Bradley S, Hughes N, Price J, Welsby D, Ekins P (2020) An equitable redistribution of unburnable carbon. Nat Commun 11(1):3968
35. Rabe W, Kostka G, Smith Stegen K (2017) China's supply of critical raw materials: risks for Europe's solar and wind industries? Energy Policy 101:692–699
36. Ravishankar E, Booth RE, Saravitz C, Sederoff H, Ade HW, O'Connor BT (2020) Achieving net zero energy greenhouses by integrating semitransparent organic solar cells. Joule 4(2):490–506
37. Sahoo SK (2016) Renewable and sustainable energy reviews solar photovoltaic energy progress in India: A review. Renew Sustain Energy Rev 59:927–939
38. Serrano-Lujan L, Espinosa N, Abad J, Urbina A (2017) The greenest decision on photovoltaic system allocation. Renew Energy 101:1348–1356. https://doi.org/10.1016/j.renene.2016.10.020. <GotoISI>://WOS:000388775700121, type: Journal Article
39. Shishlov I, Bellassen V (2016) Review of the experience with monitoring uncertainty requirements in the clean development mechanism. Clim Policy 16(6):703–731
40. Shishlov I, Morel R, Bellassen V (2016) Compliance of the Parties to the Kyoto Protocol in the first commitment period. Clim Policy 16(6):768–782
41. SHS (1999) Universal Technical Standard for Solar Home Systems. Tech. rep., Directorate-General for Energy (European Commission). https://op.europa.eu/en/publication-detail/-/publication/d97b36dd-2431-4c57-b525-74106bb4a1be, iSBN 92-828-5825-1
42. Toledo C, Scognamiglio A (2021) Agrivoltaic systems design and assessment: a critical review, and a descriptive model towards a sustainable landscape vision (Three-Dimensional Agrivoltaic Patterns). Sustainability 13(12). https://doi.org/10.3390/su13126871
43. UNDP (2018) Human Development Indices and Indicators. Statistical Update. Tech. rep., United Nations Development Programme. http://hdr.undp.org/sites/default/files/2018_human_development_statistical_update.pdf
44. UNFCCC (1997) Kyoto Protocol to the United Nations Framework Convention on Climate Change. Tech. Rep. FCCC/CP/1997/L.7/Add.1, United Nations Framework Convention on Climate Change, Kyoto, Japan. https://unfccc.int/documents/2409
45. UNFCCC (2015) The Paris Agreement. Tech. Rep. FCCC/CP/2015/10/Add.1, United Nations Framework Convention on Climate Change, Paris. https://unfccc.int/sites/default/files/resource/docs/2015/cop21/eng/10a01.pdf

46. UNFCCC (2021) Nationally determined contributions under the Paris Agreement. Synthesis report by the secretariat. Tech. Rep. FCCC/PA/CMA/2021/2. https://unfccc.int/documents/268571
47. Welsby D, Price J, Pye S, Ekins P (2021) Unextractable fossil fuels in a 1.5°C world. Nature 597(7875):230–234. https://doi.org/10.1038/s41586-021-03821-8, https://doi.org/10.1038/s41586-021-03821-8

Conclusions

The energy transition which slowly started since the oil crisis in the 70s has been accelerated by the need to displace fossil fuels and generate energy by using renewable sources. The initial shock produced by the sudden reduction of oil supply lead to investment in photovoltaic technology, which was then seriously considered as a renewable energy technology with a large potential to become a main supplier of electricity; the huge research and development advances that were produced in the final decades of the twentieth century made possible the massive deployment of photovoltaic modules that initiated an exponential growth since 15 years ago.

Simultaneously, the growing evidence that climate change had been mainly produced by greenhouse gas emissions originated by anthropogenic activities has created a huge pressure to increase renewable energy generation as a tool to mitigate climate change. The recently released report of Working Group I of the Intergovernmental Panel on Climate Change (part of the Sixth Assessment Report, AR6) is focused on the physical basis of climate change and has declared that *"It is unequivocal that human influence has warmed the atmosphere, ocean and land"*. The importance of the AR6 report is not only that it contains the scientific consensus about climate change, but also that it is endorsed by the signatory governments, which recognize and accept its conclusions. Therefore, the governments are compelled to act decisively with policies aimed to tackle climate change. The Paris Agreement approved in 2015 provides the tool to coordinate internationally policies aimed at the reduction of emissions, and thus, many countries have declared a set of Nationally Determined Contributions (NDCs) which set specific objectives to reduce greenhouse gas emissions that should be accompanied by specific policies and roadmaps to achieve the declared objectives.

The conviction that renewable energies will need to supply a large part of primary energy and most of the electricity that is consumed worldwide before 2050 in order to mitigate climate change has substituted the initial motivation due to the "oil peak", that is, the threat that fossil fuel supply had reached its peak and was going to

A. Urbina, *Sustainable Solar Electricity*, Green Energy and Technology,
https://doi.org/10.1007/978-3-030-91771-5

be reduced because of a lack of resources. Renewable energy technology benefits from inexhaustible resources (by definition), but needs large investments in research and development, design and construction of generators, maintenance and operation during lifetime and end-of-life treatment, including recycling or landfilling.

In particular, photovoltaic technology will be one of the main contributors to the future world energy supply. The progress of photovoltaic technology is a long story which started almost two hundred years ago, with the discovery by Edmund Becquerel of the photovoltaic effect in 1839; the understanding of the physical phenomena underlying electricity generation from light needed the development of the quantum theory and its application to solid-state electronic devices; it took half a century to advance from the concept of light *quanta* proposed by Max Planck in 1900 to the fabrication of the first solar cell with power conversion efficiency above 5% in 1954. Since then, technological advancements for the construction of photovoltaic modules and a deeper understanding of photogeneration phenomena make photovoltaic electricity generation from Sunlight simultaneously a mature commercial technology and a scientific subject still under investigation. Solar electricity is competing with any other energy technology to achieve the cheapest generation costs, but it is also a technology still under intense research effort in laboratories around the world to propose and realize new concepts in emerging photovoltaic technologies that are constantly producing breakthroughs in power conversion efficiency or in the lifetime of devices. Photovoltaic technology is an energy supplier which is competing in the electricity market with cheaper cost per generated unit of electricity than any other technology and at the same time it is one of the more exciting topics in fundamental research in laboratories worldwide. The two ends of the technological readiness level of solar electricity are immersed, respectively, in fierce competitions to make solar electricity the best option to supply sustainable electricity.

Still the contribution of solar electricity to the actual world supply is small; it reached more than 700 Giga Watts (GW) of cumulative installed capacity worldwide, including 140 GW of new capacity added in 2020, thus indicating that the Tera Watt (TW) milestone is within reach in the coming years. The International Energy Agency projections for energy supply that are in line with the objectives of reducing to near zero the greenhouse gas emissions in 2050 indicate that 5 TW of cumulative photovoltaic capacity will be installed by 2030, 10 TW by 2040 and almost 15 TW by 2050 which will provide a third of total electricity generation worldwide in 2050. This steady growth, of more than 10% annually on average, will demand the use of a large quantity of materials and energy inputs required to manufacture a huge quantity of photovoltaic modules that will be deployed in a large variety of photovoltaic systems of different sizes. The environmental, human health and socioeconomic impacts of such a large deployment require a careful evaluation to demonstrate the sustainability of solar electricity at all scales. This electricity can be generated by a variety of different photovoltaic technologies that can be roughly grouped in crystalline silicon (comprising both mono- and multi-crystalline), thin film technologies (comprising different materials but sharing the reduced thickness of the solar cells and therefore the reduced amount of semiconducting materials that can be cadmium telluride, CdTe, copper indium gallium (di)selenide, CIGS or semiconducting alloys from groups III and V of the periodic table just to mention a few examples), both already in the market, or emerging technologies that still have not reached the market (comprising either inorganic kesterite, organic bulk heterojunctions or hybrid dye-sensitized or perovskite solar cells). Although the technical parameters used to describe the performance of photovoltaic modules are the same for all technologies, the working principles of photocurrent generation and the manufacturing processes or embedded materials are different and must be analysed separately from a sustainability point of view.

The best methodological tool to evaluate the impacts of the photovoltaic technology is Life Cycle Assessment (LCA); it is a robust methodology regulated by ISO standards (14040 and 14044) that enable a fair comparison of all photovoltaic technologies between themselves, either available in the market or in a research stage, and also with other energy technologies (renewable or non-renewable). The experience of many years of work on LCA applied to photovoltaics by several research groups has provided a large amount of evidence that quantifies the impacts in many categories of the manufacture, operation and decommissioning of photovoltaic technologies. It also evaluates the risks that a massive deployment of solar electricity may face in the future. And also many groups go beyond the standard LCA approach and are proposing extended tools with the aim to evaluate the sustainability of solar electricity with a broader scope; proposals such as carbon footprint, or methods to integrate the socioeconomical evaluation and associated risks are still under development and subject of methodological discussions.

The LCA methodology proceeds carefully by selecting a functional unit (FU) whose impacts are going to be evaluated. Several FUs can be selected for an LCA study of solar electricity: one square meter of photovoltaic module, the nominal power in standard conditions, or the unit of supplied electricity (either direct current or alternating current, including the electronic devices required for power conversion); a clear scope of the study is declared in each study, which may range from cradle to gate (the manufacture of modules) or include all life stages of the photovoltaic system, that is, manufacture, operation, decommissioning, recycling and landfilling in a full cradle to grave study.

The first deliverable of an LCA study is the inventory of materials and energy inputs, which allows a first evaluation of sustainability of the photovoltaic technology under investigation. From inventory analysis, the first conclusion emerges: photovoltaic technologies do not have high risks associated with materials supply. The most important technology in the market is based on crystalline silicon; in the form of oxides and silicates, silicon is the most abundant element on the Earth's crust; but the technology also depends on other materials that could face supply tensions, like silver or tin, but its replacements are being investigated and can be available at competitive prices with similar performance (carbon-based electrodes or metal grids with lower metal contents). Thin film technologies in the market may face higher risks in materials supply; in particular, tellurium and indium are strategic elements whose supply in the future may be at risk. They pose an additional burden to CdTe and CIGS technologies, which are struggling in the market to recover some of their old share, but they now are below 5% of global new annual capacity installations, around

7 GW out of a total of 140 GW in 2020. Similarly, some emerging technologies rely on indium supply for transparent conducting oxides, but good replacements have been demonstrated (fluor-doped tin oxide, conjugated polymers, or semitransparent metal grids). Gallium is also a scarce material that may affect CIGS and III-V technologies, although its supply could be increased in the near future because known reserves are larger than in the previous cases. Regarding energy input, the discussion about the energy balance of photovoltaic technology and the doubts that were cast in the early 80s saying that the energy input required to manufacture a PV module was larger than the energy supplied along its lifetime is no more an issue of discussion: the energy balance of solar electricity is always positive, and the energy payback time (EPBT, a parameter out of the scope of LCA but very useful for energy balance discussions) is between 1.2 and 3 years for crystalline silicon technology (depending on the performance ratio and the environmental parameters of the geographical location where the system was operating); this figure must be compared with the lifetime of any commercial photovoltaic technology, which is 25 or 30 years, defined as the time that it takes to lose 20% of initial power output; nevertheless, operational lifetimes can be much longer. Emerging technologies, such as organic or hybrid solar cells have demonstrated EPBTs as low as a few days or weeks, although their lifetime is still low in comparison to inorganic technologies and therefore, the extension of lifetime is a major research topic in the field of emerging photovoltaic technologies.

Once the inventory of materials and energy has been compiled, the next step in the LCA study is to calculate the impacts that they produce in several impact categories which are carefully described and calculated by using an up-to-date scientific knowledge that is constantly revised; there are more than 25 categories that are commonly used. For these purposes, several life cycle impact assessment methods have been proposed; they use different approaches to analyse the path linking inventories to impact categories, which are then aggregated in fewer end-point broader impact groups. The obtained results are similar, but depending on the LCIA method used they may present variations and therefore sensitivity and robustness analysis are commonly carried out. If the most important fifteen categories are chosen for a fair comparison of PV technologies, the results show that all of them deliver impact quantitative values that are of the same order of magnitude, and between one and three orders of magnitude lower than impacts of fossil fuel technologies. These categories are Climate change, Ozone depletion, Human toxicity (cancer and non-cancer effects), Particulate matter, Ionizing radiation, Photochemical ozone formation, Acidification, Eutrophication (terrestrial, fresh water and marine), Ecotoxicity, Land use and Resource depletion (water, mineral and fossil). The only exception is land occupation, where ground photovoltaic systems require more land than other energy alternatives (but not for the case of roof-top systems). When the comparison is made between different commercial photovoltaic technologies, the conclusion is that crystalline silicon is above the other technologies in all categories, but with similar values, always in the same order of magnitude as indicated above (with one exception: in the mineral and fossil resource depletion category, CIGS has higher impacts). When emerging technologies are considered, the results of LCA are varied because very different manufacturing routes are available and still not a standard procedure is clearly established; nevertheless, the LCA studies for organic and hybrid technologies show a large potential for reducing impacts in most of the categories, a fact mainly due to the low energy consumption processing routes: printing methods *versus* evaporation methods, or annealing temperatures much lower than those required for crystallization of silicon; the emerging technologies since they are thin film technologies require a reduced amount of materials per square meter of module (or per peak power output). For the LCA study of emerging technologies, the reduced lifetimes make compulsory considering one or more replacements when their impacts are compared to commercial technologies. When LCA of complete photovoltaic systems are considered, the contribution of Balance of System (BoS) components is important, and while the contribution of the module has been reduced along time, the BoS share of impacts becomes important and will require more studies in the future to recommend ways to reduce its contribution in many categories.

It is becoming of paramount importance the issue of end of life of photovoltaic modules, of those already deployed and those to be deployed in the future. Although the lifetime of modules is large, at least 25 years, many photovoltaic systems installed at the end of the 90s are reaching their end of life, which together with early replacement of defective modules (or simply revamping systems with new modules to repower them) are already producing a large amount of photovoltaic panels that must be treated as electronic waste. According to the International Renewable Energy Agency estimations, around 8 hundred thousand tonnes in 2020 and more than 7 billion tonnes in 2030 of PV modules will reach their end of life. Despite this pressing issue, still a few recycling facilities have been designed and constructed worldwide, and some of them are still pilot plants. Landfilling of modules will produce impacts that can be strongly reduced if a large share of modules is recycled. Also, the recovery of materials from recycled modules will reduce their impact on resource depletion categories. The economic return of recovered silicon wafers or recycled poly-silicon or other elements are still non-profitable when compared to the purchase of original materials, but could become profitable if new technological advances are achieved and up-scaled at the industrial level since some efficient recycling methods have already been demonstrated at the laboratory scale. Regulatory measures and their enforcement will increase the rate of recycled modules in the future and create incentives for more research and development investment for recycling technologies. Still, the regulations framework worldwide is very poor and although some countries are advancing in this regard (European Union), there is a risk of creating an additional problem with exports of electronic waste to third countries where regulations are more permissive or non-existent.

As part of a more broader approach to sustainability, the Life Cycle Assessment methods must be complemented with socioeconomic considerations. The economic cost of manufacturing modules and cost of ownership of final systems (installed and operational) have been strongly reduced in the past years, thanks to potent drivers such as availability of poly-silicon, module manufacture optimization and reduction of Balance of System costs. Total photovoltaic system costs fell below 1 USD/W_p in 2020, with module costs below 0.4 USD/W_p. In many cases, the reduction of these hardware costs has been accompanied by a reduction in soft costs: engineering, workforce, taxes or permits. The hardware cost reduction is very robust and occurs worldwide, while soft costs are more dependent on country regulations and they are the main source of differences of total capital costs of photovoltaic systems in different locations. When the solar electricity cost is analysed, the influence of the geographical location where the system has been installed is large due to the direct dependence on irradiance and temperature of the produced electricity. As a world average, the levelized cost of solar electricity was 0.057 USD/kWh in 2020, with reductions in most countries larger than 80% in the past 10 years. Solar electricity has become the cheapest option in many countries, and yet there is room for further reduction of costs.

Other socioeconomic impacts are more difficult to evaluate. The contribution of energy access to human development index (HDI) has been clearly demonstrated many years ago, with initial access to electricity in rural livelihoods producing a large impact up to average world HDI values and reducing its impact in countries reaching high HDI values (almost 9 for 200 Giga Joule per year of primary energy consumption per capita) and then showing that additional per capita energy consumption seems useless for improving human development. Access to electricity in developing countries with programmes of rural electrification with solar home systems has proven to be effective for human development. Despite the continued effort of electrification, still almost one billion people lack access to electricity, and photovoltaic solar home systems combined with microgrids will continue to be an important application of photovoltaic systems which at the same time contributes to the reduction of greenhouse gas emissions because they mainly substitute diesel generators.

Land occupation is an impact which depends strongly on the size of the PV systems; the same total capacity can be installed with a large set of small to medium size photovoltaic systems (roof-top or ground-mounted but occupying reduced amounts of land) or with a few large plants (occupying hundreds of hectares, in some cases competing with food production), with much higher impact on land occupation. Since the economic return for large scale solar plants is higher, incentives for roof-top systems, complemented with electricity self-consumption schemes, will need to be kept in order to fully exploit the roof-top and building integrated applications market. Additionally, the competition for fertile land between agriculture and cattle on one side and large photovoltaic utility plants on the other side is growing and it can be considered a conflict between food versus electricity production. The impacts of these large PV plants on landscapes and ecosystems are growing and must be carefully assessed. In some cases, this competition has lead to social unrest. Agri-voltaic applications, where photovoltaic systems and crops share the same land, are a possible solution to this dilemma, and the design of photovoltaic systems compatible with agriculture is one of the most fastest growing market areas for photovoltaics (either integrated in greenhouses or in open fields with different rack-mounting structures). These are only two examples that illustrate the complexity of sustainability evaluation, where Life Cycle Assessment approaches need to be complemented with additional tools.

Solar electricity is a fundamental part of the energy transition that is already occurring. Electricity generated by photovoltaic modules will supply a third of worldwide electricity consumption in 2050 according to some scenarios of the International Energy Agency. Such a massive deployment of solar modules is feasible, but its sustainability must be carefully assessed. The Life Cycle Assessment studies carried out so far indicate that solar electricity produces much lower impacts than any other fossil fuel alternative, and has reduced impacts in several categories even when compared to other renewable energy sources such as wind or hydro. The risk arising from material scarcity is low and affects only some technologies, thus indicating that large-scale manufacture of solar modules can be achieved to supply even the scenarios with higher photovoltaic contribution to electricity generation. Nevertheless, environmental and human health impacts still have room for improvement and more investment is necessary to develop new technologies with lower impacts, and importantly, to develop recycling pathways for the massive amount of photovoltaic modules that will reach their end of life in the coming decades. The strong cost reduction of solar electricity (both cost of ownership of the systems and levelized costs of electricity) is a driver for technological improvement, but recycling may require additional regulations and worldwide coordination to ensure minimal impacts on the environment and human health.

Index

A. Urbina, *Sustainable Solar Electricity*, Green Energy and Technology,
https://doi.org/10.1007/978-3-030-91771-5